New Results and Advances in PGE Mineralogy in Various Ni-Cu-Cr-PGE Ore Systems

New Results and Advances in PGE Mineralogy in Various Ni-Cu-Cr-PGE Ore Systems

Special Issue Editors

Andrei Y. Barkov
Federica Zaccarini

MDPI • Basel • Beijing • Wuhan • Barcelona • Belgrade

MDPI

Special Issue Editors

Andrei Y. Barkov
Research Laboratory of
Industrial and Ore Mineralogy,
Cherepovets State University
Russia

Federica Zaccarini
Department of Applied
Geosciences and Geophysics,
University of Leoben
Austria

Editorial Office
MDPI
St. Alban-Anlage 66
4052 Basel, Switzerland

This is a reprint of articles from the Special Issue published online in the open access journal *Minerals* (ISSN 2075-163X) from 2018 to 2019 (available at: https://www.mdpi.com/journal/minerals/special_issues/PGM).

For citation purposes, cite each article independently as indicated on the article page online and as indicated below:

LastName, A.A.; LastName, B.B.; LastName, C.C. Article Title. *Journal Name* **Year**, *Article Number*, Page Range.

ISBN 978-3-03921-716-8 (Pbk)
ISBN 978-3-03921-717-5 (PDF)

Contents

About the Special Issue Editors

Andrei Y. Barkov is the Head of Research Laboratory of Industrial and Ore Mineralogy at the Cherepovets State University of Russia. He has a Ph.D. degree (Univ. of Oulu, Finland) and a Dr. Sci. degree (IGEM RAS, Moscow) in geology and mineralogy. His research interests are broad in the areas of platinum-group mineralogy, ore mineralogy, ore genesis, geochemistry and petrology of layered intrusions and ultramafic-mafic complexes, and involve investigations of various Cr-Ni-Cu-PGE deposits and PGE-Au-bearing placers.

Federica Zaccarini from Modena, Italy obtained an Italian-European habilitation as full Professor in Mineralogy, Petrology, Geochemistry, Volcanology and Ore Deposits. She is a Senior Scientist at the University of Leoben (Austria) and Head of the E.F. Stumpfl Electron Laboratory. Her main research fields include characterization of opaque mineral species, particularly using optical and electron microscopes, and electron microprobe analysis, investigation and characterization of accessory minerals, analytical techniques for the noble metals and their geochemistry, mineralogy and origin of ore deposits related with mafic-ultramafic rocks, platinum-group minerals in chromitites from ophiolite, stratiform complexes and Ural-Alaskan type intrusion, and sulfides deposits and sulphide deposits in ophiolitic complexes. Her work has been recognized by the international scientific community and the International Mineralogical Association has adopted a new mineral name in her honor (Zaccarinite)

Editorial

Editorial for the Special Issue "Platinum-Group Minerals: New Results and Advances in PGE Mineralogy in Various Ni-Cu-Cr-PGE Ore Systems"

Andrei Y. Barkov [1] and Federica Zaccarini [2,*

[1] Research Laboratory of Industrial and Ore Mineralogy, Cherepovets State University, 5 Lunacharsky Avenue, 162600 Cherepovets, Russia; ore-minerals@mail.ru

[2] Department of Applied Geological Sciences and Geophysics, University of Leoben, Peter Tunner Str. 5, A-8700 Leoben, Austria

* Correspondence: federica.zaccarini@unileoben.ac.at

Received: 10 June 2019; Accepted: 11 June 2019; Published: 17 June 2019

The platinum-group minerals (PGM) consist of a group of accessory minerals that concentrate the six platinum-group elements (PGE): osmium (Os), iridium (Ir), ruthenium (Ru), rhodium (Rh), platinum (Pt), and palladium (Pd). Recently, the PGE have gained tremendous importance due to their application in many modern and advanced technologies. However, with a concentration of about 10^{-6} to 10^{-7} % in the Earth's crust, the PGE are numbered among the ultratrace elements. The PGM occur naturally, as alloys, native elements, or combinations with other elements, mainly S, As, Te, Bi, Sb, Se, and, rarely, O. Moreover, the PGM are rare, representing only less than 3% of the approved minerals by the Commission on New Minerals and Mineral Names of the International Mineralogical Association (IMA). Concentration of PGE of economic importance is present in mafic–ultramafic layered intrusions, in continental flood basalts, in the Alaskan-type complexes, and in the placer deposits derived by their erosion [1]. Sub-economic concentration of PGE has been reported from the organic-rich black shales of the Zechstein deposits in Poland [2] as well as from several chromitites associated with ophiolite [1]. Regarding their genesis, the PGM are divided into two main groups: 1) the primary PGM that crystallized at high temperature during the magmatic stage and 2) the secondary PGM that precipitated or have been altered and reworked at low temperature.

This Special Issue of *Minerals* aimed to bring together recent studies on the mineralogical aspect of the PGE with special regards on how and where the PGM formed. Occurrence and description of PGM from China, Russia, South Africa, USA, and Zimbabwe are reported.

In his contribution, Oberthür [3] reported on the presence of PGM found in the layered intrusions of the Bushveld and Great Dyke complexes, discussing in great detail, the PGM genetical evolution from their lode deposits to the placers. According to this author, the final PGM suite which survived the weathering alteration from sulfide ores via oxidized ores into placers results from the continuous elimination of unstable PGM and the dispersion of soluble PGE. The final alluvial PGM assemblage consists of a great number of residual, detrital grains accompanied by minor and rare authigenic PGM that crystallized in the supergene environment.

The study by Stepanov et al. [4] demonstrated that the degree of alteration of PGM in placers associated with the Ural-Alaskan type Svetloborsky massif is linked to the transport distance. In particular, these authors had shown that at a distance of more than 10 km, the degree of PGM mechanical attrition becomes significant, and the primary morphological features, characteristic of lode platinum, are rarely preserved.

The origin of magmatic PGM in chromite deposits associated with ophiolites and Alaskan-type complexes of the Urals had been discussed in the paper by Zaccarini et al. [5]. Crystallization of Os, Ir, and Ru minerals in ophiolitic chromitites took place during and after primary chromite precipitation, under variable sulfur fugacity and temperature. The dominant PGM in the Alaskan-type chromitites are Pt–Fe alloys that crystallized at 1300 and 1050 °C. On-cooling equilibration to about 900 °C may produce lamellar unmixing of different Pt–Fe alloys and osmium. Precipitation of the Pt–Fe alloys locally is followed by an increase of sulfur fugacity leading to the formation of rare PGE sulfides.

PGM and PGE–Au phases found in alluvium along the River Bolshoy Khailyk, in the western Sayans, Russia, have been described by Barkov et al [6]. Three groups of alloy are reported and an order of crystallization is suggested: 1) Os–Ir–Ru, 2) Pt$_3$Fe, and 3) Pt–Au–Cu alloys, which likely crystallized in the sequence from Au–(Cu)-bearing platinum, Pt(Au,Cu), Pt(Cu,Au), and PtAuCu$_2$, to PtAu$_4$Cu$_5$. Many of the Os–Ir–Ru and Pt–Fe grains have porous and altered rims that contain secondary PGM, gold, and rare Cu-rich bowieite and a Se-rich sulfarsenide of Pt. Barkov et al. [6] argued that the alloys precipitated in a highly reducing environment. Late assemblages indicate the presence of an oxygenated local environment leading to Fe-bearing Ru–Os oxide and seleniferous minerals. A primitive ultramafic rock was the source of the studied alluvial PGM and associated minerals.

A careful mineralogical study of PGM grains from alluvial placers of the Gornaya Shoria of Russia [7] allowed to postulate the conditions of their formation and alteration. The original sub-graphic and layered texture pattern of PGM and gold indicate that they are the result of solid solution and eutectic decompositions occurred at magmatic stages, including the result of the interaction of Pt$_3$Fe with a sulfide melt enriched with Te and As.

In their review, Barkov and Cabri [8] summarized the compositional variations of major and minor elements in Pt–Fe alloys from both lode and placer occurrences found in layered intrusions, Alaskan-Uralian-(Aldan)-type and alkaline gabbroic complexes, ophiolitic chromitites from Canada, USA, Russia, and other localities worldwide. Typically, Ir, Rh, Pd, and minor Cu, Ni are incorporated into a compositional series (Pt,PGE)$_{2-3}$(Fe,Cu,Ni) in the lode occurrences. In contrast, the distribution of Ir, Rh, and Pd is fairly chaotic in placer Pt–Fe grains. These authors noticed that the Pt–Fe alloys from placers are notably larger in size compared to those from lode deposits. However, based on their observation of a large dataset of Pt–Fe alloys from numerous origins, Barkov and Cabri [8] concluded that they show compositional overlaps that are too large to be useful as reliable index minerals.

Yang et al. [9] reported on the presence of PGM such as laurite, Os–Ir–Ru alloys, and minor Pd–Te, anduoite, and irarsite in dunite and chromitite of the Xiadong Alaskan-type complex located in southern Central Asia. These authors noticed that the PGM described are very different compared to the typical PGM assemblage of the Alaskan-type complexes worldwide. The highest PGE concentrations (112 ppb) was detected in a dunite sample. Therefore, the authors suggested that the Xiadong complex has potential for PGE exploration.

The contribution by Koerber and Thakurta [10] is dealing with the investigation of a gabbro from the midcontinent rift Echo Lake intrusion in northern Michigan, USA. The authors found that the investigated rocks are characterized by a PGE-enrichment in a 45 m thick magnetite-ilmenite-bearing olivine gabbro unit with grades up to 1.2 g/t Pt + Pd and 0.3 wt % Cu, related to the presence of disseminated pyrrhotite and chalcopyrite.

The occurrence of a rare Pt-enriched tetra-auricupride was documented by Barkov et al. [11] from an ophiolite-associated placer at Bolshoy Khailyk, western Sayans, Russia. The authors, based on the chemical composition and a synchrotron micro-Laue diffraction study, concluded that tetra-auricupride can incorporate as much as ~30 mol % of a "PtCu" component, apparently without relevant modification of the unit cell.

A mineralogical investigation of PGM nuggets composed mainly of Os–Ir–(Ru) and minor Pt–Fe alloys from the Sisim Placer, eastern Sayans, Russia, was carried out by Barkov et al. [12]. The Os–Ir–(Ru) contains several inclusions such as PGE-rich monosulfide, PGE-rich pentlandite, and Ni–Fe–(As)-rich laurite. The authors suggested that the studied nuggets derived by the erosion of a chromitite of the Lysanskiy mafic–ultramafic complex. The presence of localized fluids was recognized and the fluids were responsible for the crystallization of the unique association of laurite with monazite-(Ce) found in the Sisim Placer.

Last but not least, this Special Issue hosts one paper in which Vymazalova et al. [13] reported the discovery of a new PGM. The new PGM, $Pd_9Ag_2Bi_2S_4$, is named thalhammerite, and it was discovered in galena-pyrite-chalcopyrite and millerite-bornite-chalcopyrite vein-disseminated ores from the Komsomolsky mine of the Talnakh and Oktyabrsk deposits, Noril'sk region, Russia. Due to the small size of thalhammerite in the natural sample, its crystal structure was solved and refined from the single-crystal X-ray-diffraction data of synthetic $Pd_9Ag_2Bi_2S_4$. The mineral is tetragonal, space group $I4/mmm$, with a 8.0266(2), c 9.1531(2) Å, V 589.70(2) Å^3, and $Z = 2$.

Overall, we hope that this Special Issue will contribute to a better understanding of the origin of the PGM and will promote future investigation of tiny and rare minerals such as the PGM.

Finally, we would like to thank the authors, referees, and editorial staff of *Minerals* for their precious effort that contributes to the success of this Special Issue.

Author Contributions: A.B and F.Z. wrote this editorial.

Conflicts of Interest: The authors declare no conflict of interest.

References

1. Pohl, W. *Economic Geology: Principles and Practice*; John Wiley & Sons: Hoboken, NJ, USA, 2011; pp. 1–680.
2. Kucha, H. Platinum-group metals in the Zechstein copper deposits, Poland. *Econ. Geol.* **1982**, *77*, 1578–1591. [CrossRef]
3. Oberthür, T. The fate of platinum-group minerals in the exogenic environment—From sulfide ores via oxidized ores into placers: Case studies bushveld complex, South Africa, and Great Dyke, Zimbabwe. *Minerals* **2018**, *8*, 581. [CrossRef]
4. Stepanov, S.Y.; Palamarchuk, R.S.; Kozlov, A.V.; Khanin, D.A.; Varlamov, D.A.; Kiseleva, D.V. Platinum-group minerals of Pt-placer deposits associated with the Svetloborsky Ural-Alaskan type massif, Middle Urals, Russia. *Minerals* **2019**, *9*, 77. [CrossRef]
5. Zaccarini, F.; Garuti, G.; Pushkarev, E.; Thalhammer, O. Origin of platinum group minerals (PGM) inclusions in chromite deposits of the Urals. *Minerals* **2018**, *8*, 379. [CrossRef]
6. Barkov, A.Y.; Shvedov, G.I.; Silyanov, S.A.; Martin, R.F. Mineralogy of platinum-group elements and gold in the ophiolite-related placer of the River Bolshoy Khailyk, Western Sayans, Russia. *Minerals* **2018**, *8*, 247. [CrossRef]
7. Nesterenko, G.V.; Zhmodik, S.M.; Belyanin, D.K.; Airiyants, E.V.; Karmanov, N.S. Micrometric inclusions in platinum-group minerals from Gornaya Shoria, Southern Siberia, Russia: Problems and genetic significance. *Minerals* **2019**, *9*, 327. [CrossRef]
8. Barkov, A.Y.; Cabri, L.J. Variations of major and minor elements in Pt–Fe alloy minerals: A review and new observations. *Minerals* **2019**, *9*, 25. [CrossRef]
9. Yang, S.H.; Su, B.X.; Huang, X.W.; Tang, D.M.; Qin, K.Z.; Bai, Y.; Sakyi, P.A.; Alemayehu, M. Platinum-group mineral occurrences and platinum-group elemental geochemistry of the Xiadong alaskan-type complex in the Southern Central Asian orogenic belt. *Minerals* **2018**, *8*, 494. [CrossRef]
10. Koerber, A.J.; Thakurta, J. PGE-enrichment in magnetite-bearing olivine gabbro: New observations from the midcontinent rift-related echo lake intrusion in Northern Michigan, USA. *Minerals* **2019**, *9*, 21. [CrossRef]
11. Barkov, A.Y.; Tamura, N.; Shvedov, G.I.; Stan, C.V.; Ma, C.; Winkler, B.; Martin, R.F. Platiniferous tetra-auricupride: A case study from the Bolshoy Khailyk placer deposit, Western Sayans, Russia. *Minerals* **2019**, *9*, 160. [CrossRef]
12. Barkov, A.Y.; Shvedov, G.I.; Martin, R.F. PGE–(REE–Ti)-rich micrometer-sized inclusions, mineral associations, compositional variations, and a potential lode source of platinum-group minerals in the Sisim placer zone, Eastern Sayans, Russia. *Minerals* **2018**, *8*, 181. [CrossRef]
13. Vymazalová, A.; Laufek, F.; Sluzhenikin, S.F.; Kozlov, V.V.; Stanley, C.J.; Plášil, J.; Zaccarini, F.; Garuti, G.; Bakker, R. Thalhammerite, $Pd_9Ag_2Bi_2S_4$, a new mineral from the Talnakh and Oktyabrsk deposits, Noril'sk Region, Russia. *Minerals* **2018**, *8*, 339. [CrossRef]

minerals

MDPI

Article

Origin of Platinum Group Minerals (PGM) Inclusions in Chromite Deposits of the Urals

Federica Zaccarini [1,*], Giorgio Garuti [1], Evgeny Pushkarev [2] and Oskar Thalhammer [1]

1 Department of Applied Geosciences and Geophysics, University of Leoben, Peter Tunner Str.5,
A 8700 Leoben, Austria; giorgio.garuti1945@gmail.com (G.G.); oskar.thalhammer@unileoben.ac.at (O.T.)
2 Institute of Geology and Geochemistry, Ural Branch of the Russian Academy of Science, Str. Pochtovy per. 7,
620151 Ekaterinburg, Russia; pushkarev@igg.uran.ru
* Correspondence: federica.zaccarini@unileoben.ac.at; Tel.: +43-3842-402-6218

Received: 13 August 2018; Accepted: 28 August 2018; Published: 31 August 2018

check for
updates

Abstract: This paper reviews a database of about 1500 published and 1000 unpublished microprobe analyses of platinum-group minerals (PGM) from chromite deposits associated with ophiolites and Alaskan-type complexes of the Urals. Composition, texture, and paragenesis of unaltered PGM enclosed in fresh chromitite of the ophiolites indicate that the PGM formed by a sequence of crystallization events before, during, and probably after primary chromite precipitation. The most important controlling factors are sulfur fugacity and temperature. Laurite and Os–Ir–Ru alloys are pristine liquidus phases crystallized at high temperature and low sulfur fugacity: they were trapped in the chromite as solid particles. Oxygen thermobarometry supports that several chromitites underwent compositional equilibration down to 700 °C involving increase of the Fe3/Fe2 ratio. These chromitites contain a great number of PGM including—besides laurite and alloys—erlichmanite, Ir–Ni–sulfides, and Ir–Ru sulfarsenides formed by increasing sulfur fugacity. Correlation with chromite composition suggests that the latest stage of PGM crystallization might have occurred in the subsolidus. If platinum-group elements (PGE) were still present in solid chromite as dispersed atomic clusters, they could easily convert into discrete PGM inclusions splitting off the chromite during its re-crystallization under slow cooling-rate. The presence of primary PGM inclusions in fresh chromitite of the Alaskan-type complexes is restricted to ore bodies crystallized in equilibrium with the host dunite. The predominance of Pt–Fe alloys over sulfides is a strong indication for low sulfur fugacity, thereby early crystallization of laurite is observed only in one deposit. In most cases, Pt–Fe alloys crystallized and were trapped in chromite between 1300 and 1050 °C. On-cooling equilibration to ~900 °C may produce lamellar unmixing of different Pt–Fe phases and osmium. Precipitation of the Pt–Fe alloys locally is followed by an increase of sulfur fugacity leading to crystallize erlichmanite and Ir–Rh–Ni–Cu sulfides, occurring as epitaxic overgrowth on the alloy. There is evidence that the system moved quickly into the stabilization field of Pt–Fe alloys by an increase of the oxygen fugacity marked by an increase of the magnetite component in the chromite. In summary, the data support that most of the primary PGM inclusions in the chromitites of the Urals formed in situ, as part of the chromite precipitation event. However, in certain ophiolitic chromitites undergoing annealing conditions, there is evidence for subsolidus crystallization of discrete PGM from PGE atomic-clusters occurring in the chromite. This mechanism of formation does not require a true solid solution of PGE in the chromite structure.

Keywords: chromitite; platinum group minerals; primary inclusions; ophiolite; Alaskan-type complex.; Urals; Russia

1. Introduction

Ever since pioneer geochemical surveys of ultramafic rocks [1–13], chromitite has been recognized as a potential concentrator of platinum-group elements (PGE = Os, Ir, Ru, Rh, Pt, Pd). These metals are mainly carried in minute grains (<20 μm) of specific platinum-group minerals (PGM) that occur enclosed in chromite crystals, at the crystal rims or in the interstitial silicate gangue of the chromitite. Because of their textural position the PGM not included in chromite are exposed to alteration and can be remobilized at a small scale by the action of low-temperatures hydrous fluids [8,12]. The PGM included in fresh chromite are preserved by alteration and generally are considered to have crystallized at high temperature. Since the beginning, it was established that some chromitites preferentially concentrate IPGE (Os, Ir, Ru), and others are dominated by PPGE (Rh, Pt, Pd), depending on PGE fractionation during crystallization of the parent melt, and partial melting of the mantle source. Apart from this question, there has been considerable debate concerning the mechanism by which discrete PGM crystals are enclosed in the chromite. Some authors have interpreted the PGM to have exsolved from the oxide host at some subsolidus stage, and actually experimental works have demonstrated that some PGE have crystal-chemical compatibility for the spinel structure, being incorporated as a true solid solution in the oxide [9]. However, some factual observations strongly argue against this model as the sole mechanism for the formation of PGM in chromite. For example, dissolution-exsolution cannot explain the mineralogical diversity of PGM included in a single chromite crystal, neither can it account for the compositional similarity of PGM found included in chromite and co-crystallizing mafic silicates [4–6,10]. There is now a general consensus that the primary PGM are pristine liquidus phases mechanically trapped in chromite and mafic silicates precipitating from the magma at high temperature would appear the most likely. This model may easily account for the mineralogical variability of PGM-bearing composite inclusions reported from a number of chromitites, and does not require crystallographic substitution of PGE in the framework of chromite and mafic silicates [6,11–13]. The crystallization of PGM at high temperature is somehow supported by the so called "clusters theory" [14]. These authors suggested that in natural magmas the PGE do not occur as free cations or any other molecular species, but form disordered clusters consisting of a few hundred atoms, suspended in the melt. Because of their physical and chemical properties the clusters will tend to coalesce with decreasing temperature to form specific PGM alloys, sulfides, or compounds with other ligands (e.g., As, Te, Bi, Sb). The "clusters theory" has received much attention from the students since it can explain many characters of the primary PGM inclusions in chromitites. Nevertheless, the magmatic association between chromite and PGE at high temperature still presents some unresolved questions which need further investigation [15].

The chromite deposits of the Urals offer a unique opportunity to study genetic mechanisms of the chromite-PGE mineralization associated with ophiolites and Alaskan-type ultramafic complexes. These chromitites formed by a sequence of magmatic events marking the geodynamic evolution of the Uralian Ocean, from its opening in Ordovician pre-Palaeozoic to closure in Upper Devonian-Carboniferous Permo-Triassic times.

Preliminary overviews have revealed that chromitites associated with ophiolites and Alaskan-type intrusions have distinct Os–Ir–Ru or Ir–Rh–Pt–Pd geochemical specialization, respectively [7,16], providing a useful guideline for the interpretation of the primary source of PGE nuggets in alluvial placer deposits of the Urals [7,16,17]. Mineralogical investigations carried out between 1996 and 2016 have described in detail the diverse PGM assemblages of the chromitites, thereby establishing a close consistency between the geochemistry and mineralogy of the PGE [18–45]. The results of these studies, however, did not lead to a conclusive model able to adequately explain the primary origin of PGM inclusions in the chromite. Several aspects remain unsolved, for example, the physical state of the PGM at the time of entrapment (liquid, solid crystals, solid + liquid), or the effects of chromite re-equilibration in the subsolidus. The introduction of the "cluster theory" [14] clearly provides further argument for debate making the role of the re-adjustment of PGM assemblages in the subsolidus stage more likely. In order to contribute to the discussion, we have examined approximately 2500 microprobe analyses of PGM grains, partly taken from the above literature and partly provided by new analyses carried out in the course of this overview. A survey of electron-microscope images of primary PGM inclusions in the fresh chromite is provided, in order to illustrate crystallization relations among the different PGM phases. Comparison of PGE-PGM data with chromite composition allows us to explore the possible correlation between the type of PGM mineralization and conditions invoked for the precipitation of chromite at high magmatic temperature, or the influence of chromite on-cooling equilibration in the subsolidus stage.

2. Geological Setting and Sample Provenance

Ural orogenic belt extends over 2500 km along the 60° East Meridian (Figure 1A). Ophiolites and Alaskan-type complexes hosting the chromite deposits examined in this overview occur distributed along the axial zone of the Ural belt (Figure 1B). Geologic classification of the chromite deposits [46] includes: (1) Mantle-hosted ophiolitic chromitite (Ray-Iz, Alapaevsk, Kluchevsk, Nurali, Kraka, Kempirsai Main Ore Field); (2) Banded chromitite in supra-Moho cumulate of ophiolites (Nurali, Kempirsai Batamshinsk, Tagashasai, Stepninsk); (3) Chromitite lenses in Alaskan-type zoned intrusions (Kytlym, Kachkanar, Nizhny Tagil, Uktus) (Table 1). Type-1 chromitites are characterized by high-Cr, low-Ti composition and are hosted in harzburgite or lherzolite type ophiolites; most deposits have economic size. Type-2 chromitites have high-Al composition, but their Ti content varies from low to high according to the type of host-rock, dunite, pyroxenite, troctolite, clinopyroxenite. Type-3 chromitites have high-Cr, high-Ti composition. Type 2 and 3 chromitites have sub economic size as a source of chromite ore. The geometrical relationships of type-3 deposits indicate "syngenetic" deposition of chromitite in equilibrium with host dunite, or late "epigenetic" emplacement of chromitite within solid dunite [47]. PGM in the syngenetic type chromitite mostly form small inclusions in chromite crystals. PGM in the epigenetic chromitites fill interstitial space between chromite grains and form huge aggregates which are the main source for the famous Uralian placers.

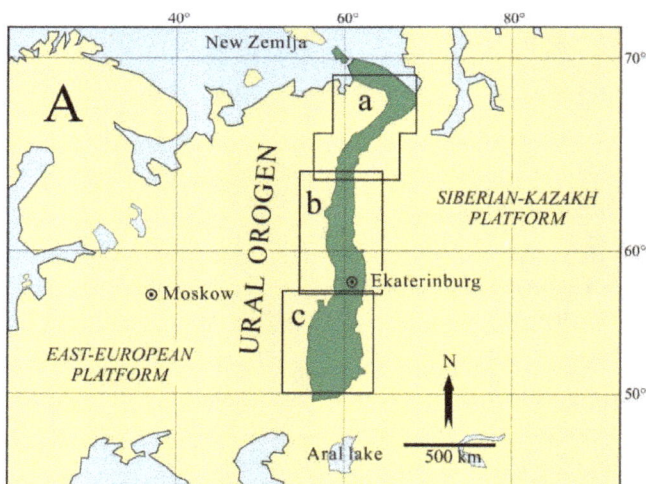

Figure 1. *Cont.*

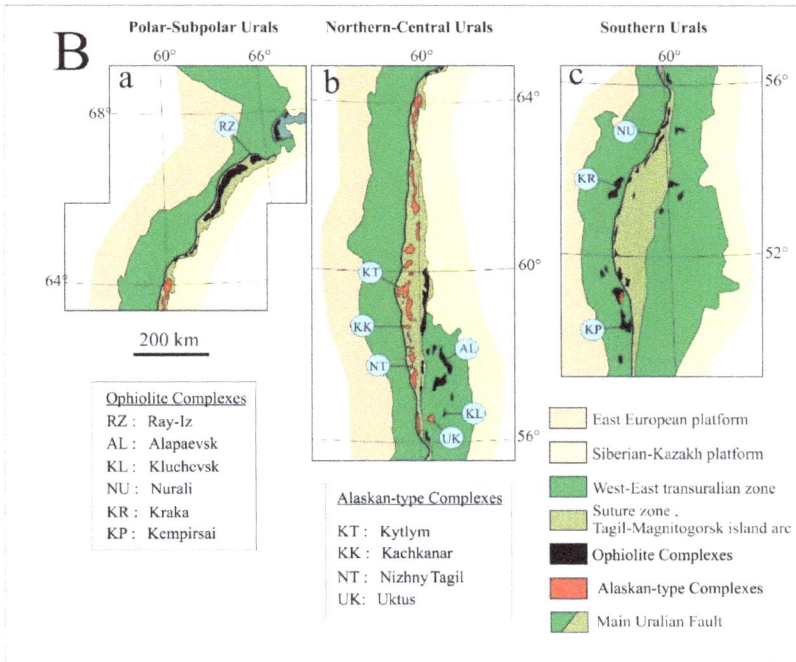

Figure 1. (**A**) The Ural orogen. (**B**) Simplified geological map of the Polar, Central, and Southern Urals, showing the location of the complexes hosts the studied chromitites.

Type-3 chromite compositions reported in Table 1 refer to "syngenetic" chromitites which better reflect conditions of chromitite-dunite co-precipitation from the Alaskan primitive magma, at high temperature. Overview of the PGE geochemistry [24,28,39,40,44,48,49] indicates that the chromitites are characterized by distinctive chondrite-normalized [50] PGE patterns, showing alternative predominance of Os–Ir–Ru (Figure 2A), and Rh–Pt–Pd (Figure 2B,C) in podiform and banded chromitites associated with ophiolites, or a marked Pt–Ir specialization with Ru negative anomaly in those within Alaskan-type intrusions (Figure 2D).

Table 1. Geologic classification of chromite deposits of the Urals.

	TiO$_2$	Al$_2$O$_3$	Cr$_2$O$_3$	#Cr	#Fe^2	#Fe^3	Host Rock	Magma Type	Geological Setting
Type-1, Mantle hosted ophiolitic chromitite									
Ray-Iz high-Cr (8)	0.10	9.69	59.19	0.80	0.33	0.05	Hz-D	IA-bon	SSZ
Kempirsai MOF high-Cr (21)	0.16	9.76	59.69	0.80	0.30	0.05	Hz-D	IA-bon	SSZ
Kraka high-Cr (5)	0.17	12.84	55.76	0.74	0.31	0.06	L-H-(D)	picritic-thol	MOR/BA?
Kluchevsk high-Cr (7)	0.19	11.62	55.94	0.77	0.38	0.06	D-tz	IA-bon	SSZ
Alapaevsk high-Cr (5)	0.21	10.55	57.78	0.79	0.38	0.07	Hz-D	IA-bon	SSZ
Alapaevsk high-Al (9)	0.27	22.42	44.63	0.57	0.31	0.06	L-H-(D)	MORB	MOR?
Type-2, Banded chromitite in supra-Moho cumulates of ophiolites									
Kempirsai BAT (9)	0.08	26.67	40.99	0.51	0.29	0.05	D	MORB	MOR
Kempirsai TAG (3)	0.30	25.68	38.60	0.50	0.37	0.07	D	MORB	MOR
Kempirsai STEP (7)	0.53	28.50	36.33	0.46	0.36	0.06	D-T	MORB	MOR
Nurali low-Ti (7)	0.08	28.70	36.88	0.46	0.35	0.06	D-Wr	picritic-thol	CM
Nurali high-Ti (5)	0.77	19.73	33.22	0.53	0.56	0.19	Wr-Cpx	Fe-thol	CM
Type-3, Chromitite lenses in Alaskan-type zoned intrusions									
Kachkanar (5)	0.50	7.64	50.60	0.82	0.47	0.17	D	ankar	IA
Nizhny Tagil (18)	0.44	7.45	51.30	0.82	0.44	0.17	D	ankar	IA
Kytlym (8)	0.77	10.57	42.13	0.73	0.47	0.23	D	ankar	IA
Kytlym Butyrin vein (6)	1.84	6.30	30.12	0.76	0.71	0.43	Cpx	IPB?	IA
Uktus S-dunite (10)	0.61	12.91	48.26	0.72	0.42	0.12	D	ankar	CM?
Uktus W-dunite (2)	0.55	11.45	50.70	0.75	0.50	0.11	D	ankar	CM?
Uktus N-dunite (7)	1.00	11.66	40.53	0.69	0.54	0.22	D	ankar	CM?

Note: Column 1: Chromitite type, locality, number of samples (n). MOF = Main Ore Field, BAT = Batamshinsk, TAG = Tagashasai, STEP = Stepninsk; #Cr = Cr/(Cr + Al), #Fe^2 = Fe^2/(Fe^2 + Mg), #Fe^3 = Fe^3/(Fe^3 + Cr + Al), at %; host rock: H = harzburgite; D = dunite; L = lherzolite; Wr = wherlite; Px = pyroxenite; Cpx = clinopyroxenite; T = troctolite; magma type: IA-bon = island arc boninite; thol = tholeiite; ankar = ankaramite; IPB = intra-plate basalt; setting (inferred from host-rock petrology): SSZ = supra-suduction zone; MOR = mid oceanic ridge; FA = fore arc; IA = island arc; BA = back arc; CM = continental margin.

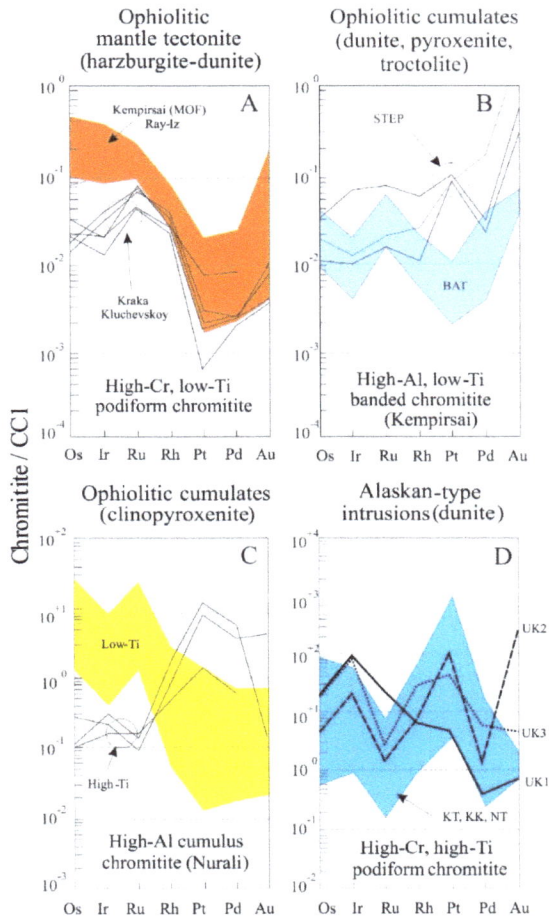

Figure 2. Platinum-group elements (PGE) chondrite [50] normalized patterns for selected chromitites of the Urals. (**A**) PGE distribution in mantle hosted ophiolitic chromitites, MOF = main ore field. (**B**) PGE distribution in banded chromitites from Kempirsai ophiolite, STEP = Stepninsk deposit, BAT = Batamshinsk deposit. (**C**) PGE distribution in cumulus chromitites from the Nurali complex. (**D**) PGE distribution in Alaskan-type chromitites from Uktus (UK), Kytlym (KT), Kachkanar (KK) and Nizhny Tagil (NT) complexes. See the text for the data source.

3. Distribution and Mineralogy of the PGM Inclusions

The amount of PGM inclusions varies greatly in both the ophiolitic and Alaskan-type chromitites of the Urals, displaying extremely irregular distribution even at the scale of single hand samples. For example, in [27] it was reported that about 30% of the polished sections investigated from the mantle-hosted chromitites of Kempirsai contain relatively abundant PGM, whereas only a few grains were observed in the remainder. About 100 PGM grains were identified in 23 samples from three chromite deposits of the Ray-Iz mantle tectonite, however only 10% of the samples showed significant PGM enrichment [22]. Most samples from Kraka, Kluchevsk, Alapevsk and the banded chromitite in the supra-moho cumulates of Kempirsai display very low contents of PGM, with a maximum frequency of 4–5 grains per 6.5 cm^2 [27,38,40,44]. In contrast, more than 400 PGM grains have been found in chromitite bands in the supra-moho cumulates of the Nurali ophiolite, with a maximum frequency of 8–10 grains per 6.5 cm^2 [39]. About 50% of the chromitite samples investigated from the Alaskan-type complexes were found to contain disseminated PGM. Chromitites from Nizhny-Tagil (18 samples) and Kachkanar (4 samples) contain PGM with maximum contents of 50 and 8 grains, respectively [19]. About 420 PGM grains were analyzed in 15 samples from the Uktus and Kytlym zoned intrusions, with a maximum frequency of 12 grains per 6.5 cm^2 in the richest samples of Kytlym [23,24,41]. The mineralogy of PGM in the Ural chromitites includes alloys, sulfides, sulfarsenides, arsenides, tellurides, antimonides, and accessory base metal (BM) minerals (sulfides, arsenides, oxides, alloys) containing minor amounts of PGE. Accessory Au and Ag phases have also been reported from various chromitites. A list of known and unknown PGM identified in the chromitites of the Urals is given in Table 2.

Table 2. Platinum Group Minerals, and accessory PGE-bearing base-metal minerals identified in chromitites of the Urals.

Mineral Species	Ideal Composition
Sulfides	Laurite (RuOsIr)S$_2$, Erlichmanite (OsRuIr)S$_2$, Kashinite Ir$_2$S$_3$, Bowieite Rh$_2$S$_3$, Cuproiridsite CuIr$_2$S$_4$, Cuprorhodsite CuRh$_2$S$_4$, Malanite CuPt$_2$S$_4$, Cooperite (PtPdNi)S, Braggite (PtPdNi)S, Vysotskite (PtPtNi)S, unknown Ir–Rh–Ni–Fe thiospinels and monosulfides;
PGE-bearing Base Metal Sulfides	Ru-Pentlandite (NiFeRu)$_9$S$_8$, Rh-Pentlandite (NiFeRh)$_9$S$_8$, Pt-pyrrhotite (FePt)$_{1-x}$S, Ir–Rh–Heazlewoodite (NiIrRh)$_3$S$_2$
Sulfarsenides, Arsenides	Irarsite IrAsS, Osarsite OsAsS, Ruarsite RuAsS, Hollingworthite RhAsS, Platarsite PtAsS, Omeiite OsAs$_2$, Ruthenarsenite (RuNi)As, Cherepanovite RhAs, Zaccariniite RhNiAs, Sperrylite PtAs$_2$, unknown Ir-Rh-Os arsenides
PGE-bearing Base Metal Arsenides	Rh-Orcelite (NiIrRh)$_{5-x}$As$_2$, Rh-Maucherite (NiIrRh)$_{11}$As$_8$
Alloys	Osmium, Iridium, Ruthenium, Rutheniridosmine (OsIrRu), Platinum, Isoferroplatinum Pt$_3$Fe, (Pt$_{2.5}$(FeNiCu)$_{1.5}$, Tetraferroplatinum PtFe, Pt(FeNiCu), Ferronickelplatinum Pt$_2$FeNi, Tulameenite Pt$_2$FeCu, Potarite HgPd, unknown Pt–Cu, Pt–Pd–Cu–Ni–Fe, Ru–Ir–Os–Fe–Ni
PGE-bearing Base Metal Alloys	Ru-Awaruite NiFeRu, Garutiite NiFeIr
Tellurides, Antimonides	Merenskyite PdTe$_2$, Tolvkite IrSbS, Geversite PtSb$_2$, Stibiopalladinite Pd$_{5+x}$Sb$_{2-x}$, unknown Ir–Sb, Pt–Fe–Sb, Rh–Te, Rh–Sb
PGE-bearing oxides	Unknown Ru–Os–Ir–Fe–Ni–O

Most authors have divided the PGM into two genetically distinct categories, based on their textural relations: (1) the "primary" PGM occurring enclosed in fresh chromite far from cracks and alteration zones, and (2) the "secondary" PGM being invariably associated with low-temperature assemblages, either included in the ferrianchromite rim of chromite grains, or in the interstitial silicate matrix (serpentine, chlorite, talc). In this overview we have focused our attention on the paragenetic characters of the primary PGM occurring enclosed in fresh chromite thereby reflecting high temperature conditions of formation, having been preserved from low-temperature alteration. The ophiolitic chromitites, as a whole, display predominance of Os–Ir–Ru minerals (IPGM) over Rh–Pt–Pd (PPGM) (Figure 3A), in agreement with the dominant negative trend of the PGE profiles (Figure 2A–C). The IPGM population mainly consists of Ru–Os disulfides of the laurite-erlichmanite series, cuproiridsite, and various Ir–Ni thiospinels and monosulfides occurring in primary inclusions. The sulfides are accompanied by decreasing amounts of Os–Ir alloys, As-bearing phases (irarsite), and BM minerals containing detectable amounts of PGE (mainly Ru-pentlandite) (Figure 3B), which may be found in both primary and secondary assemblages.

The small proportion (~6%) of PPGM reflects the occurrence of Pt and Pd phases (sperrylite, stibiopalladite, tetraferroplatinum, potarite, unknown Pt–Pd–Cu–Ni–Fe alloys) in chromitites from the uppermost cumulus layers of Nurali and Kempirsai [39,45], which distinguish for nearly flat to markedly positive chondritic profiles. Particularly numerous is the category "Ox" including rare

PGE oxydes and a great variety of unknown O-bearing compounds consisting of native Ru–Os–Ir intermixed with Fe-oxide and relicts of sulfide. These grains are interpreted to have derived from partial to complete desulfidation of primary PGM during serpentinization [21]. The Alaskan-type chromitites contain high proportion (81%) of PPGM in front of 19% IPGM. Consistent with the saw-like trend typical of most PGE profiles (Figure 2D), the PGM assemblage is characterized by the predominance of Pt- and Ir-phases, and paucity of Ru and Pd PGM [15,24,41,48,50]. PGE alloys are by far the most abundant PGM phase, accompanied by decreasing amounts of sulfides, As-PGM, BM minerals, and Te-, Sb-, and Hg-based PGM (Figure 3B). The Pt–Fe alloys isoferroplatinum and tetraferroplatinum, along with iridium, osmium, erlichmanite, kashinite, cooperite-braggite, Ir–Rh–Pt thiospinels (cuproiridsite, cuprorhodsite, malanite), and unknown Ir–Ni sulfides (Table 2) are the major components of primary inclusions in fresh chromite. Euhedral laurite crystals have also been reported as primary inclusion in chromitites of the Uktus complex. Other PGM such as tulameenite, Cu–Pd–Pt alloys, potarite, irarsite, geversite, tolovkite, unknown Rh–Te (Table 2) are found exclusively as replacement of primary PGM, or associated with alteration assemblages (ferrianchromite, chlorite, serpentine) indicating that they formed in a low-temperature post-magmatic stage [23,24].

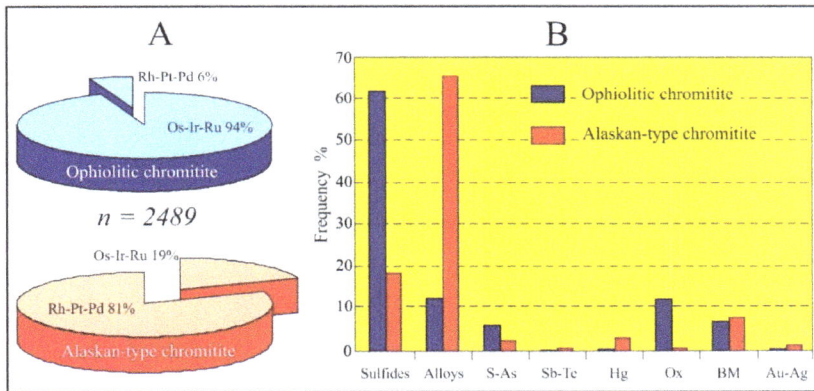

Figure 3. Platinum group minerals (PGM) abundance in the ophiolitic and Alaskan-type chromitites of the Urals. (**A**) Distribution of the PGM based on the dominant PGE. (**B**) Frequency of the PGM according to their mineralogical species.

4. Primary PGM in Ophiolitic Chromitites

4.1. Paragenetic Assemblages of PGM as Function of Sulfur Fugacity

The primary PGM inclusions in ophiolitic chromitites are generally less than 20 µm in size, and may occur either as solitary, polygonal crystals, or composite grains consisting of two or more PGM, with or without BM sulfide and silicate (Figure 4). Representative compositions of the PGM are given in Table 3. The paragenesis and composition of these PGM can be modelled according to a sequence of crystallization events controlled by relative stability of PGE alloys and sulfides as function of sulfur fugacity, $f(S_2)$, and temperature, T °C (Figure 5 [28,51,52]). In the mantle-hosted chromitites of Kempirsai and Ray–Iz, as well as in the cumulus chromitite of Nurali, primary PGM disulfides, with compositions ranging continuously between laurite (RuS_2) and erlichmanite (OsS_2), coexist with primary Os–Ir alloys characterized by Ru-poor compositions [20,28,39,53]. The paragenesis indicates that the $f(S_2)$ was initially close to the threshold for the formation of laurite (Ru + S_2 = RuS_2) coexisting with Os-Ir alloys. Increase of $f(S_2)$ with decreasing temperature results in the progressive stabilization of erlichmanite, various Ir–Ni sulfides, and Ni–Fe sulfides (Figure 5). Since an increment in $f(S_2)$ favours entering of Os in laurite [12], early laurite coexisting with Os–Ir alloys will have relatively high

Ru/Os ratio comprised between Ru_{100} and $Ru_{72}Os_{28}$, the latter corresponding to the unfractionated chondritic composition [53]. Further increase in $f(S_2)$ will cause progressive decrease of the Ru/Os ratio well below the chondritic value [53], and the composition of laurite will enter the erlichmanite field (Figure 4E–G). At Nurali, several laurite grains are zoned showing Os enrichment in the rim (Figure 4D). This has been interpreted as evidence supporting that the increase in f(S2) also accompanies fractional crystallization of chromitite in the cumulus pile of ophiolites [39]. The sulfarsenides and arsenides frequently appear as small particles attached to the external border of the Ir–Ni sulfides (Figure 4H), or overgrowing primary Os–Ir alloy crystals (Figure 4I). Although representing very accessory members of the primary assemblage of PGM inclusions, the crystallization of primary As-based PGM requires a relative increase of arsenic-fugacity in the latest stage of PGM precipitation at high temperature. Based on morphological considerations it would appear that the PGE alloys and sulfides were trapped in crystallizing chromite as solid particles or, at least in part, as liquid droplets adherent to solid crystals, for example, the Ru-arsenide enveloping the Os–Ir alloy (Figure 4I). This observation requires that the whole sequence of PGM crystallization alloy-sulfide-arsenide must have taken place in high thermal range between the crystallization temperature of Os-poor laurite (~1100 °C), and the solidus of chromite as the lower limit (see arrow T1 in Figure 5). However, some authors [27] have presented $f(S_2)/T$ °C lines trending along a much wider thermal gradient (see arrow T2 in Figure 5), well below the chromite solidus, implying that certain PGM might have split off the chromite in the subsolidus under long-lasting annealing conditions.

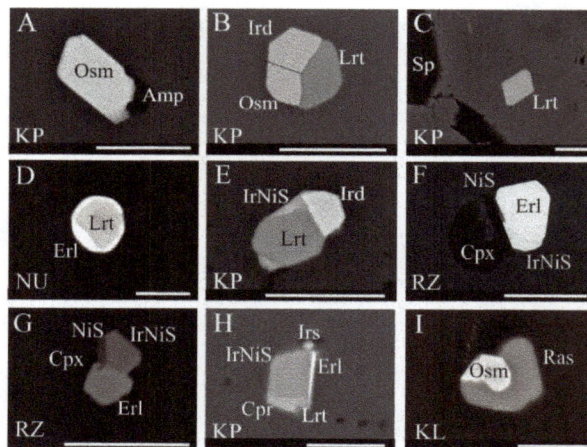

Figure 4. Back-scattered electron images of selected PGM included in chromite from the ophiolitic chromitites of the Urals. (**A**) Osmium associated with amphibole. (**B**) Bi-phase PGM composed of osmium and laurite. (**C**) single phase laurite. (**D**) Laurite rimmed by erlichmanite. (**E**) Complex PGM composed of laurite, iridium and an unnamed Ir and Ni sulfide. (**F**) Composite grain of erlichmanite, unnamed Ir and Ni sulfide, Ni sulfide in contact with clinopyroxene. (**G**) PGM consisting of erlichmanite, unnamed Ir and Ni sulfide and clinopyroxene. (**H**) Complex grain of laurite, erlichmanite, cuprorhodsite, unnamed Ir and Ni sulfide, and irarsite. (**I**) Bi-phase PGM composed of osmium and ruthenarsenide. Scale bar = 10 μm. Abbreviations: Osm = osmium, Amp = amphibole, Ird = iridium, Lrt = laurite, Sp = serpentine, Erl = erlichmanite, IrNiS = unnamed Ir and Ni sulfide, Cpx = clinopyroxene, NiS = Ni sulfide, Cpr = cuproiridsite, Irs = irarsite, Ras = ruthenarsenide. Abbreviations for the studied chromitites see Figure 1.

Minerals **2018**, *8*, 379

Table 3. Electron microprobe analyses (wt %) of PGM in ophiolitic chromitites of the Urals.

Sample Locality	PGM	Figure	Os	Ir	Ru	Rh	Pt	Pd	Fe	Ni	Cu	S	As	Tot
Kempirsai	osmium	4A	89.43	5.51	1.51	0.01	0.00	0.00	1.27	1.02	0.17	0.41	1.20	100.53
Kempirsai	iridium	4B	32.08	64.53	0.56	0.47	0.74	0.09	0.41	0.11	0.00	0.00	0.00	98.99
Kempirsai	osmium	4B	53.99	42.98	0.99	0.08	0.00	0.00	0.29	0.00	0.00	0.08	0.00	98.41
Kempirsai	laurite	4B	15.91	8.30	43.11	0.00	0.00	0.20	0.12	0.00	0.00	33.71	0.00	101.35
Kempirsai	laurite	4C	7.39	6.91	46.58	1.12	0.00	0.69	0.71	0.27	0.18	34.14	1.17	99.16
Kempirsai	laurite	4C	7.35	6.87	45.61	1.18	0.08	0.44	0.67	0.30	0.15	37.16	1.05	100.86
Kempirsai	iridium	4E	34.19	37.28	8.66	0.25	0.00	0.00	6.44	0.00	0.54	7.51	3.58	98.45
Kempirsai	cuproiridsite	4E	0.06	44.99	0.28	1.46	0.00	0.29	0.00	0.29	10.79	20.39	0.00	98.55
Kempirsai	laurite	4E	24.77	9.44	28.30	0.75	0.00	0.15	4.29	0.00	0.16	30.23	1.53	99.62
Kempirsai	laurite	4E	26.98	8.96	32.29	0.00	0.00	0.36	0.70	0.00	0.00	29.98	0.00	99.27
Kempirsai	Ir–Ni monosulfide	4H	0.00	43.16	0.00	0.34	0.00	0.13	6.81	18.31	6.81	25.51	0.00	101.07
Kempirsai	Ir–Ni monosulfide	4H	0.00	43.29	0.00	0.36	0.00	0.11	6.49	18.02	6.96	24.74	0.00	99.97
Kempirsai	Ir–Ni monosulfide	4H	0.00	45.89	0.00	0.46	0.00	0.08	5.77	14.28	7.78	24.82	0.00	99.08
Ray-Iz	laurite		14.51	7.57	39.21	1.13	1.37	0.30	0.04	0.00	0.00	34.86	0.00	98.99
Ray-Iz	laurite		4.94	8.17	50.23	0.86	0.00	0.00	0.31	0.23	0.13	35.45	0.00	100.32
Ray-Iz	laurite core		1.72	2.98	57.63	0.02	0.00	0.21	0.00	0.06	0.29	35.79	0.00	98.70
Ray-Iz	laurite rim		14.51	7.57	39.21	1.13	1.37	0.30	0.04	0.00	0.00	34.86	0.00	98.99
Ray-Iz	erlichmanite	4F	27.88	12.87	26.83	0.68	0.00	0.00	0.02	0.18	0.16	30.98	0.00	99.60
Ray-Iz	Ir–Ni sulfide	4F	0.04	44.35	0.00	1.22	0.96	0.21	3.93	18.26	4.52	26.33	0.12	99.94
Ray-Iz	erlichmanite	4G	45.11	10.71	13.42	0.06	0.00	0.04	0.13	0.11	0.05	31.50	0.50	101.63
Ray-Iz	Ir–Ni thiosp	4G	0.16	43.18	0.05	0.37	0.45	0.00	5.25	16.56	5.03	28.84	0.04	99.93
Ray-Iz	Ir–Ni thiospin		0.00	43.56	0.00	3.85	0.08	0.24	5.12	11.87	6.46	29.08	0.02	100.28
Ray-Iz	cuproiridsite		0.00	55.97	0.04	5.95	1.63	0.11	0.00	0.22	10.52	26.26	0.02	100.72
Kraka	laurite		14.26	6.61	39.13	0.83	0.00	0.57	0.00	0.00	0.08	36.54	0.99	99.01
Kraka	laurite		16.40	6.15	41.65	0.12	0.00	0.00	0.20	0.24	0.02	34.87	0.00	99.65
Nurali	laurite core	4D	19.10	5.41	39.92	0.00	0.00	0.00	0.36	0.35	0.00	34.88	0.00	100.02
Nurali	erlichmanite rim	4D	64.18	2.01	9.01	0.21	0.00	0.00	0.09	0.02	0.00	26.10	0.00	101.62
Kluchevsk	rutheniridosmine	4I	54.55	16.02	24.76	0.62	0.00	0.29	0.00	2.29	0.00	0.00	1.88	100.41
Kluchevsk	ruthenarsenite	4I	7.49	1.29	40.18	1.41	0.57	0.54	0.28	5.91	0.00	0.02	41.32	99.01

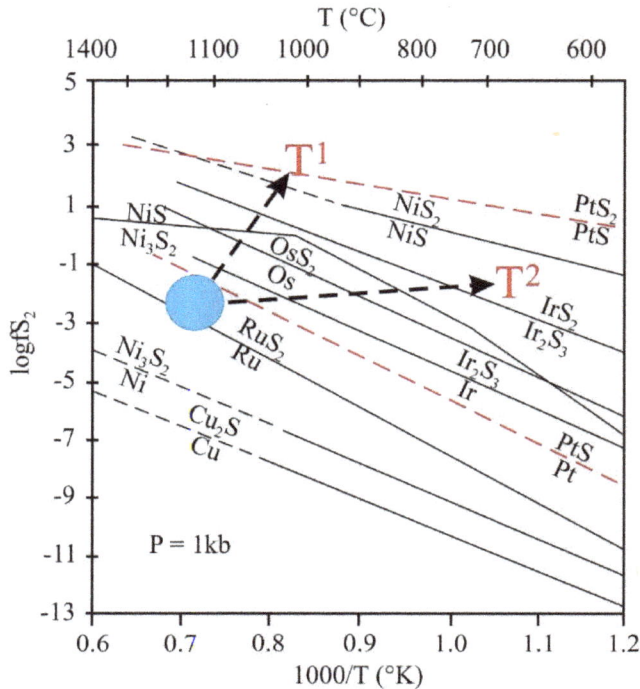

Figure 5. Metal-sulfide equilibrium curves for Ru, Os, Ir, Pt, Ni, and Cu as function of sulfur fugacity and temperature. Two different possible trends of the PGM precipitation are given [28,52].

4.2. Relationships with Chromite Composition

The oxygenthermobarometry [46,54] based on the olivine-spinel equilibrium [55] shows that chromitites of the Urals equilibrated in a wide range from 1510 to 590 °C. Discarding unrealistic values above 1400 °C and below 700 °C, most samples (e.g., Kempirsai) plot between 1400 and 700 °C, following a trend of increasing $Fe^3/(Fe^3 + Fe^2)$ and $\Delta logf(O_2)$ with decreasing temperature (Figure 6A) suggesting that the increase of fO_2 continued in the subsolidus [27]. Adjustment of the $Fe^3/(Fe^3 + Fe^2)$ ratio occurs by oxidation of Fe^2 inside the chromite itself as indicated by the negative correlation between $Fe^3/(Fe^3 + Fe^2)$ and $Fe^2/(Fe^2 + Mg)$ (Figure 6B).

The assemblage of PGM inclusions indicates that the $f(S_2)$ was forced to increase up to values suitable for the stabilization of high-energy PGM sulfides [56], and/or sulfidation of those PGE that were still residing in the chromite as atomic clusters [27]. The temperatures calculated for chromitites of Kraka are sensibly higher than those of Kempirsai (1310–1090 °C). In these chromitites, laurite shows limited Ru–Os substitution, and occurs as isolated euhedral crystals, sometimes associated with primary mafic silicates, mainly clinopyroxene. The Ir–Ni sulfides are conspicuously absent, indicating precipitation of the PGM at lower sulfur fugacity compared with Kempirsai and Ray–Iz [40]. Composition and textural characters of these laurites are totally similar to the laurite grain in Figure 4C, supporting that these grains formed in the magma prior to chromite crystallization, and escaped re-equilibration at low temperature. In conclusion, the data presented here support that laurite can precipitate from the magma at relatively low fS_2, and is enclosed in the chromite as a solid particle. Under a high cooling-rate (line T1 in Figure 5) the fS_2 increases rapidly, stabilizing the expected sequence of PGM sulfides prior to the crystallization of chromite. In the lack of annealing, there will be no reequilibration-adjustment of the chromite composition and limited chance for the formation of PGM inclusions in the subsolidus.

On the contrary, post-magmatic equilibration under slow cooling-rate, as it was the case in the Kempirsai chromite deposits, will favour formation of a high number of PGM inclusions characterized by a variegate composition.

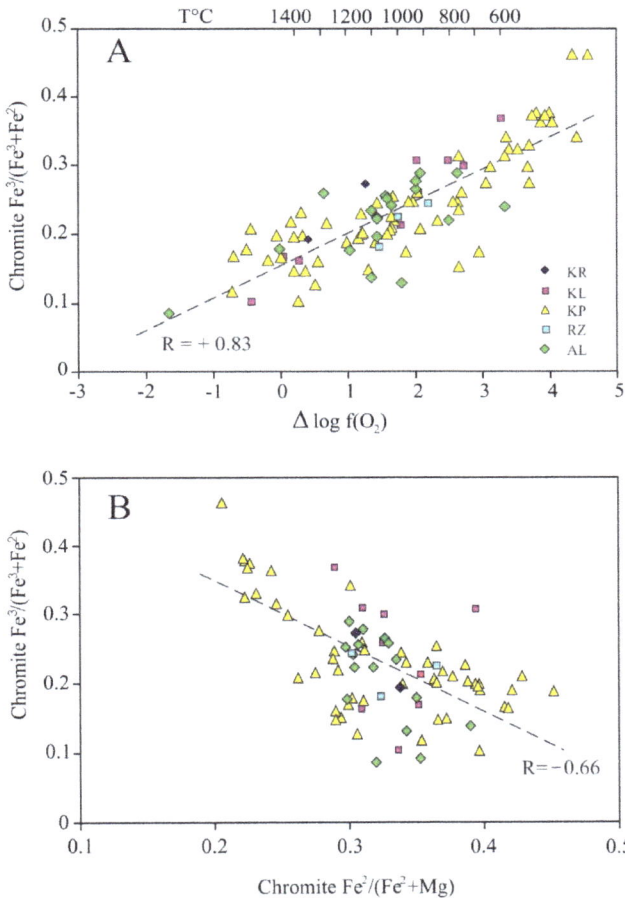

Figure 6. (**A**) Variation of the chromite oxidation ratio [$Fe^3/(Fe^3 + Fe_2)$] as function oxygen fugacity and temperature. Oxygen fugacity is expressed as deviation from the fayalite-magnetite-quartz (FMQ) buffer. The temperature scale is based on the equation T (°C) = [Δlog(O_2) − 5.6231]/−0.0039. See Figure 1 for the abbreviation of names of ophiolite and Alaskan-type complexes. (**B**) Negative correlation between the oxidation ratio and the #Fe^2 number of chromite. See text for explanation.

5. Primary PGM in Alaskan-Type Chromitites

5.1. Primary PGM and Sulfur Fugacity in Alaskan-Type Chromitites

Primary PGM in "syngenetic" chromitites of the Urals occur as minute disseminated grains (1–35 µm) enclosed in fresh chromite, as either single-phase crystals, or forming composite aggregates of more PGM, sometimes with associated silicates.

Most common mineral assemblages and compositions of the PGM inclusions are given in Figure 7 and Table 4, respectively.

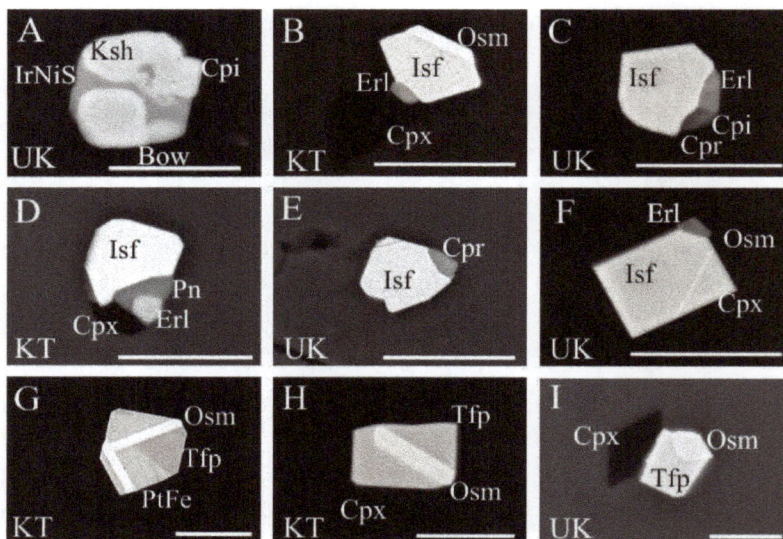

Figure 7. Back-scattered electron images of representative PGM included in chromite from the Alaskan-type chromitites of the Urals. (**A**) Composite crystal composed of kashinite rimmed by bowieite, cuproiridsite, and an unnamed Ir and Ni sulfide. (**B**) PGM consisting of isoferroplatinum, erlichmanite, and osmium in contact with clinopyroxene. (**C**) Complex grain of isoferroplatinum, cuprorhodsite, cuproiridsite and erlichmanite. (**D**) Isoferroplatinum in contact with erlichmanite, pentlandite, and clinopyroxene. (**E**) Bi-phase grain of isoferroplatinum and cuprorhodsite. (**F**) PGM composed of isoferroplatinum, osmium, and erlichmanite in contact with clinopyroxene. (**G**) Complex crystal of tetraferroplatinum, osmium, and a Pt–Fe alloy. (**H,I**) Bi-phase grains of tetraferroplatinum and osmium in contact with clinopyroxene. Scale bar = 10 µm. Ksh = kashinite, Cpi = cuproiridsite, Bow = bowieite, IrNiS = unnamed Ir and Ni sulfide, Isf = isoferroplatinum, Osm = osmium, Erl = erlichmanite, Cpx = clinopyroxene, Cpr = cuprorhodsite, Pn = pentlandite, Tfp = tetraferroplatinum, PtFe = Pt–Fe alloy. Abbreviations for the studied chromitites see Figure 1.

The predominance of Pt–Fe alloys over sulfides provides evidence for the crystallization of primary PGM under fS_2 as low as to prevent formation of Pt sulfides. Primary precipitation of sulfides at high temperature is reported exclusively from the Uktus chromitite where laurite with low Os content and kashinite with Ir–Ni–sulfide (Figure 7A) occur included in fresh chromite. The study of Kytlym and Uktus chromitites [23] allows identification of three groups of primary Pt–Fe alloys (Figure 8). The most abundant alloys consist of isoferroplatinum Pt_3Fe and tetraferroplatinum PtFe containing less than 3 at % Ni + Cu. These alloys, mainly Pt_3Fe, are preferentially accompanied by a sulfide-rich assemblage, erlichmanite, Ir–Rh–Cu–Ni thiospinels, and pentlandite occurring as epitaxic overgrowth at the alloy's boundary (Figures 7B–F). This indicates that the alloys were a stable phase under relatively high fS_2 capable of stabilizing the suite of PGM sulfides reported in the pictures. In contrast, isoferroplatinum and tetraferroplatinum both enriched in Ni and Cu, would appear to have formed at relatively low fS_2 as suggested by failure to crystallize Ni–Cu Ir-thiospinels and erlichmanite (Figure 7G–I).

Table 4. Electron microprobe analyses (wt %) of PGM in ophiolitic chromitites of the Urals.

Locality	PGM	Figure	Os	Ir	Ru	Rh	Pt	Pd	Fe	Ni	Cu	S	As	Tot
Uktus	Laurite		8.12	5.46	47.50	2.16	0.00	0.00	0.74	0.04	0.00	35.60		99.62
Uktus	Laurite		23.20	7.91	34.80	0.60	0.00	0.00	0.62	0.01	0.00	31.90		99.04
Uktus	kashinite	7A	1.54	52.40	0.30	18.40	0.28	0.00	1.40	0.37	0.18	22.90		97.77
Uktus	Ir–Ni–S	7A	0.32	34.50	0.01	10.10	0.43	0.00	6.43	14.50	3.84	25.20		95.33
Uktus	Pt$_3$Fe	7C	0.23	1.49	0	0.79	84.90	0	8.50	0.28	0.36	0.00	0.02	96.57
Uktus	cuprorhodsite	7E	0.21	32.20	0.00	26.30	0.92	0.00	0.48	0.98	9.36	26.90	0.04	97.39
Uktus	Pt$_3$Fe	7E	0	0.96	0	0.36	88.70	0.33	8.90	0.28	0.21	0.00	0	99.74
Uktus	Pt$_3$Fe	7F	0.18	1.33	0.01	0.88	84.10	0.17	8.40	0.35	0.23	0.00	0.13	95.78
Uktus	Osmium	7I	96.90	1.29	0.18	0.13	2.00	0.00	0.02	0.04	0.00	0.00	0.01	100.57
Uktus	Pt$_3$Fe	7I	0.13	0.80	0.05	1.19	84.56	0.02	10.27	0.28	0.39		0.04	97.68
Kytlym	Pt$_3$Fe	7B	2.5	2.04	0.09	1.35	84.76	0.19	8.56	0.23	0.19	0.00	0.04	99.95
Kytlym	erlichmanite	7B	42.01	3.55	16.28	2.31	6.77	0.23	0	0.09	0.01	26.04	0	97.29
Kytlym	Pt$_3$Fe	7D	0.18	0.24		0.62	87.30	0.86	9.00	0.39	0.45			99.04
Kytlym	erlichmanite	7D	53.33	3.16	5.53	5.08	0.00	0.65	0.00	2.11	0.28	26.96	0.00	97.10
Kytlym	Rh-Ir sulfide	7D	11.90	12.40	0.63	13.30	6.61	0.00	11.90	9.76	4.04	26.30	0.02	96.86
Kytlym	PtFe	7G	0.11	0.87	0.04	1.29	83.10	0.33	12.00	2.98	1.09			101.81
Kytlym	Osmium	7H	87.00	6.24	0.40	0.44	2.98	0.07	0.29	0.19	0.00		0.05	97.66
Kytlym	PtFe	7H	0.18	1.01	0.00	1.04	75.14	0.33	14.52	4.84	2.37	0.00	0.11	99.54

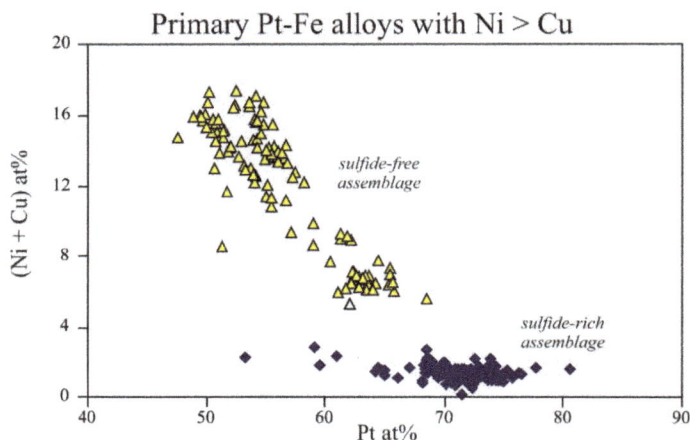

Figure 8. Variation of the (Ni + Cu) content in primary Pt–Fe alloys occurring in sulfide-free or sulfide-rich PGM assemblages from the syngenetic chromitites of Kytlym and Uktus. Under low sulfur fugacity (sulfide free assemblage) Ni and Cu cannot form independent sulfides but are forced to enter the alloy structure substituting for platinum as supported by the Pt-(Ni + Cu) negative correlation (R = −0.9).

According to the authors [23], the fS_2 exerts strong influence on the composition and paragenetic assemblage of primary Pt–Fe alloys crystallizing at high temperature. At low fS_2 both Ni and Cu are forced to enter the alloy structure in substitution for Pt, as indicated by the Pt/(Ni + Cu) negative correlation in Figure 8. At relatively higher fS_2 Ni and Cu tend to form independent sulfides with Ir and Rh overgrowing the Pt–Fe alloys. An Os-alloy is usually present as exsolved lamellae in the Pt–Fe alloys. Its textural relations support the inferred variation of fS_2 during crystallization of primary inclusions. The osmium lamellae are very small or absent in Pt–Fe alloys formed at high fS_2, in which Os is mainly carried in primary erlichmanite. On the contrary, large osmium lamellae are almost ubiquitous in Ni–Cu rich Pt–Fe alloys occurring in the sulfide free PGM assemblage (Figure 7). The composition of primary laurite incusions of the Uktus chromitites remain confined to Os concentrations lower than about 10 at%, therefore indicating crystallization at sulfur fugacities $\log fS_2$ close to −1.3 and a temperature of the order of 1300 °C, for the near-end-member laurite [24]. The Figure 5 shows that these conditions are comparable with those obtained for initial PGM precipitation within upper mantle chromitites of the ophiolites [22,53]. The presence of erlichmanite in the PGM paragenesis indicates that sulfur fugacity reached the Os–OsS$_2$ reaction line at about $\log fS_2$ = 1.0 and 1100 °C, but did not increase futher-above this limit.

5.2. The Role of Oxygen Fugacity and Temperature

Results of the olivine-spinel thermobarometer indicate that the crystallization-equilibration of syngenetic chromitite and accessory chromite in dunite follows a unique trend (Figure 9A) indicating that oxygen fugacity fO_2 was increasing during fractionation of dunite to massive chromitite [24]. At the same time, the data support that massive chromitite equilibrated at higher temperature compared with the accessory chromite in dunite, possibly due to higher olivine/chromite mass ratio in the latter. In [24] it was shown that the oxidation ratio $Fe^{3+}/(Fe^{3+} + Fe^{2+})$ in the Uktus chromitites correlates positively with the increase of oxygen fugacity in the system and, significantly, it displays distinct correlation with variation of the assemblage of primary PGM inclusions in chromite (Figure 9B). The magnesiochromite with low oxidation ratio $[Fe^{3+}/(Fe^{3+} + Fe^{2+}) = 0.23–0.25]$ characterize the chromitites in the lowermost dunite body of the complex, where the PGM inclusions display high concentration of Ru–Os–Ir sulfides (laurite, kashinite, cuproiridsite) and the extreme paucity or absence of Pt-minerals. With proceeding differentiation, the oxidation ratio of the chromite increases up to $Fe^{3+}/(Fe^{3+} + Fe^{2+}) = 0.44$, and abundant Pt–Fe alloys (isoferroplatinum, tetraferroplatinum) start to crystallize with the sulfide rich assemblage described above. The most differentiated chromitite characterizes for a Fe-rich composition and high oxidation ratio [$Fe^{3+}/(Fe^{3+} + Fe^{2+}) = 0.59$] resembling "chromian titanomagnetite". The PGM are mainly composed of Pt and Pd as expected for the common trend of PGE magmatic fractionation, although the minerals mainly occupy the interstitial space among chromite grains and occur in typical secondary PGM assemblage characterized by abundance of As- Sb-, Te-phase [24]. The isoferroplatinum and sulfide PGM assemblage of Kytlym occurs in chromitite with an oxidation ratio of $Fe^{3+}/(Fe^{3+} + Fe^{2+}) = 0.45$. In contrast, syngenetic chromites with a relatively high oxidation ratio [$Fe^{3+}/(Fe^{2+} + Fe^{3+}) = 0.52$] contain Ni–Cu rich Pt–Fe alloys but no sulfide (Figure 9B), indicating that fO_2 was increasing concomitant with a significant depression in fS_2 [23].

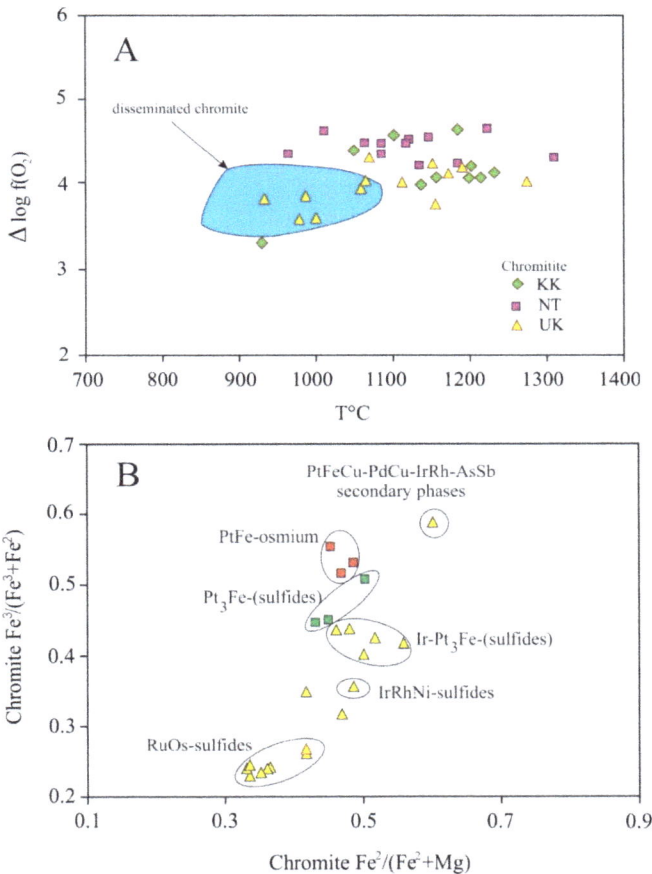

Figure 9. (**A**) Variation of oxygen fugacity as function of temperature in syngenetic chromitite and accessory chrome spinel disseminated in the host dunite. (**B**) The PGM assemblage in the syngenetic chromitites of Uktus (triangle) and Kytlym (square) evolves from enriched in Ru–Os and Ir sulfides into Pt–Fe alloy dominated with increasing oxydation ratio of the chromite.

5.3. Thermodynamic Conditions for Precipitation of Primary Pt–Fe Alloys

Considering the results of olivine-chromite thermobarometry in the Alaskan-type syngenetic chromitites of the Urals [24,33,46,54] and the maximum stability temperature of Pt–Fe alloys in a S-poor system, we may assume that the Pt–Fe alloys occurring as primary inclusions were trapped in the chromite as solid crystals, at temperatures in the range of 1050–1300 °C, and oxygen fugacity from +2.1 to +4.9 deltalog(O_2) above the FMQ buffer. These emphasizes the anomaluos behavior of Pt that co-precipitates with the refractory Ir, and is not removed from the melt together with the companion Rh and Pd as a result of the segregation of a magmatic sulfide liquid [24,57]. The discrepancy with chromitites from other geological settings (i.e., the continental layered intrusions) is evident, and has received diverse explanations. Among others, one possible model assumes that the increase in $f(O_2)$ required for the crystallization of chromite might have been responsible for the sharp drop of Pt solubility in the silicate melt, causing precipitation of the Pt–Fe alloys [58,59]. However, reversing this cause-effect order, [24] proposed that it was the strong tendency of Pt to combine with Fe to form Pt–Fe alloys that caused the Pt-solubility falling down. According to this mechanism, the extensive stabilization of Pt–Fe alloys at high temperature may actually reflect the anomalous increase of the FeO and Fe_2O_3 activity in the magma parent to Alaskan-type chromitites, that is a major consequence expected from the SiO_2-undersaturation condition of these melts. The effect on chromite composition would be the incorporation of larger amounts of magnetite component ($FeOFe_2O_3$) in the chromite structure, thus simulating an increase of $f(O_2)$ in the system [60]. For this reason, primary precipitation of Pt–Fe occurs preferentially in mafic magmas having olivine (not orthopyroxene) and chromite with relatively high oxidation ratio, $Fe^{3+}(Fe^{2+} + Fe^{3+})$, on the liquidus: the Alaskan-type magmas.

6. Conclusions

The review based on examination of more than 2500 analyses of PGM associated with ophiolitic and Alaskan-type chromitites of the Urals reveals that mineralogy of PGM crystallizing at high temperature is controlled by: (1) the nature of the parent melt and relative concentrations of PGE; (2) the presence of melt-soluble clusters of PGE in the parent melt; and (3) by the chemical-physical conditions such as temperature, sulfur- and oxygen-fucagity, prevailing during their precipitation.

The mineralogy of PGM inclusions in ophiolitic and Alaskan-type chromitites of the Urals is consistent with reported whole-rock PGE concentrations. In particular, the ophiolitic chromitites contain abundant Os–Ir–Ru minerals and rare Rh–Pt–Pd phases, whereas the most abundant PGM in the Alaskan-type chromitites are Pt–Fe alloys accompanied by minor Ir–Os–Rh phases.

The most important factors controlling the precipitation of PGM in ophiolitic chromitites are temperature and sulfur fugacity. The mineralogical assemblage shows that the chromitites formed in a narrow range of temperature (1310–1090 °C) are characterized by the presence of PGM, such as laurite and alloys in the Os–Ir–Ru system, that require relatively low sulfur fucagity to precipitate. The chromitites that have suffered compositional re-equilibration in a wide thermal range (e.g., 1400–700 °C) contain a great number of PGM, including erlichmanite and Ir–Ni–sulfides. This observation suggests that the post-magmatic equilibration under slow cooling-rate was responsible for the increasing of the sulfur fugacity and the formation of a volatile rich fluid, thus promoting precipitation of a variegate suite of PGM comprising alloys, sulfides, sulfarsenides, and arsenides. In this post-magmatic stage, all the PGE that were present in solid chromite as dispersed atomic clusters could easily be converted into discrete PGM inclusions splitting off the chromite structure.

The predominance of Pt–Fe alloys over sulfides in the Alaskan-type chromitites indicates that the crystallization of magmatic PGM occurred under fS2 as low as to prevent formation of Pt sulfides but was high enough to crystallize erlichmanite and Ir–Ru–Ni–Cu sulfides. The key factor for the precipitation of abundant Pt–Fe alloys in the Alaskan-type chromitites is the SiO_2-undersaturation condition of the parent melt and the oxygen fugacity that was increasing during fractionation of dunite to massive chromitite. The estimated temperatures suggest that the range of crystallization of PGM in the primary magmatic stage was comprised between 1050 and 1300 °C.

In summary, the whole scenario provides further support to the conclusion that the majority of primary PGM inclusions in chromite formed in situ as part of the chromite precipitation event, and eventually was modified during post-magmatic, slow cooling conditions. Only a few high refractory PGM might have formed during partial melting in the deep mantle source, being transported as suspended solid particles to the site of chromite deposition

Author Contributions: F.Z. and G.G. wrote and organized the paper. E.P. collected some of the studied samples and provided the geological and field information. O.T. revised the English. All the authors contributed to the interpretation of the data and to organize the manuscript.

Funding: E.P is grateful to the grant by the Russian Foundation of Basic Research (RFBR grant No. 16-05-00508-a).

Acknowledgments: The Editorial staff of *minerals* is thanked for the efficient assistance. The comments of two anonymous referees improved the manuscript. The University Centrum for Applied Geosciences (UCAG) is thanked for the access to the E. F. Stumpfl electron microprobe laboratory. This manuscript is dedicated to German Fershtater, Ural Branch of the Russian Academy of Science.

References

1. Barnes, S.J.; Naldrett, A.J.; Gorton, M.P. The origin of the fractionation of Platinum-group elements in terrestrial magmas. *Chem. Geol.* **1985**, *53*, 303–323. [CrossRef]

2. Cabri, L.J.; Naldrett, A.J. The nature, distribution, and concentrations of platinum-group elements in various geological environments. In Proceedings of the 27th International Geological Congress, Moscow, Russia, 4–14 August 1984; pp. 10–27.

3. Crocket, J.H. Geochemistry of the Platinum-Group Elements. In *Platinum-Group Elements: Mineralogy, Geology, Recovery*; Cabri, L.J., Ed.; Geology Division of CIM: Ottawa, ON, Canada, 1981; pp. 47–64.

4. Legendre, O.; Augé, T. Mineralogy of platinum-group mineral inclusions in chromitites from different ophiolite complexes. In *Metallogeny of Basic and Ultrabasic Rocks*; Gallagher, M.J., Ixer, R.A., Neary, C.R., Prichard, H.M., Eds.; Springer: Amsterdam, The Netherlands, 1986; pp. 361–372.

5. Page, N.J.; Cassard, D.; Haffty, J. Palladium, platinum, rhodium, ruthenium and iridium in chromitites from the Massif du Sud and Tiebaghi massif, New Caledonia. *Econ. Geol.* **1982**, *77*, 1571–1577. [CrossRef]

6. Talkington, R.W.; Watkinson, D.M.; Whittaker, P.J.; Jones, P.C. Platinum-group minerals and other solid inclusions in chromite of ophiolitic complexes: Occurrence and petrological significance. *Tschermaks Mineralogische Petrogr. Mitt.* **1984**, *32*, 285–301. [CrossRef]

7. Rudashevskiy, N.S. Origin of various types of platinoid mineralization in ultramafic rocks. *Int. Geol. Rev.* **1987**, *29*, 465–480. [CrossRef]

8. Garuti, G.; Zaccarini, F. In situ alteration of platinum-group minerals at low temperature: evidence from serpentinized and weathered chromitite of the Vourinos Complex, Greece. *Can. Mineral.* **1997**, *35*, 611–626.

9. Capobianco, C.J.; Drake, M.J. Partitioning of ruthenium, rhodium, and palladium between spinel and silicate melt and implications for platinum-group element fractionation trends. *Geochim. Cosmochim. Acta* **1990**, *54*, 869–874. [CrossRef]

10. Garuti, G.; Gazzotti, M.; Torres-Ruiz, J. Iridium, rhodium, and platinum sulfides in c chromitites from the ultramafic massifs of Finero, Italy, and Ojen, Spain. *Can. Mineral.* **1995**, *33*, 509–520.

11. Constantinides, C.C.; Kingston, G.A.; Fisher, P.C. The occurrence of platinum group minerals in the chromitites of the Kokkinorotsos chrome mine, Cyprus. In *Ophiolites, Proceedings of the International Ophiolite Symposium, Cyprus 1979*; Panayiotou, A., Ed.; Geological Survey Department: Nicosia Cyprus, 1980; pp. 93–101.

12. Stockman, H.W.; Hlava, P.F. Platinum-group minerals in Alpine chromitites from southwestern Oregon. *Econ. Geol.* **1984**, *79*, 491–508. [CrossRef]

13. Augé, T.; Johan, Z. Comparative study of chromite deposits from Troodos, Vourionos, North Oman and New Caledonia ophiolites. In *Mineral Deposits within the European Community*; Spec. publication n. 6, of the Society for Geology Applied to Mineral Deposits; Boissonnas, J., Omenetto, P., Eds.; Springer: Berlin/Heidelberg, Germany, 1988; pp. 267–288.

14. Tredoux, M.; Lyndsay, N.M.; Devis, G.; MacDonald, I. The fractionation of platinum-group elements in magmatic systems, with the suggestion of a novel causal mechanism. *S. Afr. J. Geol.* **1995**, *98*, 157–167.

15. Garuti, G. Chromite-platinum-group element magmatic deposits. In *Geology, Encyclopedia of Life Support Systems (EOLSS) UNESCO*; De Vivo, B., Stäwe, K., Grasemann, B., Eds.; Eolss Publisher: Oxford, UK, 2004.

16. Razin, L.V. Geologic and genetic features of forsterite dunites and their platinum-group mineralization. *Econ. Geol.* **1976**, *71*, 1371–1376. [CrossRef]

17. Makeyev, A.B.; Kononkova, N.N.; Kraplya, E.A.; Chernukha, F.P.; Bryanchaninova, N.I. Platinum Group Minerals in alluvium of the Northern Urals and Timan: The key to primary sources of platinum. *Trans. Acad. Sci. Earth Sci. Sect.* **1997**, *353*, 181–184.

18. Anikina, Ye.V.; Pushkarev, E.V.; Garuti, G.; Zaccarini, F.; Cabella, R. *The Evolution of Chrome Spinel Composition and PGE Minerals in the Dunite of the Uktus Massif (Middle Urals)*; Institute of Geology and Geochemistry Ekaterimburg: Ekaterimburg, Russia, 1996; p. 9. (In Russian)

19. Augé, T.; Genna, A.; Legendre, O.; Ivanov, K.S.; Volchenko, Y.A. Primary platinum mineralization in the Nizhny Tagil and Kachkanar ultramafic complexes, Urals, Russia: A genetic model for PGE concentration in chromite-rich zones. *Econ. Geol.* **2005**, *100*, 707–732. [CrossRef]

20. Distler, V.; Kryachko, V.V.; Yudovskaya, M.A. Ore petrology of chromite-PGE mineralization in the Kempirsai ophiolite complex. *Mineral. Petrol.* **2008**, *92*, 31–58. [CrossRef]

21. Garuti, G.; Zaccarini, F.; Cabella, R.; Fershtater, G.B. Occurrence of unknown Ru–Os–Ir–Fe oxide in the chromitites of the Nurali ultramafic complex, southern Urals, Russia. *Can. Mineral.* **1997**, *35*, 1431–1440.

22. Garuti, G.; Zaccarini, F.; Moloshag, V.; Alimov, V. Platinum-group minerals as indicators of sulfur fugacity in ophiolitic upper mantle: An example from chromitites of the Ray–Iz ultramafic complex, Polar Urals, Russia. *Can. Mineral.* **1999**, *37*, 1099–1116.

23. Garuti, G.; Pushkarev, E.; Zaccarini, F. Composition and paragenesis of Pt-alloys from chromitites of the Ural-Alaskan-Type Kitlim and Uktus complexes, northern and central Urals, Russia. *Can. Mineral.* **2002**, *40*, 1127–1146. [CrossRef]

24. Garuti, G.; Pushkarev, E.; Zaccarini, F.; Cabella, R.; Anikina, E. Chromite composition and platinum-group mineral assemblage in the Uktus Uralian-Alaskan-type complex (Central Urals, Russia). *Mineral. Deposita* **2003**, *38*, 312–326. [CrossRef]

25. Grieco, G.; Diella, V.; Chaplygina, N.L.; Savelieva, G.N. Platinum group elements zoning and mineralogy of chromitites from the cumulate sequence of the Nurali massif (Southern Urals, Russia). *Ore Geol. Rev.* **2007**, *30*, 257–276. [CrossRef]

26. Ivanov, O.K. *Zoned Ultramafic Complexes of the Urals (Mineralogy, Petrology, Genesis)*; Uralian University Publishing House: Ekaterinburg, Russia, 1997; p. 488. (In Russian)

27. Melcher, F.; Stumpfl, E.F.; Simon, G. Platinum-group minerals and associated inclusions in chrome spinel of the Kempirsai ultramafic massif, Southern Urals, Kazakhstan. In *Mineral Deposits*; Pašava, K.Z., Ed.; Balkema: Rotterdam, The Netherlands, 1995; pp. 153–156.

28. Melcher, F.; Grum, W.; Simon, G.; Thalhammer, T.V.; Stumpfl, E.F. Petrogenesis of the ophiolitic giant chromite deposits of Kempirsai, Kazakhstan: a study of solid and fluid inclusions in chromite. *J. Petrol.* **1997**, *38*, 1419–1458. [CrossRef]

29. Melcher, F. Base metal-platinum group element sulfides from the Urals and the Eastern Alps: Characterization and significance for mineral systematics. *Mineral. Petrol.* **2000**, *68*, 177–211. [CrossRef]

30. Moloshag, V.P.; Smirnov, S.V. Platinum mineralization of the Nurali mafic-ultramafic massif (Southern Urals). *Notes Rus. Mineral. Soc. Part.* **1996**, *125*, 48–54. (In Russian)

31. Pašava, J.; Knésl, I.; Vymazalová, A.; Vavřín, I.; Ivanovna Gurskaya, L.; Ruslanovich Kolantsev, L. Geochemistry and mineralogy of platinum-group elements (PGE) in chromites from Centralnoye I, Polar Urals, Russia. *Geosc. Front.* **2011**, *2*, 81–85. [CrossRef]

32. Pushkarev, E.V. *Petrology of the Uktus Dunite-Clinopyroxenite-Gabbro Massif (the Middle Urals)*; Russian Academy of Sciences, Ural Branch, Institute of Geology and Geochemistry: Ekaterinburg, Russia, 2000; pp. 1–296. (In Russian)

33. Pushkarev, E.V.; Anikina, E.V. Low temperature origin of the Ural-Alaskan type platinum deposits: Mineralogical and geochemical evidence. In Proceedings of the 9th International Platinum Symposium, Billings, Montana, MT, USA, 21–25 July 2002; pp. 387–390.

34. Pushkarev, E.V.; Anikina, E.V.; Garuti, G.; Zaccarini, F. Chromium-Platinum deposits of Nizhny-Tagil type in the Urals: Structural-substantial characteristic and a problem of genesis. *Litosfera.* **2007**, *3*, 28–65. (In Russian)

35. Sedler, I.K.; Wipfler, E.L.; Anikina, E.V. Platinum group minerals and associated chrome-spinels of the Alaskan-type Nizhny Tagil massif, Middle Urals. In *Mineral Deposits: Processes to Processing*; Stanley, C.J., Ed.; Taylor & Francis: Rotterdam, The Netherlands, 1999; pp. 787–790.

36. Smirnov, S.V.; Moloshag, V.P. Two types of Platinum deposits of the Nuraly ultramafic pluton (South Urals). In Proceedings of the VII International Platinum Symp, Moskow, Russia, 1–4 August 1994.

37. Thalhammer, T.V. The Kempirsai Ophiolite Complex, South Urals. Petrology, Geochemistry, Platinum-Group Minerals, Chromite Deposits. Ph.D. Thesis, University of Leoben, Leoben, Austria, 1996.

38. Zaccarini, F. Comparative Study of Platinum-Group Minerals in Podiform Chromitites from Mesozoic, Paleozoic, and Precambrian Ophiolite Complexes: Examples from the Mediterranean Area, the Ural Chain, and the Egyptian Eastern Desert. Ph.D. Thesis, University of Bologna, Bologna, Italy, 1999; p. 135. (In Italian)

39. Zaccarini, F.; Pushkarev, E.; Fershtater, G.B.; Garuti, G. Composition and mineralogy of PGE-rich chromitites in the Nurali Lherzolite-gabbro complex, southern Urals, Russia. *Can. Mineral.* **2004**, *42*, 545–562. [CrossRef]

40. Zaccarini, F.; Pushkarev, E.; Garuti, G. Platinum-group element mineralogy and geochemistry of chromitite of the Kluchevskoy ophiolite complex, central Urals (Russia). *Ore Geol. Rev.* **2008**, *33*, 20–30. [CrossRef]

41. Zaccarini, F.; Garuti, G.; Pushkarev, E. V. Unusually PGE-rich chromitite in the Butyrin vein of the Kytlym Uralian-Alaskan complex, Northern Urals, Russia. *Can. Mineral.* **2011**, *49*, 52–72. [CrossRef]

42. Zaccarini, F.; Garuti, G.; Bakker, R.J.; Pushkarev, E.V. Electron microprobe and Raman Spectroscopy investigation of an oxygen-bearing Pt–Fe–Pd–Ni–Cu compound from Nurali chromitite (Southern Urals, Russia). *Microsc. Microanal.* **2015**, *21*, 1070–1079. [CrossRef] [PubMed]

43. Zaccarini, F.; Bindi, L.; Pushkarev, E.V.; Garuti, G.; Bakker, R.J. Multi-analytical characterization of minerals of the bowieite-kashinite series from Svetly Bor complex, Urals (Russia) and comparison with worldwide occurrences. *Can. Mineral.* **2015**, *54*, 461–473. [CrossRef]

44. Zaccarini, F.; Pushkarev, E.V.; Garuti, G.; Kazakov, I. Platinum-Group Minerals and other accessory phases in chromite deposits of the Alapaevsk ophiolite, Central Urals, Russia. *Minerals* **2016**, *6*, 108. [CrossRef]

45. Zoloev, K.K.; Volchenko, Yu.A.; Koroteev, V.A.; Malakhov, I.A.; Mardirosyan, A.N.; Khripov, V.N. *Platinum Ores in Different Complexes of the Urals*; Ural State University Press: Ekaterinburg, Russia, 2001; p. 199. (In Russian)

46. Garuti, G.; Pushkarev, E.; Thalhammer, O.A.R.; Zaccarini, F. Chromitites of the Urals (Part 1): Overview of chromite mineral chemistry and geo-tectonic setting. *Ofioliti* **2012**, *37*, 27–53.

47. Betekhtin, A.G. Mikroskopische Untersuchungen an Platinerzen aus dem Ural. *N. Jb. Miner. Abh.* **1961**, *97*, 1–34.

48. Garuti, G.; Pushkarev, E.V.; Zaccarini, F. Diversity of chromite-PGE mineralization in ultramafic complexes of the Urals. In Proceedings of the Platinum-Group Elements—From Genesis to Beneficiation and Environmental Impact: 10th International Platinum Symposium, Oulu, Finland, 8–11 August 2005; Geological Survey of Finland: Esbo, Finland, 2005.

49. Kojonen, K.; Zaccarini, F.; Garuti, G. Platinum-Group-Elements and gold geochemistry and mineralogy in the Ray–Iz ophiolitic chromitites, Polar Urals. In *Mineral Exploration and Sustainable Development*; Eliopoulos, D.G., Ed.; Millpress: Rotterdam, The Netherlands, 2003; pp. 599–602.

50. Naldrett, A.J.; Duke, J.M. Platinum metals in magmatic sulfide ores. *Science* **1980**, *208*, 1417–1424. [CrossRef] [PubMed]

51. Wood, S.A. Thermodynamic calculation of the volatility of the platinum group elements (PGE): The PGE content of fluids at magmatic temperatures. *Geochim. Cosmochim. Acta.* **1987**, *51*, 3041–3050. [CrossRef]

52. Ferrario, A.; Garuti, G. Platinum-group mineral inclusions in chromitites of the Finero mafic-ultramafic complex (Ivrea-Zone, Italy). *Mineral. Petrol.* **1990**, *41*, 125–143. [CrossRef]

53. Garuti, G.; Zaccarini, F.; Economou-Eliopoulos, M. Paragenesis and composition of laurite from chromitites of Othrys (Greece): Implication for Os–Ru fractionation in ophiolitic upper mantle of the Balkan Peninsula. *Mineral. Deposita.* **1999**, *34*, 312–319. [CrossRef]

54. Chashchukhin, I.S.; Votyakov, S.L.; Uimin, S.G. Oxygen thermometry and barometry in chromite-bearing ultramafic rocks: An exmple of ultramafic massis on the Urals. II. Oxidation state of ultramafics and the composition of mineralizing fluids. *Geochem. Int.* **1998**, *36*, 783–791.

55. Ballahaus, C.; Berry, R.F.; Green, D.H. High-pressure experimental calibration of the olivine-orthopyroxene.spinel geobarometer: Implications for the oxidation state of the upper mantle. *Contr. Mineral. Petrol.* **1991**, *107*, 27–40. [CrossRef]

56. Westland, A.D. Inorganic chemistry of the platinum group elements. In *Platinum Group Elements: Mineralogy, Geology, Recovery*; Cabri, L.J., Ed.; Geology Division of CIM: Ottawa, ON, Canada, 1981; pp. 5–18.

57. Garuti, G.; Fershtater, G.B.; Bea, F.; Montero, P.; Pushkarev, E.; Zaccarini, F. Platinum-group elements as petrological indicators in mafic-ultramafic complexes of the central and southern Urals. *Tectonophysics* **1997**, *276*, 181–194. [CrossRef]

58. Amossé, J.; Allibert, M.; Fisher, W.; Piboule, M. Experimental study of the solubility of platinum and iridium in basic silicate melts-implications for the differentiation of platinum group elements during magmatic processes. *Chem. Geol.* **1990**, *81*, 45–53. [CrossRef]

59. Nixon, G.T.; Cabri, L.J.; Laflamme, J.H.G. Platinum group elements mineralization in lode and placer deposits associated with the Tulameen Alaskan type complex, British Columbia. *Can. Mineral.* **1990**, *28*, 503–535.

60. Irvine, T.N. Chromian spinel as a petrogenetic indicator. Part 2. Petrologic applications. *Can. J. Earth Sci.* **1967**, *4*, 71–103. [CrossRef]

minerals

MDPI

Article

Platinum-Group Mineral Occurrences and Platinum-Group Elemental Geochemistry of the Xiadong Alaskan-Type Complex in the Southern Central Asian Orogenic Belt

Sai-Hong Yang [1,2,*], Ben-Xun Su [2,3], Xiao-Wen Huang [4], Dong-Mei Tang [2], Ke-Zhang Qin [2,3], Yang Bai [2,3], Patrick Asamoah Sakyi [5,*] and Melesse Alemayehu [1,6]

[1] State Key Laboratory of Lithospheric Evolution, Institute of Geology and Geophysics, Chinese Academy of Sciences, P.O. Box 9825, Beijing 100029, China; melesse555@yahoo.com
[2] Key Laboratory of Mineral Resources, Institute of Geology and Geophysics, Chinese Academy of Sciences, Beijing 100029, China; subenxun@mail.iggcas.ac.cn (B.-X.S.); tdm@mail.iggcas.ac.cn(D.-M.T.); kzq@mail.iggcas.ac.cn (K.-Z.Q.); by@mail.iggcas.ac.cn (Y.B.)
[3] University of Chinese Academy of Sciences, Beijing 100049, China
[4] State Key Laboratory of Ore Deposit Geochemistry, Institute of Geochemistry, Chinese Academy of Sciences, Guiyang 550081, China; huangxiaowen@vip.gyig.ac.cn
[5] Department of Earth Science, University of Ghana, P.O. Box LG 58, Legon-Accra 00233, Ghana
[6] School of Applied Natural Science, Department of Applied Geology, Adama Science and Technology University, P.O. Box 1888, Adama 00251, Ethiopia
* Correspondence: shyang@mail.iggcas.ac.cn (S.-H.Y.); pasakyi@ug.edu.gh (P.A.S.); Tel.: +86-010-8299-8592 (S.-H.Y.); Fax: +86-010-62010846 (S.-H.Y.)

check for updates

Received: 7 August 2018; Accepted: 23 October 2018; Published: 1 November 2018

Abstract: Alaskan-type complexes commonly contain primary platinum-group element (PGE) alloys and lack base-metal sulfides in their dunite and chromite-bearing rocks. They could therefore host PGE deposits with rare sulfide mineralization. A detailed scanning electron microscope investigation on dunites from the Xiadong Alaskan-type complex in the southern Central Asian Orogenic Belt revealed: various occurrences of platinum-group minerals (PGMs) that are dominated by inclusions in chromite grains containing abundant Ru, Os, S and a small amount of Pd and Te, indicating that they mainly formed prior to or simultaneously with the crystallization of the host minerals; A few Os–Ir–Rurich phases with iridium/platinum-group element (IPGE) alloy, anduoite (Ru,Ir,Ni)(As,S)$_{2-x}$ and irarsite (IrAsS) were observed in chromite fractures, and as laurite (RuS$_2$) in clinopyroxene, which was likely related to late-stage hydrothermal alteration. The rocks in the Xiadong complex display large PGE variations with ∑PGE of 0.38–112 ppb. The dunite has the highest PGE concentrations (8.69–112 ppb), which is consistent with the presence of PGMs. Hornblende clinopyroxenite, hornblendite and hornblende gabbro were all depleted in PGEs, indicating that PGMs were likely already present at an early phase of magma and were mostly collected afterward in dunites during magma differentiation. Compared with the regional mafic–ultramafic intrusions in Eastern Tianshan, the Xiadong complex show overall higher average PGE concentration. This is consistent with the positive PGE anomalies revealed by regional geochemical surveys. The Xiadong complex, therefore, has potential for PGE exploration.

Keywords: Alaskan-type complex; Central Asian Orogenic Belt; PGM; PGE mineralization

1. Introduction

Platinum-group element (PGE) mineralization in Alaskan-type complexes is invariably present in dunite units [1,2], occasionally in the vein-type settings associated with bodies of magnetite-rich clinopyroxenite in the dunite or peridotite core, in pegmatitic and micaceous rocks [3]. Previous studies revealed general enrichments in Pt, Pd and Rh relative to Ru, Ir and Os in Alaskan-type complexes [4,5]. Platinum-rich minerals, mainly Pt–Fe alloys are thought to form in the early stages of magmatic evolution and are associated with chromite formation [6]. Palladium-rich minerals on the other hand form at later stages and are associated with Cu–Fe–V–Ti metal formation [4,7]. As the Alaskan-type complexes originated from highly oxidizing and low-sulfur magmatism [8], they commonly occur with the presence of PGE alloys and lack base-metal sulfides [8]. However, sulfide mineralization coupled with chromite and/or PGE mineralization was explored in a few Alaskan-type complexes (e.g., Turnagain, British Columbia; and Duke Island, Alaska) [9,10]. This discrepancy is not well understood. Further investigation into the occurrences and geochemical histories of chromites, platinium-group minerals (PGMs) and sulfides is required in these Alaskan-type complexes.

Geological survey mapping of ultralow density using chemical spectrometry analysis of Pt, Pd and Au from floodplain overbank sediment samples revealed PGE anomalies in the Eastern Tianshan orogen for a long time [11]. Platinum-group element deposits are yet to be found in this area, although an increasing number of Ni–Cu sulfide deposits were identified in Permian mafic–ultramafic intrusions [12–14]. The Xiadong mafic–ultramafic complex in the Mid-Tianshan Terrane, the southern part of the Eastern Tianshan orogen, differs from the Ni–Cu sulfide deposit-hosting mafic–ultramafic intrusions in terms of its geochronology, petrology, mineralogy and geochemistry [15–20]. It is composed of dunite, hornblende clinopyroxenite, hornblendite and hornblende gabbro. These rocks are characterized by cumulate textures, high Mg contents, low trace–element abundances, flat rare-earth element (REE) and arc-magma-type trace-element patterns, and considerable Mg isotopic variations at both mineral and whole-rock scales [15,17–20]. The Xiadong complex was formed following multiple magmatic pulses according to zircon U–Pb ages (479, 477, 379, and 313 Ma) of four hornblende gabbros [17]. It was been identified as a typical Alaskan-type complex showing potential chromite mineralization [15]. A more recent study reported the discovery of base-metal minerals in ultramafic rocks of the Xiadong Alaskan-type complex [18]. In order to investigate potential PGE and metal sulfide mineralization in the Xiadong complex, we present the observations of PGMs and sulfides as well as PGE whole-rock geochemistry of Xiadong intrusive rocks. The Xiadong complex shows variable but relatively higher PGE concentrations when compared with regional mafic–ultramafic intrusions. The complex, thus, has potential for PGE exploration. The PGM and sulfide segregations in this Alaskan-type complex are also discussed.

2. Geological Background

The Eastern Tianshan and Beishan orogens are located in the southern part of the Central Asian Orogenic Belt (CAOB) (Figure 1A). Numerous mafic–ultramafic intrusions are widespread in the orogens and they host the major Ni–Cu sulfide deposits in China [14,21,22] (Figure 1B). These mafic–ultramafic intrusions formed mostly at the post-orogenic extensional setting in the early Permian [12–15]. Results of geochemical exploration revealed that the positive PGE anomalies (0.4–0.6 ppb Pt and 0.4–1.0 ppb Pd) are spatially related with the mafic–ultramafic intrusions in the Eastern Tianshan and Beishan [11,23]. However, progress is yet to be made on PGE deposit exploration in this area.

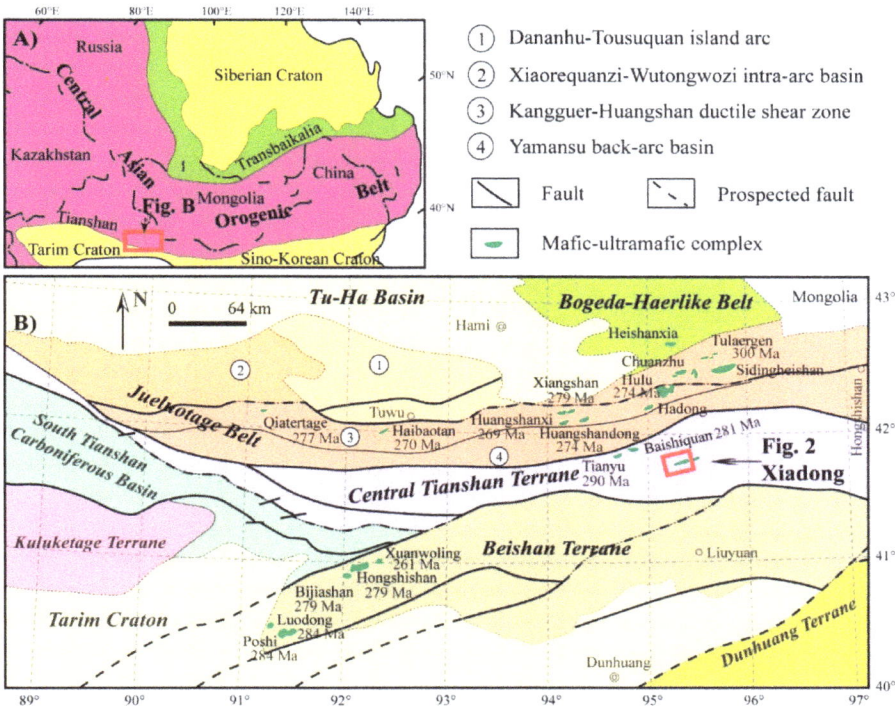

Figure 1. (**A**) Location map of the study area in the Central Asian Orogenic Belt. (**B**) Geological map of the Eastern Tianshan and Beishan orogens showing the distribution of Paleozoic mafic–ultramafic complexes (modified from References [15,17]). The reported numbers represent the ages of known intrusions.

The Xiadong mafic–ultramafic complex is situated in the Mid-Tianshan Terrane (Figure 1B), which is interpreted as a continental arc from the Ordovician to the Carboniferous. The complex was formed by multiple pulses of magma via high-degree melting of the lithospheric mantle, most likely related to subduction of the South Tianshan ocean and subsequent subduction of the Junggar ocean. The detailed geology was summarized by Tang et al. and Su et al. [17,24]. The Xiadong complex stretches east–west and covers an area measuring 7 km in length with a maximum width of 0.5 km. It has distinctive petrological and mineralogical features relative to the Permian mafic–ultramafic intrusions in Eastern Tianshan and Beishan. The Xiadong complex is composed of dunite in the core and surrounded by hornblende clinopyroxenite, hornblendite and hornblende gabbro (Figure 2), representing a complete set of rock suites compared with other Alaskan-type complexes worldwide [15,18]. These rock units are characterized by cumulate textures and display intrusive contacts with the country rocks [15].

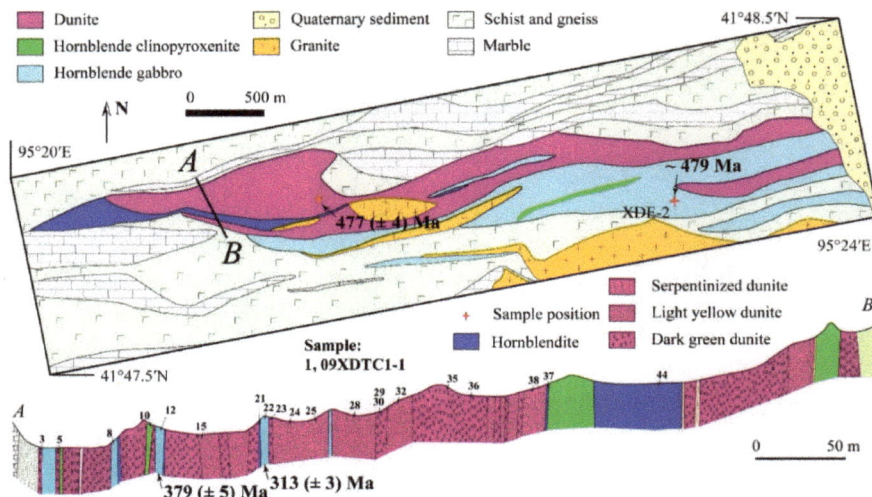

Figure 2. Geological map of the Xiadong mafic–ultramafic complex, accompanied by a horizontal profile along A to B showing its rock units and sampling positions (modified from References [15,17]). Isotopic ages are the U–Pb ages of zircons from four hornblende gabbros.

3. Materials and Analytical Methods

3.1. Sample Description

The studied rocks were mostly fresh with slight alteration. The minerals in the complex are dominated by olivine, clinopyroxene, and hornblende with accessory chromite and magnetite. Orthopyroxene is very rare in all of the rock units. Plagioclase is absent in the dunite and hornblende clinopyroxenite. The dunite is made up of olivine (90–98 vol.%) and chromite (0.5–8 vol.%) with accessory hornblende and clinopyroxene (<1–2 vol.%). The Fo content of olivine in the dunites ranges from 92 to 96.7 [15,18]. Subhedral to euhedral interstitial chromite occurs in elongated crystal or rounded shapes.

3.2. Sample Preparation

Samples were collected using a hammer before polishing thin sections for optical microscopy work. From the samples, 14 polished thin sections including eight dunite samples, two hornblende clinopyroxenite samples, two hornblendite samples, and two hornblende gabbro samples were prepared for electron microscopy work. The thin sections were then coated with the carbon with ~10 nm thickness using a Leica EM ACE200 coating equipment (Leica, Vienna, Austria). Sixteen samples extracted from the different rock units were also crushed into powder with 200 mesh for measuring PGE concentrations.

3.3. Sample Analyses

The thin sections were studied using a FEI Nova Nano450 scanning electron microscope (SEM) equipped with an energy dispersive X-ray spectrometer (EDS) from Oxford Instruments (High Wycombe, UK) at the Institute of Geology and Geophysics, Chinese Academy of Sciences. Backscattered electron (BSE) images were obtained at an accelerating voltage of 10 kV and a beam current of 3.0 nA. The acquisition time of one BSE image was ~1 min. Semi-quantitative spot analyses and EDS mapping were obtained at an accelerating voltage of 20 kV and a beam current of 8.8 nA.

Minerals **2018**, *8*, 494

The acquisition time of each spot was about 15 s, ensuring the spectrum area exceeded 2.5×10^6 counts. The mapping determination of each grain took about 5 min, with a dwell time of 5 μs for each pixel.

Whole-rock PGE concentrations were measured at the State Key Laboratory of Ore Deposit Geochemistry, Institute of Geochemistry, Chinese Academy of Sciences. The concentrations were determined by isotope dilution (ID) inductively coupled plasma (ICP)-MS using a modified digestion method [25,26]. Five grams of rock powder (>200 mesh) and appropriate amounts of ^{101}Ru, ^{193}Ir, ^{105}Pd and ^{194}Pt isotope spikes were used. The PGE measurements were made with a Perkin Elmer Elan DRC-e ICP-MS. Ir, Ru, Pt and Pd concentrations were measured by isotope dilution, and Rh concentrations were calculated using ^{194}Pt as an internal standard [27]. The analyses were monitored using the standard reference materials, WGB-1 (gabbro) and TDB-1 (diabase). Measured values were in good agreement with certified values. The total procedural blanks were lower than 0.009 ng for Ru, Rh, and Ir, and 0.030 ng for Pt and Pd.

4. Results

4.1. Occurrences of PGM and Sulfide

Fourteen samples were investigated using SEM. PGMs were only found in dunite sample. The number of identified grains per sample was different. There were about nineteen grains with sizes of less than one micron, as well as two grains of about four microns, from eight dunite samples. Limited by the resolution of SEM, some micro-scale individual grains with PGE elements detected by EDS could not be imaged. Some of the back scattered electron images and EDS analyses are shown in Figures 3–7 and Table 1.

Platinum-group minerals mostly occur as inclusions in chromite from sub-micrometer to several tens of nanometers in size (Figures 3–5 and Figure 6a). These PGM inclusions are composed of PGE sulfides and tellurides, dominated by Os–Ru–S with seven grains discovered in five samples (Figure 3a–c), Ru–S with six grains discovered in five samples (Figure 3d) and Os–Ru–Pd–Te with three grains discovered in two samples (Figure 3e,f).

Two PGMs were observed in fractures of chromite in one sample. One grain was only about ten nanometers in size, appearing irregular in shape. The larger grain forming composite grain with a maximum diameter of 4×3 microns, displays a well-defined contact with chromite (Figure 4a). The BSE image and EDX maps in Figure 4 show that the Ir and Pt distributions positively correlate with S in the grain, whilst Ru and Ni positively correlate with As distribution (Figure 4c–i). The average composition measured by EDS on the basis of a total of 3 number of atoms in one molecule or formula unit (apfu). gave $(Ru_{0.89}Ir_{0.27}Ni_{0.10}Pt_{0.08})\Sigma_{1.3}(As_{1.36}S_{0.30})\Sigma_{1.7}$ (Table 1) and $Ir_{0.77}Ru_{0.08}As_{0.76}S$, corresponding to the formula $(Ru,Ir,Ni)(As,S)_{2-x}$ likely to be anduoite [28] and IrAsS, known as irarsite, respectively. Osmium was concentrated in a local area (Figure 4c), measuring 52.9 wt.% Os, 11.6 wt.% Ir and 34.5 wt.% Ru (Table 1), indicating that it was an alloy.

An elliptical grain of laurite (about 5 μm in size) was observed in clinopyroxene (Figure 5). This grain contained S, Ru, and Se with small amounts of Os, Ir, and Ni. The average composition of the laurite from EDS was $Ru_{0.47}Fe_{0.09}Os_{0.03}Ir_{0.03}Ni_{0.02})_{0.53}(Se_{0.08}S)$ on the basis of 1 S, with an empirical formula of RuS_2.

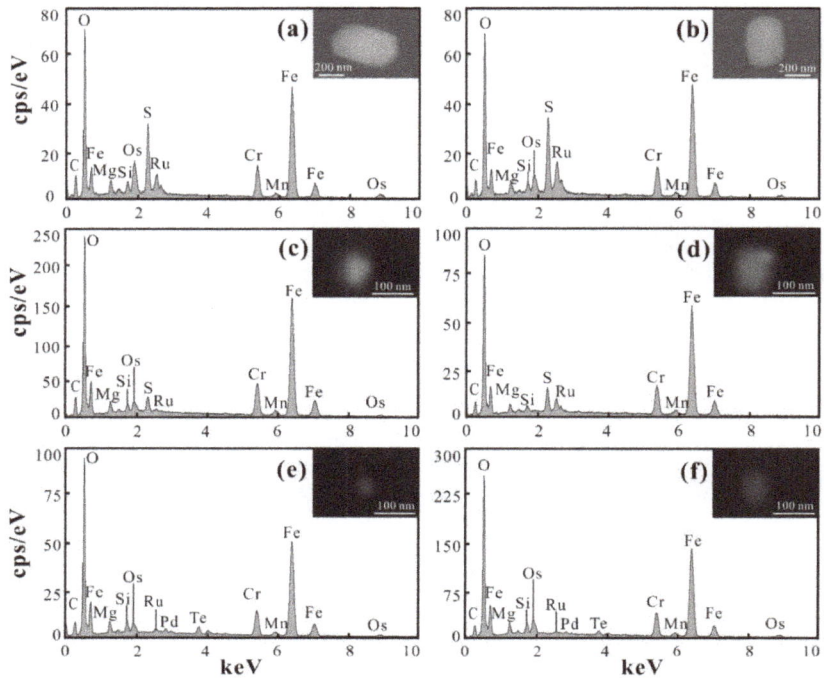

Figure 3. Energy dispersive X-ray spectrometry (EDS) and backscattered electron (BSE) images of platinum-group mineral (PGM) inclusions in chromite in the Xiadong dunites. (**a–c**) Ru–Os–S; (**d**) Ru–S; (**e,f**) Os–Ru–Pd–Te.

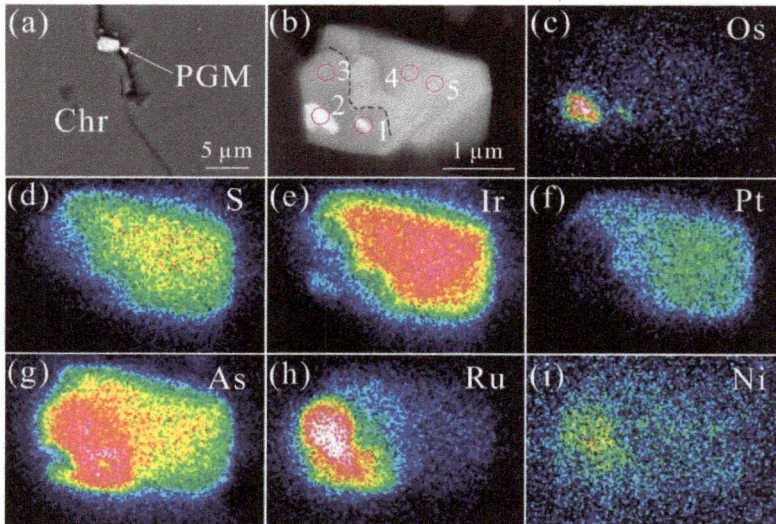

Figure 4. BSE images and S, As, Ir, Ru, Pt, Os, and Ni element maps of a three-phas PGM grain from the chromite fracture. (**a,b**) BSE images; (**c–i**) Os, S, Ir, Pt, As, Ru, and Ni element maps. Chr: chromite; numbers represent the positions of semi-quantitative spot analyses listed in Table 1.

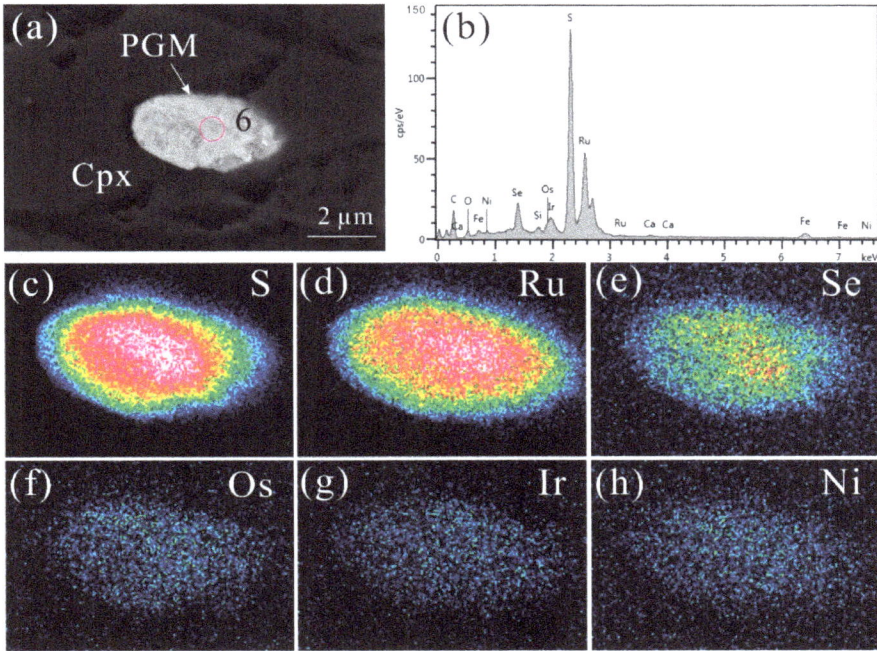

Figure 5. BSE images and S, Ru, Se, Os, Ir, and Ni element maps of a single phase of PGM grain from the silicate matrix of clinopyroxene. (**a**) BSE image; (**b**) the EDS spectrum of spot 6; (**c**–**h**) S, Ru, Se, Os, Ir, and Ni element maps. Cpx: clinopyroxene; 6: the position of semi-quantitative spot analysis of spot 6 in Table 1.

A variety of sulfides and sulfarsenide occur as inclusions within chromite (Figure 6) and serpentinized olivine (Figure 7), as well as in fractures of chromite (Figure 6). Chemically, they are maucherite ($Ni_{11}As_8$), pentlandite (($Fe,Ni)_9S_8$), millerite (NiS), heazlewoodite (Ni_3S_2), jaipurite (CoS), and galena (PbS). The largest grain size was 10 µm. Maucherite enclosed in chromite was mostly associated with heazlewoodite (Figure 6b,c), whereas those in the serpentinized olivine mostly occurred as single mineral phases (Figure 7a), occasionally associated with pentlandite (Figure 7b). Pentlandite phases also mainly occurred as single grains (Figure 6i–l) and were sometimes associated with galena in chromite (Figure 6g,h), and maucherite in serpentinized olivine (Figure 7b,d).

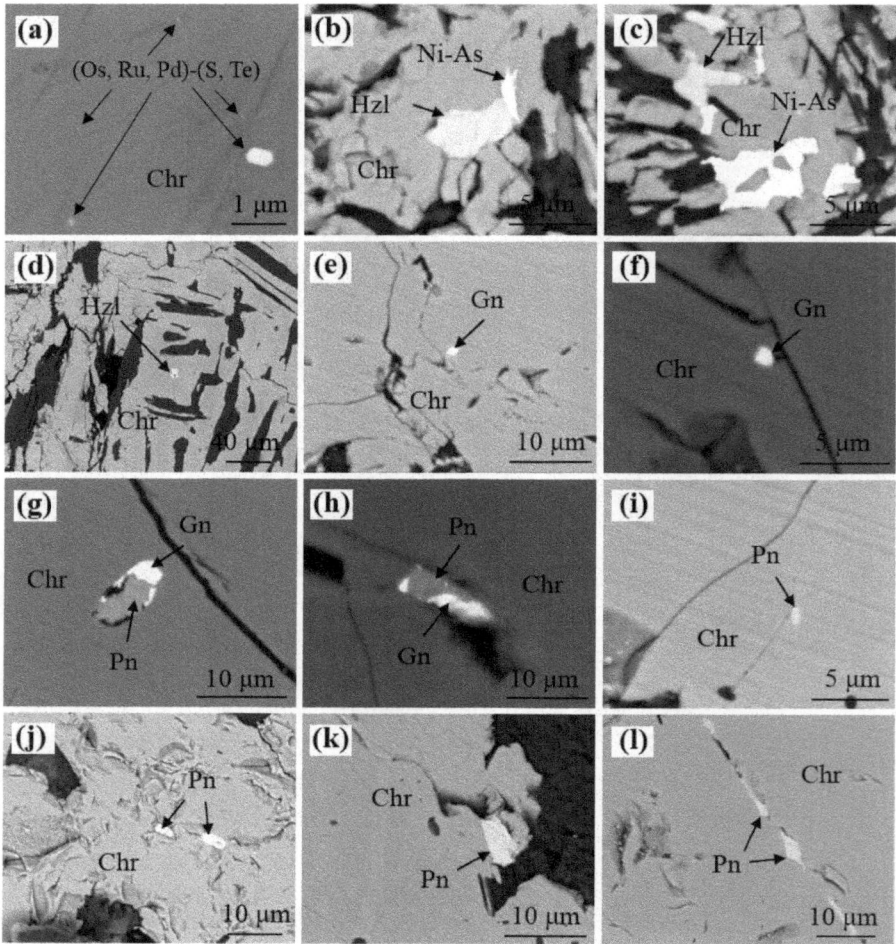

Figure 6. BSE images of base-metal minerals in chromite. (**a**) PGM of Os–Ru–Pd–S–Te; (**b,c**) heazlewoodite associated with maucherite; (**d**) heazlewoodite occurring as a single mineral phase; (**e,f**) galena occurring as a single mineral phase; (**g,h**) pentlandites associated with galena; (**i–l**) pentlandites occurring as a single mineral phase. Ni–As: maucherite; Hzl: heazlewoodite; Gn: galena; Pn:pentlandite.

Table 1. Representative energy dispersive spectrometry (EDS) analyses of platinum-group minerals from the Xiadong Alaskan-type complex.

Element/Spot	Os	Ir	Ru	Pt	Fe	Ni	Se	As	S	Total (wt.%)	Os *	Ir *	Ru *	Pt *	Fe *	Ni *	Se *	As *	S *
1	52.9	11.6	34.5	-	-	1.0	-	-	-	100	0.40	0.09	0.49	-	-	0.03	-	-	-
2	54.2	11.3	34.5	-	-	-	-	-	-	100	0.42	0.09	0.50	-	-	-	-	-	-
3	-	18.9	32.9	5.50	-	2.20	-	37.1	3.50	100	-	0.91	3.01	0.26	-	0.34	-	4.58	1.00
4	-	60.6	3.10	-	-	-	-	23.3	13.1	100	-	0.77	0.08	-	-	-	-	0.76	1.00
5	-	60.0	3.40	-	-	-	-	23.4	13.1	100	-	0.77	0.08	-	-	-	-	0.77	1.00
6	5.60	5.30	44.2	-	4.70	0.90	5.70	-	33.6	100	0.03	0.03	0.47	-	0.09	0.02	0.08	-	1.00

Note: 1 and 2—Os–Ir–Ru alloys; 3—(Ru, Ir, Pt, Ni)$_{0.81}$(As$_{4.58}$S); 4 and 5—irarsite; 6—laurite. Element (wt.%); element * (number of atoms).

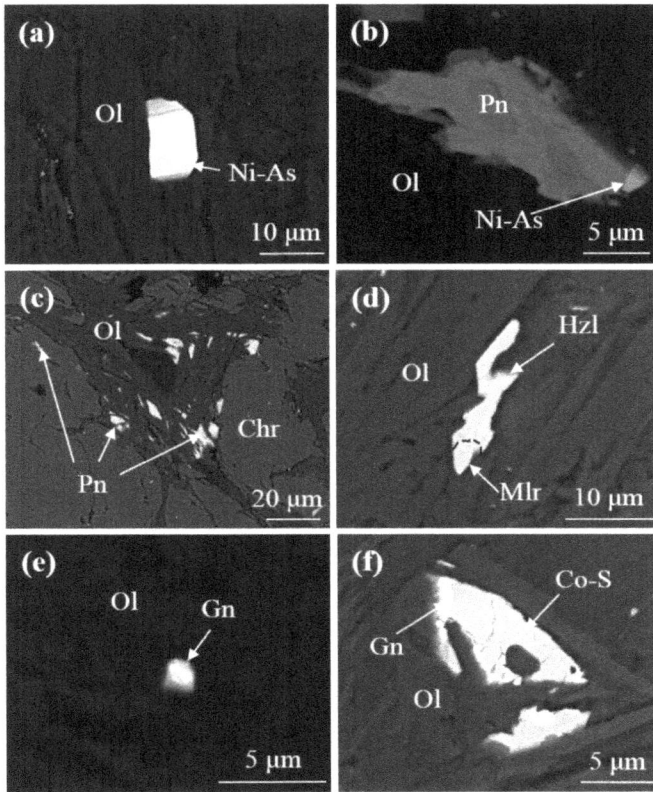

Figure 7. BSE images of base-metal minerals in serpentinized olivine. (**a**) Euhedral maucherite occurring as a single mineral phase; (**b**) pentlandites associated with maucherite; (**c**) pentlandites occurring as a single mineral phase; (**d**) heazlewoodite associated with millerite; (**e**) galena occurring as a single mineral phase; (**f**) galena occurring at the boundary of jaipurite. Ol: olivine; Mlr: millerite; Co–S: cobaltous sulfide.

4.2. Whole-Rock PGE Geochemistry

The PGE concentrations in dunite, hornblende clinopyroxenite, hornblendite, and hornblende gabbro from the Xiadong complex are reported in Table 2. These rocks show a wide range in total PGE concentrations from 8.69 to 112 ppb for dunite, 3.41 ppb for hornblende clinopyroxenite, 0.38 to 2.08 ppb for hornblendite, and 0.41 to 6.98 ppb for hornblende gabbro.

Irridium concentrations in the dunites varied between 0.80 and 5.02 ppb, whilst the other rock types had relatively low Ir concentrations (<0.1 ppb). Ruthenium concentrations of the dunites varied between 3.27 and 11.0 ppb, which was higher than those of the hornblende clinopyroxenite (2.64 ppb), hornblendites (0.03 to 0.05 ppb), and hornblende gabbros (0.04 to 0.61 ppb). Rhodium concentrations of the dunites varied between 0.63 and 3.34 ppb, whilst the other rock types had lower Rh concentrations (<0.2 ppb). The dunites had a large variation and overall higher Pt concentrations of 0.92 to 50.4 ppb than hornblende clinopyroxenite (0.32 ppb), hornblendite (0.20 to 1.63 ppb), and hornblende gabbro (0.23 to 5.89 ppb). Palladium concentrations in the dunites showed large variation from 0.16 to 42.2 ppb, whilst the other rocks showed limited Pd variations (hornblende clinopyroxenite: 0.23 ppb; hornblendite: 0.12 to 0.38 ppb; hornblende gabbro: 0.10 to 0.47 ppb).

Table 2. Platinum-group element (PGE) concentrations (ppb, except where specified) of the rocks from the Xiadong Alaskan-type complex. PPGE—Pd-PGE; IPGE—Ir-PGE.

Sample	Rock Type	Ir	Ru	Rh	Pt	Pd	Cu (ppm)	ΣPGE	IPGE	PPGE	PPGE/IPGE	Pd/Ir	Cu/Pd (×10³)
09XDTC1-35	Dunite	5.02	11.0	3.34	50.4	42.2	-	112	16.0	95.9	5.98	8.40	-
09XDTC1-36	Dunite	0.80	4.23	2.02	14.6	12.0	3.37	33.6	5.03	28.6	5.68	15.00	0.28
09XDTC1-15	Dunite	1.28	5.00	0.71	1.49	0.39	246	8.87	6.28	2.59	0.41	0.31	627
09XDTC1-28	Dunite	1.54	3.27	0.75	2.05	1.09	4.73	8.69	4.81	3.88	0.81	0.70	4.35
09XDTC1-29	Dunite	2.61	7.87	0.63	2.09	0.71	-	13.9	10.5	3.44	0.33	0.27	-
09XDTC1-24	Dunite	1.76	5.55	0.78	1.27	0.16	-	9.52	7.31	2.21	0.30	0.09	-
09XDTC1-25	Dunite	1.12	4.82	0.79	2.09	2.80	7.32	11.6	5.95	5.68	0.96	2.49	2.61
09XDTC1-32	Dunite	1.38	5.38	0.79	0.92	1.31	8.35	9.78	6.76	3.02	0.45	0.95	6.38
09XDTC1-10	Hornblende clinopyroxenite	0.07	2.64	0.14	0.32	0.23	9.63	3.41	2.71	0.69	0.26	3.17	41.7
09XDTC1-21	Hornblendite	0.01	0.05	0.04	1.63	0.34	8.91	2.08	0.06	2.02	32.3	26.5	26.2
09XDTC1-37	Hornblendite	0.02	0.05	0.01	0.24	0.38	8.54	0.70	0.07	0.64	9.75	23.4	22.4
09XDTC1-44	Hornblendite	0.01	0.03	0.01	0.20	0.12	7.31	0.38	0.04	0.33	7.50	11.9	59.1
09XDTC1-8	Hornblende gabbro	0.01	0.04	0.01	0.23	0.24	15.0	0.54	0.05	0.49	9.36	19.0	61.5
09XDTC1-12	Hornblende gabbro	0.02	0.05	0.01	0.23	0.10	88.8	0.41	0.07	0.34	4.94	6.61	890
09XDTC1-22	Hornblende gabbro	0.01	0.05	0.03	1.07	0.47	67.2	1.63	0.06	1.57	26.4	41.9	143
XDE2	Hornblende gabbro	0.09	0.61	0.19	5.89	0.20	1.81	6.98	0.70	6.28	9.01	2.27	9.00

5. Discussion

5.1. Primary and Late-Stage Generation of PGM and Sulfide

The PGMs in the Xiadong Alaskan-type complex were mainly PGE sulfides with a few iridium/platinum-group element (IPGE) alloys, PGE sulfarsenides, and tellurides. The inclusion size in chromite was much smaller, usually not exceeding 1 μm, with mainly micro-scale individual grains. Additionally, some inclusions showed euhedral crystals (Figure 3a,b). Some other base metal-minerals including maucherite, pentlandite, millerite, cobaltous sulfide, and galena were also observed. These are different from Alaskan-type complexes found worldwide that are typified by the presence of PGE alloys and the lack of base-metal sulfides [6]. The Pt–Fe alloys, including silicate minerals from Uralian–Alaskan-type intrusion, are formed after the crystallization of chromite, clinopyroxene, and albite. The formation temperature of Pt–Fe alloys is estimated at about 900 °C [5]. In most cases, the inclusions in chromite (Figures 3 and 6), clinopyroxene (Figure 5) and olivine (Figure 7) in the Xiadong Alaskan-type complex were dominated by high-temperature sulfide minerals (e.g., Os–Ru–S, Os–Ru–Pd–Te, Os alloy, and laurite), and would have crystallized from magmas prior to or simultaneously with the formation of the host minerals. A general consensus is that PGMs crystallize prior to or simultaneously with chromite, and deposit from the primary melt in the following order: IPGE alloys, followed by Ru-rich laurite, Os-rich laurite, and finally, base-metal mineral phases [29,30]. The temperature decreases and fS_2 increases [31]. The maximum thermal stability of RuS_2 is experimentally indeed high (1275 °C), consistent with a magmatic crystallization temperature [29]. This crystallization order and the occurrence of high-temperature sulfide minerals suggested for the Xiadong dunites would imply that the PGMs and chromite formed under relatively high fS_2 and temperature conditions. The model for PGE concentration in the Merensky Reef of the Bushveld Complex in South Africa indicates that a silicate melt coexisting with a PGE-enriched sulfide melt with high temperatures of 1200–1400 °C, would be oversaturated by the least soluble PGE upon cooling [32]. With the cooling of a mafic magma, silicate minerals and chromite start crystallizing at some stage and trap some of the PGE "nuggets" [32]. A combined focused ion beam and high-resolution transmission electron microscopy (FIB/HRTEM) investigation of PGE-rich samples from the Merensky Reef suggests that the PGE-rich nanophases found in the base metal sulfide might represent an early phase of magmatic PGM that formed from the silicate melt to be later collected by the sulfide melt [33]. The occurrence of PGE-rich nano inclusions in chromite (Figures 3 and 6) in our study requires that PGMs were already formed in the silicate melt and were subsequently incorporated into the growing chromite grains.

The anduoite–irarsite intergrowth in the cracks of chromite (Figure 4) could be related to late-stage hydrothermal alteration [18]. The incompatible As could have accumulated along with the late-stage PGE sulfides deposited as interstitial phases in the chromitite at a postmagmatic hydrothermal stage of crystallization [34]. Augé et al. [35] affirmed that hydrothermal laurite typically has more As content (1.01–5.97 wt.%) than magmatic laurite. The unidentified phase (Ru, Ir, Pt, Ni)$_{0.81}$(As$_{4.58}$S), as well as the occurrence of irarsite, at the margins of PGM grains (Figure 4) indicates the interaction of PGMs with late-stage As- and S-rich fluids. In addition, the laurite in the Xiadong dunite had significantly lower IrS_2 content (<5.3 mol.%; Table 1) than that in ophiolitic chromitite (ca. 15 mol.%) [36], suggesting the limited solubility of Ir in RuS_2. The laurite in clinopyroxene (Figure 5) likely reflects a process of ultimate S loss and Se incorporation into RuS_2 during a late-stage evolution of the H_2O-bearing fluid [37]. Laurite is normally depleted in Se with hundreds of ppm of Se [38].

Heazlewoodite crystallizes under a wide range of T and fS_2 conditions [39]. However, primary heazlewoodite inclusions within the chromite grains (Figure 6b–d) indicate a magmatic origin. Furthermore, the inclusion relationship texturally suggests that the heazlewoodite crystallized prior to or simultaneously with the chromite crystals. Some of the base-metal minerals such as millerite, pentlandite, and galena in the cracks of chromite (Figure 6) or serpentinized olivine (Figure 7) are medium- or low-temperature minerals, and are known to form in hydrothermal systems.

5.2. Relationship between PGM Occurrences and Bulk PGE Concentrations

The rocks of the Xiadong complex displayed large variation in PGE concentration with ∑PGE of 0.38–112 ppb, which is slightly lower than that of Ural–Alaskan-type complexes with ∑PGE of 9.1–196.1 ppb [40]. The highest PGE concentration in the dunite was 112 ppb, indicating PGE enrichment in the parental magma of the Xiadong complex. The dunite samples displayed primitive mantle-normalized PGE patterns with a weakly positive slope from Ir to Pd [41] (Figure 8). It is noteworthy that six dunite samples showed negative slopes from Rh to Pt and had a large variationin Pd (Figure 8). The PGE patterns ranged from nearly unfractionated in the dunites (Pd/Ir = 0.09–15) to mildly fractionated in the hornblende clinopyroxenite, hornblendite, and hornblende gabbro (Pd/Ir = 2.27–41.9). This distinct geochemical behavior was emphasized by the Pd/Ir ratio, considered as the "index of fractionation" of the PGE, during petrological processes [40]. Having a chalcophile and siderophile nature, PGEs are mostly concentrated in the earth's core and mantle, but are remarkably low in the crustal "Clarke" values or frequently close to the detection limits [40]. Because of various melting temperatures, PGEs are present in several sulfide phases in the mantle and are incorporated into the melts depending on the degree of partial melting in the mantle [42]. Compared to the mid-ocean range basalt (MORB) PGE concentration with the low degree of partial melting in the range of 2–15% [43], the Xiadong complex showed the higher PGE concentration (Figure 8), suggesting that both IPGEs (Os, Ir, and Ru) and palladium/platinum-group elements (PPGE: Rh, Pt, and Pd) transferred to the melt with high-degree of melting. A higher degree of partial melting can transfer IPGEs from the mantle rocks to the formed melt [44]. On the other hand, a low degree partial melting only transfers PPGEs to the formed melt due to the varying sulfide melt/silicate liquid partition coefficients [45,46]. Quantitative simulation studies indicated that the primitive magma of the Xiadong complex was derived from 24% bulk partial melting of primitive mantle [47]. Whole-rock high-Mg features indicate that they formed from a depleted mantle source under a high degree of partial melting [17].

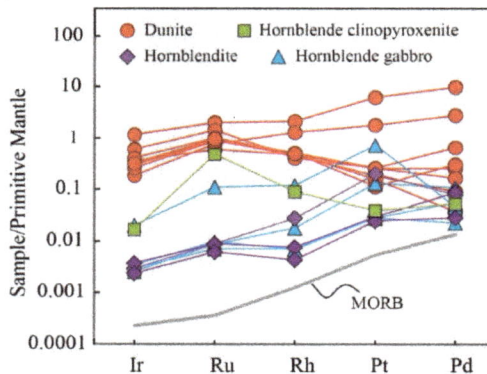

Figure 8. Primitive mantle-normalized PGE patterns for the rocks from the Xiadong Alaskan-type complex. Primitive mantle-normalized values were sourced from Reference [41]. Mid-ocean range basalt (MORB) values were sourced from Reference [43].

Platinum-group elements are commonly found in the form of discrete chromite-hosted PGMs [48]. As discussed above, PGEs are known to be mostly present as micron-sized or nanometer-sized inclusions forming prior to or simultaneously with chromite formation from the silicate magma. Dunite had the highest PGE concentrations (8.69–112 ppb) (Table 2), which is generally consistent with the presence of PGMs. Many laurite and irarsite grains were observed either as euhedral inclusions or along fractures in chromite grains, consistent with IPGE enrichment in all dunites. A few PPGMs were observed with a wide range of concentrations of Pt (0.92 to 50.4) and Pd (0.16 to 42.2) in most dunites (Table 2), suggesting that Pt and Pd were likely to be mobilized during late alteration processes resulting in differing slopes on the primitive mantle-normalized diagrams (Figure 8). The presence of abundant Pd-bearing PGM is consistent with a pronounced Pd anomaly from chromitites of the Butyrin vein compared with dunite-hosted chromitites at Kytlym, indicating that a high fugacity of sulfur and high-temperature fluids enriched in Hg, Te and Cu play an important roles during the formation of the mineralization in the Butyrin vein [49]. Hornblende clinopyroxenite, hornblendite and hornblende gabbro were all depleted in PGEs, indicating that PGEs were already formed the early solid phases prior to the crystallization of chromite and were mostly collected afterward in dunites during magma differentiation. This is consistent with the identification of only one laurite inclusion in hornblende clinopyroxene (Figure 5).

5.3. Implications on PGE Mineralization in Eastern Tianshan

The formation of PGE-bearing sulfide deposits mainly depends on two conditions: (1) the availability of PGE-rich primary magma and (2) an immiscible sulfide liquid separation and segregation during magma evolution [50]. Palladium and Cu are both considered as incompatible elements with silicate melt [51]. The Pd minimum partition coefficient of 17,000 in sulfide/silicate liquid is considerably higher than the Cu partition coefficient (1000) in sulfide/silicate [46]. Palladium has a stronger sulfide affinity than Cu when sulfide immiscibility occurs [46]. Cu/Pd ratios would remain constant in basaltic magmas during S-undersaturated fractional crystallization [51]. Therefore, the depletion in Pd relative to Cu can provide an estimation of the percentage of crystal fractionation under sulfide saturated conditions required to deplete the magma in PGEs and other chalcophile elements. The Cu/Pd ratio is an important evaluation parameter of magma evolution, and is widely used in the study of PGE deposits [52]. In magma evolution processes, sulfide separation under saturation would result in the significant depletion of Pd relative to Cu in the residual magma, subsequently giving rise to the observed Cu/Pd ratio. In the Xiadong complex, most dunite samples had Cu/Pd ratios (2815–6377) lower than that of the primary mantle (6364) [52], except for one sample having a significantly higher Cu/Pd ratio (627,000). Along with fractional crystallization, Cu/Pd ratios of clinopyroxenite, hornblendite and gabbro in the range of 9000–142,516 were clearly higher than the primary mantle value, indicating that PGE sulfide separation took place prior to or simultaneously with the formation of the dunite.

The PGEs had similar partition coefficients in both sulfide and silicate melts, with a strong preference for the sulfide melt. The crystallization of PGMs in a PGE-rich primary magma indicates that sulfur saturation occurred during the formation of chromite. In addition, the occurrence of a large number of base metal-sulfides as disseminated accessory grains in all dunites indicates that they formed prior to olivine and chromite crystallization. In the early stage of magma crystallization, enrichment of Ni in the rocks depends on the sulfur fugacity [8,18]. Meanwhile, the crystallization of large amounts of chromite and ilmenite results in the loss of Fe^{2+} in the melt leading to sulfur saturation [50]. The availability of sufficient S, As, Te, and PGEs facilitates the formation of PGE sulfides prior to silicate mineral crystallization. The country rocks of the Xiadong complex are dominated by schists, gneisses and marbles, and their occurrences may reduce the oxygen fugacity of the magmas. It was been demonstrated that an increase in the fO_2 in melt results in the increase in the degree of partitioning of Ir, Ru, and Rh into spinel [29]. Accordingly, PGEs would not be soluble and could form PGM under reducing fO_2. In addition, the generally small sizes of the Alaskan-type complexes (a few to hundreds of km^2) [53] are preferable for sulfur saturation. The Xiadong complex is relatively smaller than most Alaskan-type complexes worldwide and could, therefore, facilitate sulfur saturation.

The Eastern Tianshan is one of the most important Ni–Cu metallogenic provinces in China [21,54]. Abundant mafic–ultramafic complexes are distributed mainly along deep fractures in the Kangguer–Huangshan ductile shear zone in the Jueluotage Belt in the north of Eastern Tianshan (Figure 1B). They are mostly explored for magmatic Ni–Cu sulfide deposits. The two largest Ni–Cu sulfide deposits of Huangshandong and Tulaergen have very low PGE total concentrations

(<5 ppb) [12,55,56]. The Tudun, Hulu, and Xiangshan Ni–Cu sulfide deposits have total PGE concentrations varying from 0.99 ppb to 2.57 ppb [57], 1.98 ppb to 26.6 ppb [58], and 1.21 ppb to 3.26 ppb [55], respectively. The total PGE concentrations of the Baishiquan [59] and Tianyu [60] Ni-Cu sulfide deposits along the northern margin of the Middle Tianshan terrane are relatively high and close to that of the Xiadong complex (Figure 9). The Luodong and Poyi mafic–ultramafic intrusions in the Beishan terrane have low total PGE concentrations with a maximum of 3.18 ppb and 18.8 ppb, respectively [61,62]. In the Kalatongke Ni–Cu sulfide deposit in the northern margin of the Junggar terrane, the total PGE concentrations vary from 0.23 ppb to 43.6 ppb [63]. The Xiarihamu magmatic Fe–Ni–Cu sulfide deposit is the largest magmatic Ni–Cu deposit ever discovered in an arc setting worldwide and is hosted in a small ultramafic body in the Eastern Kunlun Orogenic Belt of the northern Tibet–Qinghai Plateau in western China. The total concentrations of PGEs in the host rocks of the Xiarihamu deposit vary from 0.09 to 1.45 ppb [64]. When compared with other regional mafic–ultramafic complexes (Figure 9), the total concentrations of PGE in the Xiadong Alaskan-type complex are significantly higher than in other intrusions. Furthermore, PGE mineralization was found in a few Cu–Ni sulfide deposits in Eastern Tianshan [65]. In addition, regional geochemical exploration led to the discovery of PGE anomalies in Middle Tianshan terrane [9], whilst sulfide was also discovered in the Xiadong complex [18]. Therefore, the Xiadong Alaskan-type complex could have potential at least the regional anomalies for PGE as well as Ni sulfide exploration.

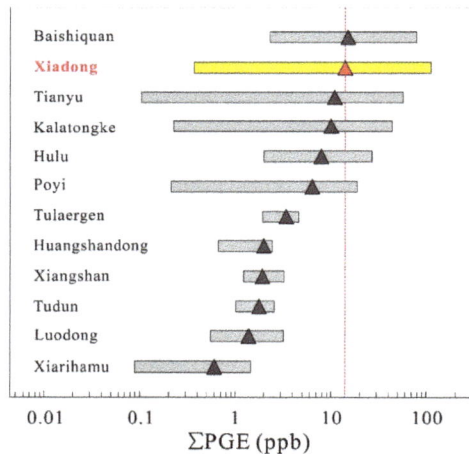

Figure 9. Comparisons of PGE concentrations in the Xiadong Alaskan-type complex and other mafic–ultramafic complexes in the southern Central Asian Orogenic belt (CAOB). Data were taken from References [55–64]. The triangle indicates the average.

6. Conclusions

1. Platinum-group minerals in the dunite from the Xiadong Alaskan-type complex are mainly PGE sulfide and sulfarsenide and occur as inclusions in chromite and clinopyroxene or as interstitial grains along fractures in chromite. The occurrence of PGE-rich inclusions such as laurite in chromite indicates that PGMs already have formed in the silicate melt and subsequently incorporated in the growing chromite grains. The occurrence of interstitial anduoite-irarsite grains along fractures of chromite and Se incorporation in laurite in clinopyroxene are ikely related to late-stage hydrothermal alteration.

2. The dunites have the highest PGE concentrations relative to other lithologies of the Xiadong complex, which is generally consistent with the presence of PGMs. Many laurite and irarsite grains were observed either as euhedral inclusions or along fractures in chromite grains, consistent

with IPGE enrichment in all dunites. A few PPGMs were observed with a wide range of Pt and Pd concentrations in most dunite samples. Hornblende clinopyroxenite, hornblendite and hornblende gabbro are all depleted in PGEs, consistent with the identification of only one laurite inclusion in hornblende clinopyroxene. They also suggest that PGEs had already formed the early solid phases prior to the crystallization of chromite and mostly afterwards were collected in dunites during magma differentiation.

3. Compared to the MORB PGE concentration with the low degree of partial melting, the Xiadong complex shows higher PGE concentration, suggesting that both IPGE and PPGE transferred to the melt with a high-degree of melting. When compared with the regional mafic–ultramafic intrusions in Eastern Tianshan, the Xiadong complex has the greatest PGE enrichment. This is consistent with the identification of PGE anomalies by regional geochemical surveys, demonstrating the potential for PGE mineralization in this region.

Author Contributions: Conceptualization, B.-X.S. and S.-H.Y.; Methodology, S.-H.Y. and X.-W.H.; Validation, B.-X.S., S.-H.Y. and P.A.S.; Formal Analysis, S.-H.Y. and P.A.S.; Investigation, Y.B.; Resources, B.-X.S.; Data Curation, S.-H.Y. and X.-W.H.; Writing-Original Draft Preparation, S.-H.Y.; Writing-Review and Editing, P.A.S., D.-M.T. and M.A.; Supervision, K.-Z.Q.; Project Administration, K.-Z.Q.; Funding Acquisition, B.-X.S. and S.-H.Y.

Funding: This research was financially supported by the National Natural Science Foundation of China (Grant 41522203), Youth Innovation Promotion Association, Chinese Academy of Sciences (Grant 2016067) and Key Laboratory of Mineral Resources, Institute of Geology and Geophysics, Chinese Academy of Sciences (Grant KLMR2017-04).

Acknowledgments: We thank Hitesh Changela for helping improve the quality of English of this manuscript.

Conflicts of Interest: The authors declare no conflict of interest.

References

1. Razin, L.V. Geologic and genetic features of forsterite dunites and their platinum-group mineralization. *Econ. Geol.* **1976**, *71*, 1371–1376. [CrossRef]

2. Cabri, L.J.; Genkin, A.D. Re-examination of Pt alloys from lode and placer deposits, Urals. *Can. Mineral.* **1991**, *29*, 419–425.

3. Barkov, A.Y.; Shvedov, G.I.; Polonyankin, A.A.; Martin, R.F. New and unusual Pd-Tl-bearing mineralization in the Anomal'nyi deposit, Kondyor concentrically zoned complex, northern Khabarovskiy Kray, Russia. *Mineral. Mag.* **2017**, *81*, 679–688. [CrossRef]

4. Johan, Z.; Ohnenstetter, M.; Slansky, E.; Barron, L.M.; Suppel, D. Platinum mineralization in the Alaskan-type intrusive complexes near Fifield, New South Wales, Australia Part 1. Platinum-group minerals in clinopyroxenites of the Kelvin Grove Prospect, owendale intrusion. *Miner. Petrol.* **1989**, *40*, 289–309. [CrossRef]

5. Nixon, G.T.; Cabri, L.J.; Laflamme, J.G. Platinum-group element mineralization in lode and placer deposits associated with the Tulameen Alaskan-type complex, British Columbia. *Can. Mineral.* **1990**, *28*, 503–535.

6. Zaccarini, F.; Singh, K.A.; Garuti, G.; Zsatyanarayanan, M. Platinum-group minerals (PGM) in the chromitite from the Nuasahi massif, eastern India: Further findings and implications for their origin. *Eur. J. Mineral.* **2017**, *29*, 571–584. [CrossRef]

7. Johan, Z. Platinum-group minerals from placers related to the Nizhni Tagil (Middle Urals, Russia) Uralian-Alaskan-type ultramafic complex: Ore-mineralogy and study of silicate inclusions in (Pt, Fe) alloys. *Miner. Petrol.* **2006**, *87*, 1–30. [CrossRef]

8. Johan, Z. Alaskan-type complexes and their platinum-group element mineralization. *Geol. Geochem. Mineral. Miner. Benef. Platin.-Group Elem.* **2002**, *54*, 669–719.

9. Clark, T. Petrology of the Turnagain ultramafic complex, northwestern British Columbia. *Can. J. Earth Sci.* **1980**, *17*, 744–757. [CrossRef]

10. Li, C.; Ripley, E.M.; Thakurta, J.; Stifter, E.C.; Qi, L. Variations of olivine Fo–Ni concentrations and highly chalcophile element abundances in arc ultramafic cumulates, southern Alaska. *Chem. Geol.* **2013**, *351*, 15–28. [CrossRef]

11. Zhang, H.; Liu, H.Y.; Chen, F.L. Regional geochemical exploration for platinum and palladium. *Geochimica* **2002**, *31*, 55–65. (In Chinese with English abstract)

12. Qin, K.Z.; Su, B.X.; Sakyi, P.A.; Tang, D.M.; Li, X.H.; Sun, H.; Liu, P.P. SIMS zircon U-Pb geochronology and Sr-Nd isotopes of Ni-Cu-bearing mafic-ultramafic intrusions in Eastern Tianshan and Beishan in correlation with flood basalts in Tarim Basin (NW China): Constraints on a ca. 280 Ma mantle plume. *Am. J. Sci.* **2011**, *311*, 237–260. [CrossRef]

13. Su, B.X.; Qin, K.Z.; Sakyi, P.A.; Li, X.H.; Yang, Y.H.; Sun, H.; Tang, D.M.; Liu, P.P.; Xiao, Q.H.; Malaviarachchi, S.P.K. U-Pb ages and Hf-O isotopes of zircons from Late Paleozoic mafic-ultramafic units in southern Central Asian Orogenic Belt: Tectonic implications and evidence for an Early-Permian mantle plume. *Gondwana Res.* **2011**, *20*, 516–531. [CrossRef]

14. Su, B.X.; Qin, K.Z.; Tang, D.M.; Sakyi, P.A.; Liu, P.P.; Sun, H.; Xiao, Q.H. Late Paleozoic mafic-ultramafic intrusions in southern Central Asian Orogenic Belt (NW China): Insight into magmatic Ni-Cu sulfide mineralization in orogenic setting. *Ore Geol. Rev.* **2013**, *51*, 57–73. [CrossRef]

15. Su, B.X.; Qin, K.Z.; Sakyi, P.A.; Malaviarachchi, S.P.K.; Liu, P.P.; Tang, D.M.; Xiao, Q.H.; Sun, H.; Ma, Y.G.; Mao, Q. Occurrence of an Alaskan-type complex in the Middle Tianshan Massif, Central Asian Orogenic Belt: Inferences from petrological and mineralogical studies. *Int. Geol. Rev.* **2012**, *54*, 249–269. [CrossRef]

16. Su, B.X.; Qin, K.Z.; Sun, H.; Tang, D.M.; Xiao, Q.H.; Liu, P.P.; Sakyi, P.A. Olivine compositional mapping of mafic-ultramafic complexes in Eastern Xinjiang (NW China): Implications for mineralization and tectonic dynamics. *J. Earth Sci.* **2012**, *23*, 41–53. [CrossRef]

17. Su, B.X.; Qin, K.Z.; Zhou, M.F.; Sakyi, P.A.; Thakurta, J.; Tang, D.M.; Liu, P.P.; Xiao, Q.H.; Sun, H. Petrological, geochemical and geochronological constraints on the origin of the Xiadong Ural-Alaskan type complex in NW China and tectonic implication for the evolution of southern Central Asian Orogenic Belt. *Lithos* **2014**, *200*, 226–240. [CrossRef]

18. Bai, Y.; Su, B.X.; Chen, C.; Yang, S.H.; Liang, Z.; Xiao, Y.; Qin, K.Z.; Malaviarachchi, S.P.K. Base metal mineral segregation and Fe-Mg exchange inducing extreme compositions of olivine and chromite from the Xiadong Alaskan-type complex in the southern part of the Central Asian Orogenic Belt. *Ore Geol. Rev.* **2017**, *90*, 184–192. [CrossRef]

19. Su, B.X.; Hu, Y.; Teng, F.Z.; Qin, K.Z.; Bai, Y.; Sakyi, P.A.; Tang, D.M. Chromite-induced magnesium isotope fractionation during mafic magma differentiation. *Sci. Bull.* **2017**, *62*, 1538–1546. [CrossRef]

20. Su, B.X.; Chen, C.; Bai, Y.; Pang, K.N.; Qin, K.Z.; Sakyi, P.A. Lithium isotopic composition of Alaskan-type intrusion and its implication. *Lithos* **2017**, *286–287*, 363–368. [CrossRef]

21. Qin, K.Z.; Zhang, L.C.; Xiao, W.J.; Xu, X.W.; Yan, Z.; Mao, J.W. Overview of major Au, Cu, Ni and Fe deposits and metallogenic evolution of the eastern Tianshan Mountains, Northwestern China. *Tecton. Evol. Metall. Chin. Altay Tianshan* **2003**, *10*, 227–248.

22. Zhang, M.J.; Li, C.S.; Fu, P.E.; Hu, P.Q.; Ripley, E.M. The Permian Huangshanxi Cu-Ni deposit in western China: Intrusive-extrusive association, ore genesis and exploration implications. *Miner. Depos.* **2011**, *46*, 153–170. [CrossRef]

23. Xie, X.J. Global geochemical mapping. *Geol. China* **2003**, *30*, 1–9. (In Chinese with English abstract)

24. Tang, D.M.; Qin, K.Z.; Sun, H.; Su, B.X.; Xiao, Q.H. The role of crustal contamination in the formation of Ni-Cu sulfide deposits in Eastern Tianshan, Xinjiang, Northwest China: Evidence from trace element geochemistry, Re-Os, Sr-Nd, zircon Hf-O, and sulfur isotopes. *J. Asian Earth Sci.* **2012**, *49*, 145–160. [CrossRef]

25. Qi, L.; Jing, H.; Gregoire, D.C. Determination of trace elements in granites by inductively coupled plasma mass spectrometry. *Talanta* **2000**, *51*, 507–513.

26. Qi, L.; Gao, J.; Huang, X.; Hu, J.; Zhou, M.F.; Zhong, H. An improved digestion technique for determination of platinum group elements in geological samples. *J. Anal. Atom. Spectrom.* **2011**, *26*, 1900–1904. [CrossRef]

27. Qi, L.; Zhou, M.F.; Wang, C.Y. Determination of low concentrations of platinum group elements in geological samples by ID-ICP-MS. *J. Anal. Atom. Spectrom.* **2004**, *19*, 1335–1339. [CrossRef]

28. Yu, T.; Chou, H. Anduoite, a new ruthenium arsenide. *Chinese Sci. Bull.* **1979**, *15*, 704–708. (In Chinese with English abstract)

29. Brenan, J.M.; Andrews, D.R.A. High–temperature stability of laurite and Ru Os–Ir alloys and their role in PGE fractionation in mafic magmas. *Can. Mineral.* **2001**, *39*, 341–360. [CrossRef]

30. Garuti, G.; Zaccarini, F.; Proenza, J.A.; Thalhammer, O.A.; Angeli, N. Platinum-group minerals in chromitites of the Niquelândia layered intrusion (Central Goias, Brazil): Their magmatic origin and low-temperature reworking during serpentinization and lateritic weathering. *Minerals* **2012**, *2*, 365–384. [CrossRef]

31. Uysal, İ.; Akmaz, R.M.; Kapsiotis, A.; Demir, Y.; Saka, S.; Avcı, E.; Müller, D. Genesis and geodynamic significance of chromitites from the Orhaneli and Harmancık ophiolites (Bursa, NW Turkey) as evidenced by mineralogical and compositional data. *Ore Geol. Rev.* **2015**, *65*, 26–41. [CrossRef]

32. Ballhaus, C.; Sylvester, P. Noble metal enrichment processes in the Merensky Reef, Bushveld Complex. *J. Petrol.* **2000**, *41*, 545–561. [CrossRef]

33. Wirth, R.; Reid, D.; Schreiber, A. Nanometer-sized platinum-group minerals (PGM) in base metal sulfides: New evidence for an orthomagmatic origin of the Merensky Reef PGE ore deposit, Bushveld Complex, South Africa. *Can. Mineral.* **2013**, *51*, 143–155. [CrossRef]

34. Barkov, A.Y.; Fleet, M.E. An unusual association of hydrothermal platinum-group minerals from the Imandra layered complex, Kola Peninsula, northwestern Russia. *Can. Mineral.* **2004**, *42*, 455–467. [CrossRef]

35. Augé, T.; Genna, A.; Legendre, Q.; Ivanov, K.S.; Volchenko, Y.A. Primary platinum mineralization in the Nizhny Tagil and Kachkanar ultramafic complexes, Urals, Russia: A genetic model for PGE concentration in chromite-rich zones. *Econ. Geol.* **2005**, *100*, 707–732. [CrossRef]

36. Cabri, L.J. The geology, geochemistry, mineralogy and mineral beneficiation of platinum-group elements. *Can. Inst. Min. Metall. Pet.* **2002**, *54*, 13–129.

37. Barkov, A.Y.; Nikiforov, A.A.; Tolstykh, N.D.; Shvedov, G.I.; Korolyuk, V.N. Compounds of Ru-Se-S, alloys of Os-Ir, framboidal Ru nanophases and laurite-clinochlore intergrowths in the Pados-Tundra complex, Kola Peninsula, Russia. *Eur. J. Mineral.* **2017**, *29*, 613–622. [CrossRef]

38. Hattori, K.H.; Cabri, L.J.; Johanson, B.; Zientek, M.L. Origin of placer laurite from Borneo: Se and As contents, and S isotopic compositions. *Mineral. Mag.* **2004**, *68*, 353–368. [CrossRef]

39. Stockman, H.W.; Hlava, P.F. Platinum–group minerals in alpine chromitites from Southwestern Oregon. *Econ. Geol.* **1984**, *79*, 491–508. [CrossRef]

40. Garuti, G.; Fershtater, G.B.; Bea, F.; Montero, P.; Pushkarev, E.; Zaccarini, F. Platinum-group elements as petrological indicators in mafic-ultramafic complexes of the central and southern Urals. *Tectonophysics* **1997**, *276*, 181–194. [CrossRef]

41. González-Jiménez, J.M.; Proenza, J.A.; Gervilla, F.; Melgarejo, J.C.; Blanco-Moreno, J.A.; Ruiz-Sanchez, R.; Griffin, W.L. High-Cr and high-Al chromitites from the Sagua de Tanamo District, Mayari-Cristal ophiolitic massif (Eastern Cuba): Constraints on their origin from mineralogy and geochemistry of chromian spinel and platinum group elements. *Lithos* **2011**, *125*, 101–121. [CrossRef]

42. Barnes, S.J.; Boyd, R.; Korneliusson, A.; Nilsson, L.P.; Often, M.; Pedersen, R.B.; Robins, B. The use of mantle normalization and metal ratios in discriminating between the effects of partial melting, crystal fractionation and sulfide segregation on platinum-group elements, gold, nickel and copper: Examples from Norway. In *Geo-Platinum 87*; Springer: Dordrecht, The Netherlands, 1988; pp. 113–143.

43. Maier, W.D.; Barnes, S.J.; Marsh, J.S. The concentrations of the noble metals in southern African flood-type basalts and MORB: Implications for petrogenesis and magmatic sulphide exploration. *Contrib. Mineral. Pet.* **2003**, *146*, 44–61. [CrossRef]

44. Lorand, J.P.; Alard, O. Platinum-group element abundances in the upper mantle: New constraints from in situ and whole-rock analyses of Massif Central xenoliths (France). *Geochim. Cosmochim. Acta* **2001**, *65*, 2789–2806. [CrossRef]

45. Leblanc, M. Platinum-group glements and gold in Gphiolitic complexes: Distribution and fractionation from mantle to oceanic floor. In *Ophiolite Genesis and Evolution of the Oceanic Lithosphere*; Petters, T.J., Nicolas, A., Coleman, R., Eds.; Springer: Dordrecht, The Netherlands, 1991; pp. 231–260.

46. Fleet, M.E.; Crocket, J.H.; Liu, M.; Stone, W.E. Laboratory partitioning of platinum-group elements (PGE) and gold with application to magmatic sulfide-PGE deposits. *Lithos* **1999**, *47*, 127–142. [CrossRef]

47. Sun, H. Ore-Forming Mechanism in Conduit System and Ore-Bearing Property Evaluation for Mafic-Ultramafic Complex in Eastern Tianshan, Xinjiang. Ph.D. Thesis, Institute of Geology and Geophysics, Chinese Academy of Sciences, Beijing, China, 2009. (In Chinese with English abstract).

48. Kiseleva, O.N.; Zhmodik, S.M.; Damdinov, B.B.; Agafonov, L.V.; Belyanin, D.K. Composition and evolution of PGE mineralization in chromite ores from the Il'chir ophiolite complex (Ospa-Kitoi and Khara-Nur areas, East Sayan). *Russ. Geol. Geophys.* **2014**, *55*, 259–272. [CrossRef]

49. Zaccarini, F.; Garuti, G.; Pushkarev, E.V. Unusually PGE-rich chromitite in the Butyrin vein of the Kytlym Uralian-Alaskan complex, Northern Urals, Russia. *Can. Mineral.* **2011**, *49*, 52–72. [CrossRef]

50. Maier, W.D. Platinum-group element (PGE) deposits and occurrences: Mineralization styles, genetic concepts, and exploration criteria. *J. Afr. Earth Sci.* **2005**, *41*, 165–191. [CrossRef]

51. Song, X.Y.; Li, X.R. Geochemistry of the Kalatongke Ni-Cu-(PGE) sulfide deposit, NW China: Implications for the format ion of magmatic sulfide mineralization in a post collisional environment. *Miner. Depos.* **2009**, *44*, 303–327. [CrossRef]

52. Barnes, S.J.; Picard, C.P. The behaviour of platinum-group elements during partial melting, crystal fractionation, and sulfide segregation: An example from the Cape Smith Fold Belt, northern Quebec. *Geochim. Cosmochim. Acta* **1993**, *57*, 79–87. [CrossRef]

53. Guillou-Frottier, L.; Burov, E.; Augé, T.; Gloaguen, E. Rheological conditions for emplacement of Ural-Alaskan-type ultramafic complexes. *Tectonophysics* **2014**, *631*, 130–145. [CrossRef]

54. Qin, K.Z.; Sun, H.; San, J.Z.; Xu, X.W.; Tang, D.M.; Ding, K.S.; Xiao, Q.H.; Su, B.X. Tectonic setting, geological features and evaluation of ore-bearing property for magmatic Cu-Ni deposits in eastern Tianshan, NW China: Proceedings of Xi'an International Ni-Cu (Pt) Deposit Symposium. *Northwest. Geol.* **2009**, *42*, 95–99.

55. Sun, H.; Qin, K.Z.; Li, J.X.; Tang, D.M.; Fan, X.; Xiao, Q.H. Constraint of mantle partial melting on PGE mineralization of mafic-ultramafic intrusions in Eastern Tianshan: Case study on Tulargen and Xiangshan Cu-Ni deposits. *Acta Petrol. Sin.* **2008**, *24*, 1079–1086. (In Chinese with English abstract) [CrossRef]

56. Qian, Z.Z.; Sun, T.; Tang, Z.L.; Jiang, C.Y.; He, K.; Xia, M.Z.; Wang, J.Z. Platinum-group elements geochemistry and its significances of the Huangshandong Ni-Cu sulfide deposit, Eastern Tianshan, China. *Geol. Rev.* **2009**, *55*, 874–884. (In Chinese with English abstract)

57. Wang, M.F.; Xia, Q.L.; Xiao, F.; Wang, X.Q.; Yang, W.S.; Jiang, C.L. Rock geochemistry and platinum group elements characteristics of Tudun Cu-Ni sulfide deposit in Eastern Tianshan Mountains of Xinjiang and their metallogenic implications. *Miner. Depos.* **2012**, *31*, 1195–1210. (In Chinese with English abstract)

58. Sun, T.; Qian, Z.Z.; Tang, Z.L.; Jiang, C.Y.; He, K.; Sun, Y.L.; Wang, J.Z.; Xia, M.Z. Zircon U-Pb chronology platinum-group elements geochemistry characteristics of Hulu Cu-Ni deposit, Eastern Xinjiang and its geological significance. *Acta Petrol. Sin.* **2010**, *26*, 3339–3349. (In Chinese with English abstract)

59. Chai, F.M.; Xia, F.; Chen, B.; Lu, H.F.; Wang, H.; Li, J.; Yan, Y.P. Platinum group elements geochemistry of two mafic-ultramafic intrusions in the Beishan block, Xinjiang, NW China. *Acta Geol. Sin.* **2013**, *87*, 474–685. (In Chinese with English abstract)

60. Tang, D.M.; Qin, K.Z.; Sun, H.; Qi, L.; Xiao, Q.H.; Su, B.X. PGE Geochemical characteristics of Tianyu magmatic Cu-Ni deposit: Implications for magma evolution and sulfide segregation. *Acta Geol. Sin.* **2009**, *83*, 680–697. (In Chinese with English abstract)

61. Chai, F.M.; Zhang, Z.C.; Mao, J.W.; Dong, L.H.; Ye, H.S.; Wu, H.; Mo, X.H. Platinum group elements geochemistry of Baishiquan mafic-ultramafic intrusions in Central Tianshan block, Xinjiang. *Acta Geol. Sin.* **2006**, *27*, 123–128. (In Chinese with English abstract)

62. Xue, S.C.; Qin, K.Z.; Li, C.S.; Tang, D.M.; Mao, Y.J.; Qi, L.; Ripley, E.M. Geochronological, petrological, and geochemical constraints on Ni-Cu sulfide mineralization in the Poyi ultramafic-troctolitic intrusion in the northeast rim of the Tarim craton, western China. *Econ. Geol.* **2016**, *111*, 1465–1484. [CrossRef]

63. Qian, Z.Z.; Wang, J.Z.; Jiang, J.G.; Yan, H.Q.; He, K.; Sun, T. Geochemistry characters of platinum group elements and its significations on the process of mineralization in the Kalatongke Cu-Ni sulfide deposit, Xinjiang, China. *Acta Petrol. Sin.* **2009**, *25*, 832–844. (In Chinese with English abstract)

64. Zhang, Z.W.; Tang, Q.Y.; Li, C.S.; Wang, Y.L.; Ripley, E.M. Sr-Nd-Os-S isotope and PGE geochemistry of the Xiarihamu magmatic sulfide deposit in the Qinghai-Tibet plateau, China. *Miner. Depos.* **2016**, *52*, 51–68. [CrossRef]

65. Qin, K.Z.; Tang, D.M.; Su, B.X.; Mao, Y.J.; Xue, S.C.; Tian, Y.; Sun, H.; San, J.Z.; Xiao, Q.H.; Deng, G. The tectonic setting, style, basic feature, relative erosion degree, ore-bearing evaluation sign, potential analysis of mineralization of Cu-Ni-bearing Permian mafic-ultramafic complexes, northern Xinjiang. *Northwest. Geol.* **2012**, *45*, 83–116. (In Chinese with English abstract)

Article

PGE-Enrichment in Magnetite-Bearing Olivine Gabbro: New Observations from the Midcontinent Rift-Related Echo Lake Intrusion in Northern Michigan, USA

Alexander James Koerber and Joyashish Thakurta *

Department of Geological and Environmental Sciences, Western Michigan University, 1903 W Michigan Ave, Kalamazoo, MI 49008, USA; alexander.j.koerber@wmich.edu
* Correspondence: joyashish.thakurta@wmich.edu; Tel.: +1-(269)-387-3667

Received: 25 November 2018; Accepted: 24 December 2018; Published: 29 December 2018

Abstract: The Echo Lake intrusion in the Upper Peninsula (UP) of Michigan, USA, was formed during the 1.1 Ga Midcontinent Rift event in North America. Troctolite is the predominant rock unit in the intrusion, with interlayered bands of peridotite, mafic pegmatitic rock, olivine gabbro, magnetite-bearing gabbro, and anorthosite. Exploratory drilling has revealed a platinum group element (PGE)-enriched zone within a 45 m thick magnetite-ilmenite-bearing olivine gabbro unit with grades up to 1.2 g/t Pt + Pd and 0.3 wt. % Cu. Fine, disseminated grains of sulfide minerals such as pyrrhotite and chalcopyrite occur in the mineralized interval. Formation of Cu-PGE-rich sulfide minerals might have been caused by sulfide melt saturation in a crystallizing magma, which was triggered by a sudden decrease in fO_2 upon the crystallization and separation of titaniferous magnetite. This PGE-enriched zone is comparable to other well-known reef-like PGE deposits, such as the Sonju Lake deposit in northern Minnesota.

Keywords: Echo Lake; Midcontinent; palladium; platinum; magnetite; gabbro

1. Introduction

Magmatism during the 1.1 Ga old Midcontinent Rift (MCR) event is associated with the origin of several metallic mineral deposits in the Great Lakes region of North America [1–5]. These mineral deposits have been classified into two groups: sulfide-poor, PGE (Platinum Group Element)-rich, layered intrusions and sulfide-rich, conduit-type, high-grade Ni-Cu sulfide deposits (Figure 1) [6,7].

In the first group, several mafic-ultramafic intrusive bodies of the Duluth Complex in northern Minnesota have been identified and studied [3,8,9], which include the Partridge River and South Kawishiwi intrusions, emplaced between 1098 and 1107 Ma [10]. These intrusions are known for basal troctolitic units which host disseminated Cu-Ni-PGE sulfide deposits with ore grades in the following ranges: 0.5–0.7 g/t Pt + Pd + Au; 0.09–0.24 wt. % Ni; 0.27–0.66 wt. % Cu [9]. The Sonju Lake intrusion in the Beaver Bay Complex in northeastern Minnesota shows a PGE-rich interval hosted in an oxide-gabbro unit with precious metallic grades up to Pd: 410 ppb; Pt: 275 ppb and Au: 1.08 ppm [11,12].

The Eagle intrusion, in the Upper Peninsula (UP) of Michigan and the Tamarack intrusion in northeastern Minnesota are the two most important members of the second group. The Eagle Ni-Cu sulfide deposit is hosted within two mafic-ultramafic intrusive bodies, called the Eagle and the Eagle East intrusions, which are associated with the Baraga dike swarm in Marquette County, Michigan [13]. The U-Pb baddeleyite age of the Eagle intrusion is 1107.2 ± 5.7 Ma [13]. The massive, semi-massive, and disseminated sulfide ore deposits in both intrusions are hosted in plagioclase-bearing peridotite, melanocratic troctolite, and olivine gabbro. The sulfide ore grades are up to 6.11 wt. % Ni and 4.15 wt. % Cu, and up to 25 ppm Pt in Cu-rich stringers [13]. The Tamarack intrusion emplaced during the early stages of the MCR at 1105.6 ± 1.2 Ma [14] is located about 80 km west of the Duluth Complex in Minnesota and it is associated with the Carlton County dike swarm [15]. The funnel-shaped intrusion is composed of peridotite, plagioclase-bearing peridotite, plagioclase-bearing pyroxenite, melanocratic troctolite, and melanocratic gabbro [16]. Massive, semi-massive, and disseminated sulfide-bearing intervals have been identified from drill-core studies with reported drill-intersections of up to 166 m with 2.33 wt. % Ni, 1.24 wt. % Cu, and 0.75 g/t Pt + Pd + Au [16].

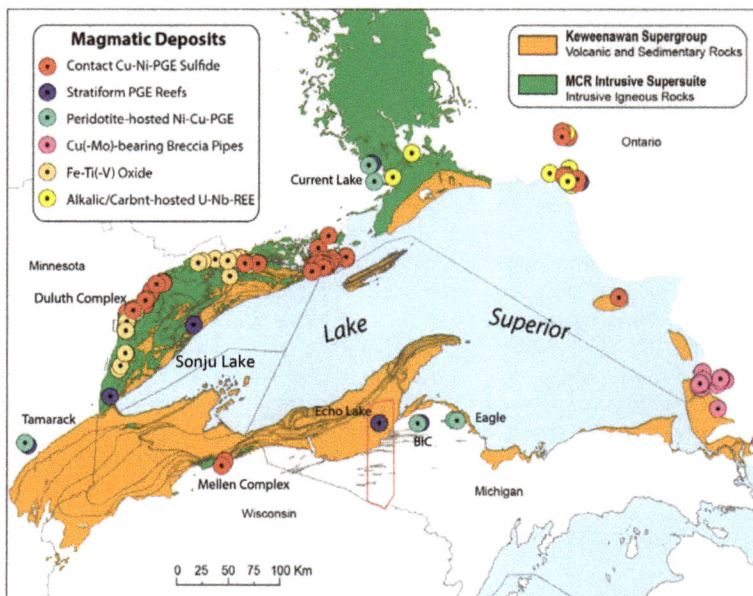

Figure 1. Magmatic ore deposits associated with the Midcontinent Rift around Lake Superior. The red box shows the location and extent of the exploration prospect zone called the Voyageur Lands [6,7].

This study is focused on the newly discovered Echo Lake intrusion, located in the Houghton and Ontonagon counties in UP Michigan (Figures 1 and 2) [17]. Exploratory drilling has identified this intrusion as a new prospect for sulfide mineralization in the Midcontinent Rift region [18]. The basement rock in the area of the intrusion is an Archean granite-gneiss [19], which is overlain by the Paleoproterozoic Michigamme Formation [20]. The latter consists of meta-greywacke, slate, metamorphosed mafic volcanics, and iron formations. The 1110 Ma Mesoproterozoic Siemens Creek Volcanics unit [1] unconformably overlies the Michigamme Formation. The Neoproterozoic Jacobsville Formation [21] overlies both the Siemens Creek Volcanics unit and the Michigamme Formation. The uppermost layer of the Echo Lake intrusion has been encountered by drill-cores at about 200 m underneath the Jacobsville Formation, and it is estimated to extend laterally for about 18 square kilometers [18]. The complete thickness of this intrusion is unknown at this stage, but a continuous succession of mafic-ultramafic rocks for more than 1 km has been reported. Being located between the Keweenaw and the Marenisco regional faults [17], the Echo Lake intrusion might have been uplifted from greater depths by displacements along these two faults. Apart from Echo Lake, there are several other small outcrops of mafic-ultramafic bodies in the area, such as the Bluff, Haystack, and Skinny intrusions (Figure 2) [22], which are mostly composed of gabbro, gabbronorite, and pyroxenite. The ages and mutual relationships of these intrusions are unknown.

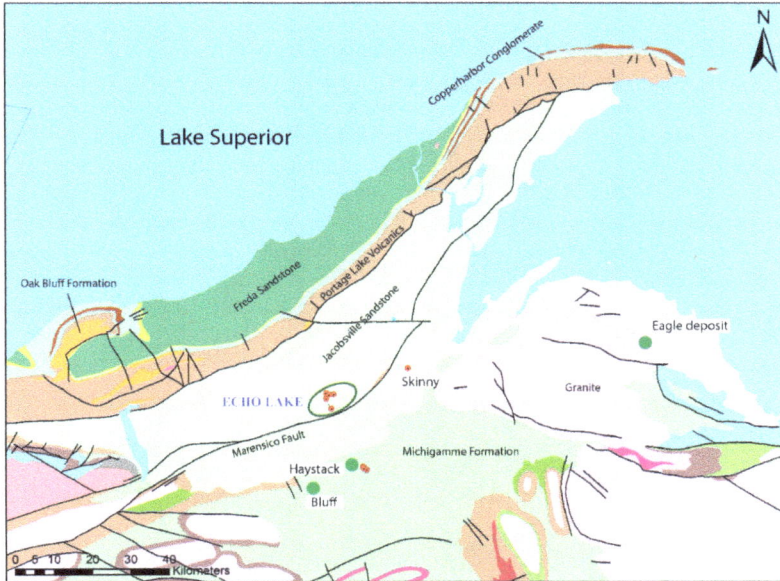

Figure 2. Location of Echo Lake prospect in the Upper Peninsula of Michigan, near the southern shore of Lake Superior [17]. Green circles indicate a few other metal prospects and mines in the area. Red circles indicate drill core locations.

The Echo Lake intrusion has been dated at 1111 Ma [23], which makes it contemporaneous with the Early Rift-forming phase of the MCR. Troctolite is the predominant rock in the intrusion, but recurrent layers of peridotite, mafic pegmatitic rock, olivine- and magnetite-bearing gabbro, and anorthosite have also been documented. This is comparable with the rhythmic layering reported from the Duluth Complex and other layered mafic-ultramafic complexes around the world, such as the Bushveld in South Africa [24] and Stillwater, in Montana, USA [25,26]. Our aims in the present study are to examine the PGE-enriched zone in the Echo Lake intrusion. Petrological analysis, along with whole rock and mineral compositional data, were used to highlight key characteristics of the Echo Lake intrusion and to compare them with the known sulfide-mineralized intrusions in the MCR.

2. Methods

For this study, one hundred and seventy samples were collected from drill cores EL-1, EL-97-05, EL-97-03, and 10EL-001 in the core repositories of Altius Resources Inc. (St. John's, NL, Canada) and the Department of Environmental Quality, Michigan.

Electron microprobe analysis (EMPA) was performed at the Department of Geoscience, University of Wisconsin-Madison, USA, using a 5-spectrometer, CAMECA SX5 electron microprobe (CAMECA, Gennevilliers, France). The probe was operated at 15 kV with a beam Faraday cup current of 20 nA and beam size of two microns. Analytical standards are both natural and synthetic silicate minerals, glasses, and oxides. Counting times were 10 s on peak. Background intensities were determined by the mean atomic number background algorithm [27].

Bulk rock chemical analyses were performed at Geoscience Laboratories in Sudbury, Ontario, Canada. Inductively Coupled Plasma Mass Spectrometer (ICP-MS) technique was used to analyze for Ni, Cu, PGE and Au. Sample analytes were prepared using fire-assay and closed vessel acid-digestion methods.

3. Results

Troctolite and olivine gabbro are the two dominant rock types in the intrusion. Troctolite samples are typically composed of 60 vol. % plagioclase, 30 vol. % olivine, 5% vol. % pyroxene, minor Fe-Ti oxides, traces of biotite and sulfide minerals such as pyrrhotite, chalcopyrite, and pentlandite (Figure 3A,B). Olivine gabbro samples are typically composed of about 50% olivine, 30% plagioclase, 10–15% pyroxene, 5–10% Fe-Ti oxides, and traces of sulfide minerals. Sulfide minerals in troctolite and olivine gabbro samples are disseminated grains of pyrrhotite, chalcopyrite, and pentlandite which form mutual intergrowths with one another (Figure 3B). Traces of other sulfide minerals, like digenite and bornite, have also been found. Sulfide minerals often form thin rims around titaniferous magnetite and ilmenite crystals (Figure 4C). Titaniferous magnetite occurs as intergrowths with biotite (Figure 4A). Titaniferous magnetite also shows delicate intergrowths and reaction textures with sulfide minerals (Figure 4B,C).

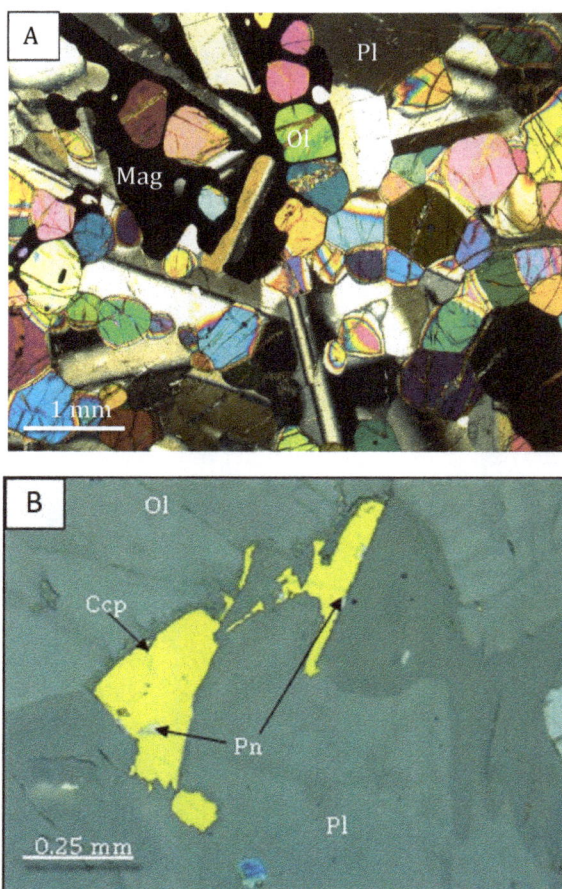

Figure 3. Troctolite showing: (**A**) Intergranular texture between plagioclase and olivine (Ol). Interstitial titaniferous magnetite (Mag) is seen surrounding grains of plagioclase (Pl) and olivine (transmitted light); (**B**) Fine intergrowths of chalcopyrite (Ccp) and pentlandite (Pn) in the interstices of olivine and plagioclase (reflected light). The blue hexagonal grain at the bottom is digenite.

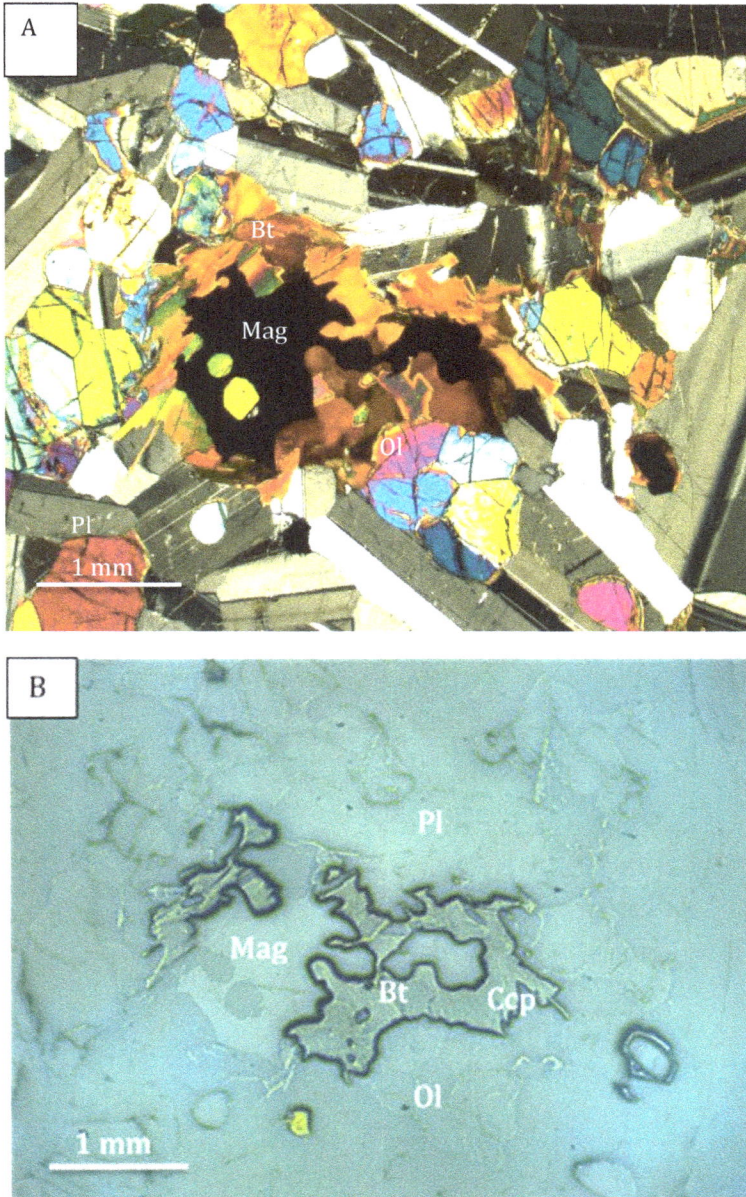

Figure 4. *Cont.*

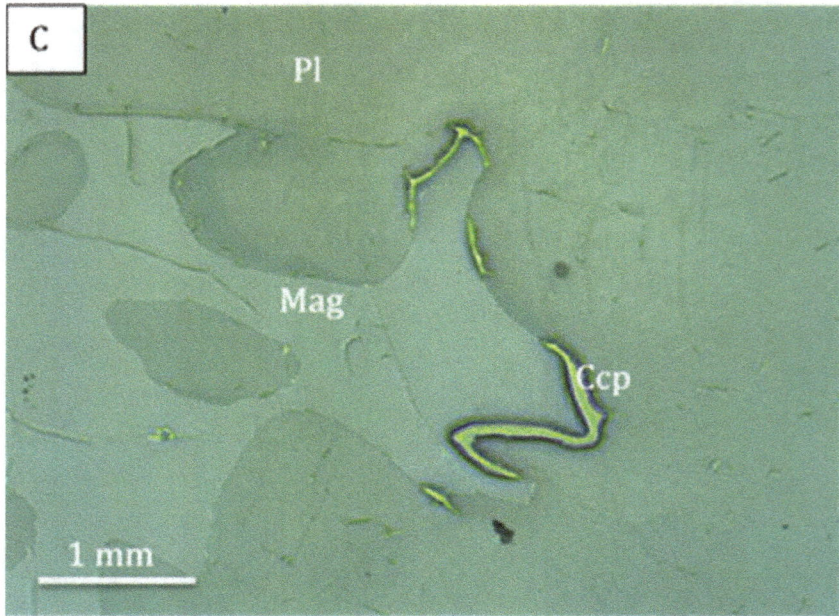

Figure 4. Magnetite-bearing olivine gabbro showing: (**A**) Cumulus texture in plagioclase and olivine with biotite (Bt)-magnetite (Mag) intergrowth in small interstitial spaces (transmitted light); (**B**) Biotite-magnetite intergrowth and fine disseminated chalcopyrite (reflected light); (**C**) Thin rim of chalcopyrite (Ccp) around titaniferous magnetite (Mag) in contact with plagioclase (Pl).

A stratigraphic sequence of the observed layered units in the Echo Lake intrusion, as revealed from drill core studies, is shown in Figures 5–7. Notable PGE-enrichment is observed within a 45 m thick layer of Fe-Ti oxide rich olivine gabbro. Titaniferous magnetite and ilmenite are the principal oxide minerals. These minerals mostly constitute 10–25 vol. % of the oxide-rich olivine gabbro unit. However, in some parts, the oxide minerals form massive zones with fine interstitial grains of silicate minerals (Figure 4). At the Fe-Ti oxide-rich zone, there is an interval with disseminated grains of sulfide minerals. Peak concentrations of Pt and Pd of 550 ppb and 634 ppb, respectively, are measured in this interval at a depth of about 998 m [18]. Pt and Pd contents show positive correlation in all studied depths of the intrusion, although Pd shows a higher concentration in most intervals (Figures 5 and 6). Stratigraphic variations of MgO, TiO_2, Cr_2O_3, and FeO (total) in the layered sequences are shown in Figure 8.

The whole-rock major element compositions of the major rock-types are shown in Table 1. Compositions of olivine, clinopyroxene, plagioclase, and titaniferous magnetite in the major rock types are shown in Tables 2–5. Olivine composition in the troctolite units of the Echo Lake intrusion shows a range between Fo_{57-62}. Clinopyroxene in these rocks plot within the compositional range of augite (Table 3). Plagioclase compositions range between An_{55-75} (Table 4). The Cr_2O_3-content of titaniferous magnetite in most intervals is about 3 wt. % (Table 5), although in some local intervals a Cr_2O_3 content above 9 wt. % is seen. The TiO_2-content in titaniferous magnetite in troctolite and olivine gabbro ranges between 7 and 10 wt. %, but in olivine-magnetite gabbro it ranges between 12 and 18 wt. %. The high TiO_2-content in magnetite is indicative of the possible existence of fine exsolution lamellae of ulvöspinel.

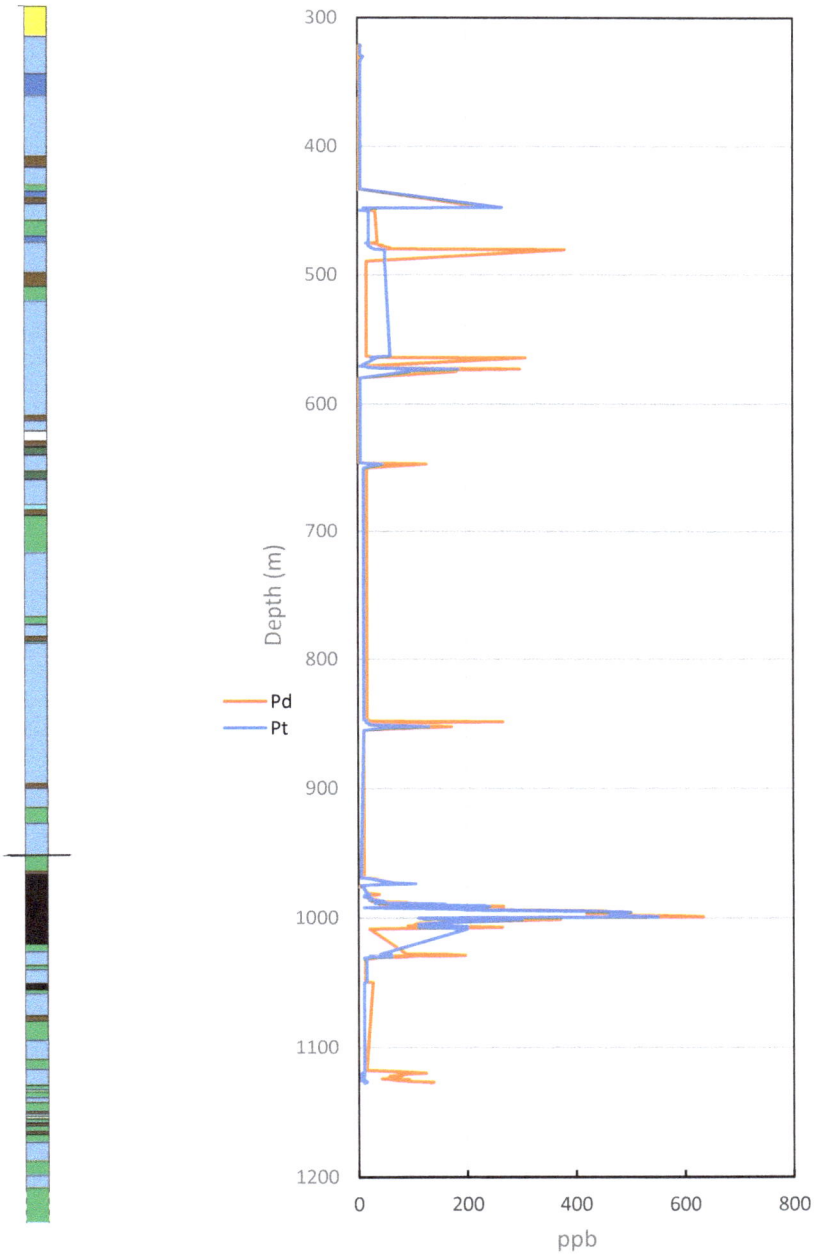

Figure 5. The layered ultramafic-mafic succession observed in the Echo Lake intrusion and locations of the zones of platinum group element (PGE)-enrichment in the intrusion. The magnetite-bearing olivine gabbro shows peak concentrations. The most highly mineralized interval between 970 m and 1050 m has been expanded in Figure 6.

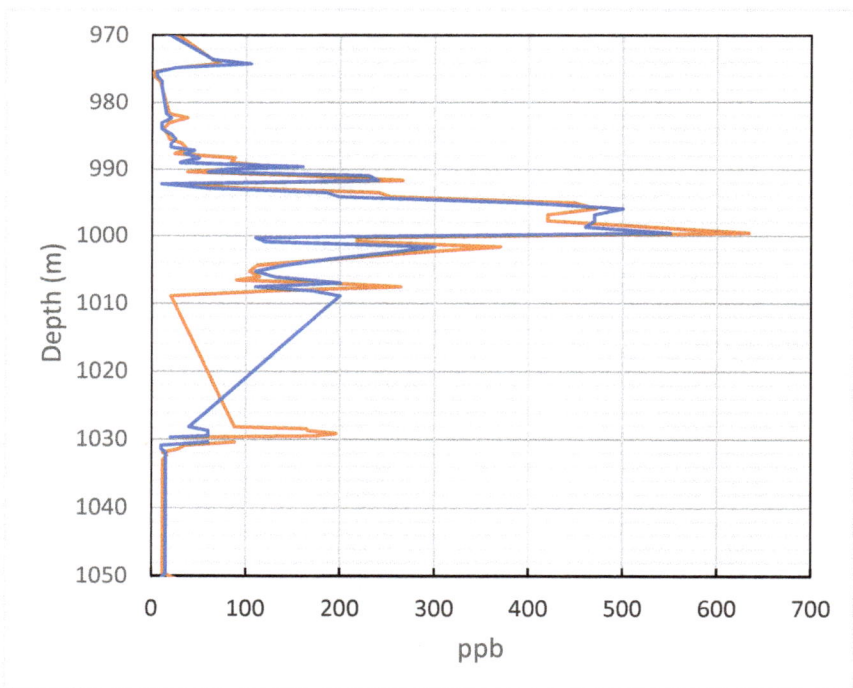

Figure 6. PGE concentrations in the mineralized zone.

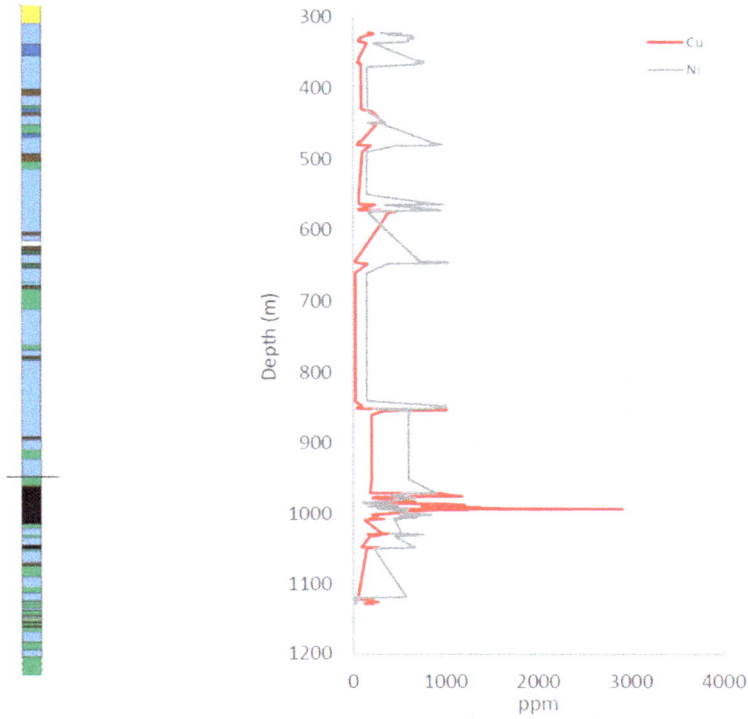

Figure 7. Concentrations of Cu and Ni in the layered succession.

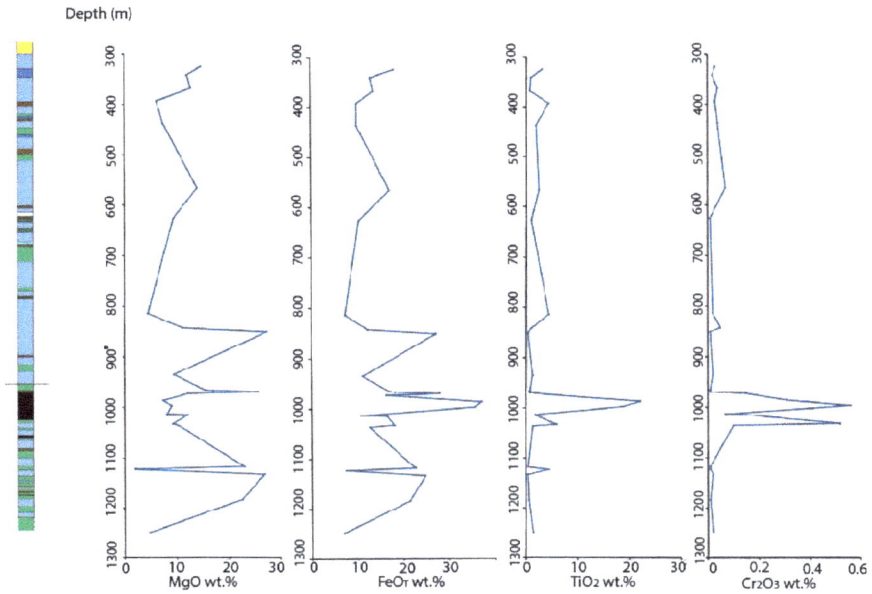

Figure 8. Stratigraphic variation plots of MgO, TiO_2, FeO_T, and Cr_2O_3.

Table 1. Major element compositions of principal rock types.

Drill Core	EL9703	10EL001	EL9703	EL9703	EL9703	EL9703	10EL001	EL9703	EL9703
Rock Type	Trt	Trt	Trt	Trt	Trt	Trt	Trt	Trt	Trt
Depth (m)	325.5	393.0	436.2	565.7	626.4	842.8	814.1	966.8	1015.0
SiO_2	43.50	48.47	46.83	43.50	47.01	45.70	49.69	42.80	43.60
TiO_2	1.45	2.15	0.90	1.21	0.47	0.44	2.20	0.28	0.93
Al_2O_3	12.30	18.75	21.55	13.90	20.60	18.10	21.23	12.60	16.35
Cr_2O_3	0.03	0.03	0.04	0.07	0.01	0.05	0.02	0.01	0.14
Fe_2O_3	19.90	11.01	11.01	18.90	11.52	13.50	8.07	20.10	18.20
FeO	17.91	9.90	9.91	17.01	10.37	12.15	7.26	18.09	16.38
MnO	0.22	0.15	0.12	0.20	0.13	0.14	0.11	0.22	0.18
MgO	14.60	6.13	7.40	13.90	9.36	11.10	4.46	16.90	12.05
CaO	6.24	9.71	9.61	6.38	9.07	8.29	11.30	5.80	7.23
Na_2O	1.79	2.74	2.79	2.00	2.70	2.34	3.10	1.70	2.38
K_2O	0.33	0.66	0.31	0.31	0.26	0.18	0.51	0.14	0.21
P_2O_5	0.15	0.27	0.07	0.14	0.06	0.03	0.07	0.03	0.02
TOTAL	99.60	100.56	100.55	99.40	100.85	99.60	100.98	99.40	100.20

Drill Core	8412	EL9703	EL9703	8413	8414	EL9703	EL9703	10EL001	EL9703
Rock Type	Mt-Ol-gab	Mt-Ol-gab	Mt-Ol-gab	Ol-gab	Ol-gab	Ol-gab	Ol-gab	Ol-gab	Ol-gab
Depth (m)	935.4	965.6	972.9	1013.8	1032.7	1036.3	1122.6	1248.3	1183.5
SiO_2	45.80	44.03	45.40	47.10	41.80	45.80	48.50	48.04	40.60
TiO_2	0.60	0.35	1.83	0.85	2.92	0.63	2.20	0.68	0.25
Al_2O_3	21.29	14.73	11.90	19.40	15.25	18.30	21.50	23.93	8.64
Cr_2O_3	0.02	0.01	0.15	0.07	0.52	0.10	0.01	0.02	0.01
Fe_2O_3	12.24	18.79	18.00	11.80	19.90	13.90	8.25	7.62	23.70
FeO	11.01	16.91	16.20	10.62	17.91	12.51	7.42	6.86	21.33
MnO	0.14	0.21	0.21	0.13	0.19	0.14	0.09	0.10	0.25
MgO	9.40	15.36	12.00	8.06	9.18	10.30	1.98	4.80	22.50
CaO	8.79	6.50	9.64	9.14	8.38	8.06	11.40	10.43	3.91
Na_2O	2.54	1.93	1.59	2.67	2.24	2.49	3.26	3.00	1.10
K_2O	0.32	0.19	0.17	0.25	0.26	0.19	0.47	0.37	0.10
P_2O_5	0.08	0.03	0.07	0.03	0.04	0.03	0.32	0.06	0.03
TOTAL	100.92	100.98	100.30	99.20	99.73	99.30	98.40	100.75	100.40

Trt = troctolite; Ol-gab = olivine gabbro; Mt-Ol-gab = magnetite-bearing olivine gabbro.

Table 2. Composition of olivine in the major rock types.

Drill Core	EL9703	EL9703	EL9703	EL9703	10EL001	EL9703	EL9703
Rock Type	Trt	Trt	Ol-gab	Ol-gab	Ol-gab	Mt-Ol-gab	Mt-Ol-gab
Depth (m)	440.2	440.2	935.5	1008.91	1158.4	992.64	985.7
SiO_2	35.09	36.44	35.85	35.62	36.68	35.60	35.79
FeO	33.20	33.25	33.43	34.55	34.48	35.46	33.36
MnO	0.45	0.43	0.46	0.45	0.47	0.44	0.38
MgO	30.29	30.41	29.91	29.30	28.76	28.72	30.43
NiO	0.13	0.09	0.12	0.12	0.07	0.11	0.03
CaO	0.06	0.05	0.03	0.06	0.04	0.10	0.08
TOTAL	99.23	100.66	99.79	100.09	100.50	100.43	100.06
Normalized to 4 oxygens							
Si	0.97	1.01	0.98	0.99	1.01	0.99	0.99
Fe(2+)	0.77	0.77	0.77	0.81	0.80	0.83	0.77
Mn	0.01	0.01	0.01	0.01	0.01	0.01	0.01
Mg	1.25	1.25	1.22	1.22	1.18	1.20	1.26
TOTAL	3.00	3.05	2.98	3.04	3.01	3.03	3.04
Fo (mol %)	61.6	61.7	61.1	59.9	59.5	58.8	61.7
Ni (ppm)	1060.7	739.9	950.1	926.0	575.0	838.5	259.0

Table 3. Composition of clinopyroxene in the major rock types.

Drill Core	EL9703	EL9703	EL9703	EL9703	EL9703	10EL001
Rock Type	Trt	Trt	Ol-gab	Mt-gab	Ol-gab	Ol-gab
Depth (m)	440.22	440.22	935.52	985.76	1008.91	1158.4
SiO_2	51.06	51.31	51.85	51.45	51.22	51.20
TiO_2	0.86	0.85	0.96	1.15	1.08	0.95
Al_2O_3	2.13	2.28	2.59	2.57	2.34	2.42
Cr_2O_3	0.11	0.13	0.07	0.13	0.11	0.06
$FeO_{(total)}$	9.26	10.07	9.45	10.96	9.91	9.57
MnO	0.24	0.25	0.23	0.28	0.24	0.24
MgO	14.80	15.66	14.67	14.12	14.35	14.33
CaO	20.73	19.02	20.69	19.72	20.78	21.31
Na_2O	0.27	0.23	0.32	0.28	0.28	0.31
TOTAL	99.49	99.83	100.84	100.66	100.32	100.43
Normalized to 6 oxygens						
Si	1.91	1.92	1.94	1.92	1.91	1.91
Al	0.09	0.10	0.11	0.11	0.10	0.11
Cr	0.00	0.00	0.00	0.00	0.00	0.00
Fe(3+)	0.01	0.06	0.16	0.06	0.06	0.09
Fe(2+)	0.28	0.25	0.14	0.28	0.25	0.21
Ti	0.02	0.02	0.03	0.03	0.03	0.03
Mn	0.01	0.01	0.01	0.01	0.01	0.01
Mg	0.82	0.87	0.82	0.79	0.80	0.80
Ca	0.83	0.76	0.83	0.79	0.83	0.85
Na	0.02	0.02	0.02	0.02	0.02	0.02
TOTAL	4.00	4.02	4.05	4.02	4.02	4.03
Wo	42.1	38.6	42.0	40.5	42.3	43.1
En	41.8	44.2	41.5	40.4	40.6	40.3
Fs	15.1	16.3	15.3	18.0	16.1	15.5
Ae	1.0	0.9	1.2	1.0	1.0	1.1

Table 4. Composition of plagioclase in the major rock types.

Drill Core	EL9703	EL9703	EL9703	EL9703
Rock Type	Trt	Trt	Mt-Ol-gab	Ol-gab
Depth (m)	440.2	440.2	985.76	1008.91
SiO_2	54.15	51.66	53.50	53.56
Al_2O_3	28.50	31.09	29.73	29.49
FeO	0.33	0.47	0.26	0.48
CaO	11.38	13.64	11.78	11.91
Na_2O	4.82	2.29	4.77	4.48
K_2O	0.41	0.39	0.33	0.45
TOTAL	99.59	99.54	100.37	100.36
Normalized to 8 oxygens				
Si	9.81	9.36	9.69	9.70
Al	6.08	6.64	6.35	6.30
Fe (2+)	0.05	0.07	0.04	0.07
Ca	2.21	2.65	2.29	2.31
Na	1.69	0.80	1.68	1.57
K	0.10	0.09	0.08	0.10
TOTAL	19.94	19.61	20.12	20.06
An	55.3	74.7	56.6	58.0
Ab	42.4	22.7	41.5	39.4
Or	2.4	2.4	1.9	2.6

Table 5. Composition of titaniferous magnetite in major rock types.

Drill Core	EL97-03	EL-97-03	EL-97-03	EL-97-03	10EL001	10EL001
Rock Type	Ol-gab	Mt-gab	Mt-gab	Trt	Ol-gab	Ol-gab
Depth (m)	980.67	985.40	984.4	440.2	968.96	1158.4
TiO_2	5.66	17.38	15.98	8.32	7.01	5.44
Al_2O_3	3.88	3.76	4.27	5.45	3.87	2.21
FeO tot	84.62	74.02	73.21	75.01	83.70	88.44
Fe_2O_3 (calc)	49.87	29.19	29.80	37.56	47.66	53.15
FeO (calc)	34.76	44.82	43.40	37.45	36.05	35.29
MnO	0.30	0.32	0.38	0.41	0.26	0.27
MgO	1.12	1.67	1.72	1.25	1.12	0.56
Cr_2O_3	3.42	2.19	3.37	9.29	3.23	2.73
TOTAL	99.00	99.34	98.93	99.73	99.18	99.65
Normalized with respect to 3 cations						
Ti	0.15	0.47	0.43	0.22	0.19	0.15
Al	0.16	0.16	0.18	0.23	0.16	0.09
Fe(3+)	1.44	0.85	0.86	1.07	1.37	1.54
Fe(2+)	1.08	1.37	1.33	1.14	1.12	1.11
Mn	0.01	0.01	0.01	0.01	0.01	0.01
Mg	0.06	0.09	0.09	0.07	0.06	0.03
Cr	0.10	0.06	0.10	0.26	0.09	0.08
TOTAL	3.00	3.00	3.00	3.00	3.00	3.00

4. Discussion

In this section the PGE-enrichment reported from the Echo Lake intrusion is discussed with reference to the geochemical principles of sulfide-melt saturation in basaltic and picritic magmas. These discussions are then placed in the context of economic mineral deposits in igneous intrusive bodies associated with the Midcontinent Rift.

4.1. Sulfide-Melt Saturation in Basaltic and Picritic Magmas

Sulfur is an incompatible element during crystallization of silicate minerals, and thus it can remain dissolved in magma until a late stage of crystallization. Sulfur concentration in magma can increase by mechanisms such as fractional crystallization and the assimilation of sulfur from external sources [28–31]. Upon the attainment of sulfide liquid saturation, an immiscible sulfide liquid is formed from the magma, which sequesters chalcophile elements such as Cu, Ni, and PGE. For most S-rich Ni-Cu deposits, such as Eagle [32] and Tamarack [33], incorporation of external sulfur is regarded as an essential requirement to attain sulfide liquid saturation [31]. However, for sulfur-poor PGE deposits, such as the Partridge River and South Kawishiwi intrusions in the Duluth Complex, prolonged fractional crystallization of magma has been proposed as an effective mechanism to induce sulfide liquid saturation [5,31].

Sulfide liquid saturation is also strongly influenced by fO_2 [34,35] and the solubility of sulfur in the magma can be inversely correlated with the Fe^{2+} content. Prolonged fractional crystallization in a closed system is one mechanism for increasing the fO_2 of the magma. Early fractionations of anhydrous minerals such as olivine, plagioclase, and pyroxene in a closed system can increase the mole fraction of dissolved H_2O in the magma [36]. In an open system, H_2O-content of the magma can be increased by assimilation of hydrous country rocks [37,38]. These factors can cause higher oxygen availability, and thus cause higher fO_2 in the magma [31,39–42]. As proposed by Ripley and Li [31], a rise in fO_2, caused by an increase in the H_2O-content of the magma changes the oxidation state of Fe from Fe^{2+} to Fe^{3+}, based on the reaction:

$$H_2O + 2FeO = Fe_2O_3 + H_2$$

This reaction causes a decrease in the effective Fe^{2+} content and thereby increases the solubility of sulfur [34]. Therefore, the formation of Fe_2O_3 during magmatic crystallization is indicative of high ambient fO_2. However, continued fractional separation of Fe_2O_3 from the magma eventually increases the relative concentration of FeO and decreases the solubility of sulfur. Sulfide liquid saturation can then be attained.

Sulfur content at sulfide saturation (SCSS) is the amount of sulfur at the time of attainment of sulfide liquid saturation in the magma [31,43]. During crystallization, the lowering of temperature decreases the solubility of sulfur and thereby lowers the value of SCSS. However, as explained above, an increase in fO_2 of the magma, as indicated by an increase in the Fe_2O_3 content of the crystallized minerals, can increase the SCSS substantially [44–47], and thereby prevent sulfide liquid saturation.

Conversely, a decrease in SCSS, caused by a sudden drop in fO_2, can induce sulfide liquid saturation and consequently can cause the separation of tiny droplets of sulfide liquid from the magma. In this case, the determinative factor for sulfide liquid separation could be the crystallization of large quantities of oxide phases, such as titaniferous magnetite and ilmenite from the magma. The separation by fractional crystallization of Fe^{3+} and Ti-oxide phases from the magma can thus reduce the SCSS, cause sulfide liquid saturation, and a rapid separation of sulfide liquid from the magma. As proposed by O'Neill and Mavrogenes [46], the excess sulfur displaces O^{2-} anions that bond with Fe^{2+} and forms a Fe-rich immiscible sulfide liquid as seen in the reaction:

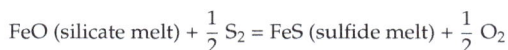

$$FeO \text{ (silicate melt)} + \frac{1}{2} S_2 = FeS \text{ (sulfide melt)} + \frac{1}{2} O_2$$

4.2. Layered and Conduit-Type Intrusions Associated with the MCR

In the MCR-related layered intrusions of northern Minnesota, PGE mineralization is hosted within sulfide-bearing horizons in troctolitic and gabbroic rocks [3,9,48]. In the Sonju Lake intrusion, PGE and Au mineralization has been reported from an Fe-Ti oxide rich gabbro unit [11]. Sulfide liquid saturation was reached after ~60% crystallization of magma by slow cooling in a closed system [8], possibly triggered by events of devolatilization and decompression.

Cu and PGE are mostly incompatible during crystallization of silicate minerals and thus, the concentrations of these metals increased in the differentiated liquid of a closed system. When the immiscible sulfide liquid formed upon sulfide liquid saturation, Cu, Pt, and Pd partitioned into the sulfide liquid and upon crystallization, the sulfide liquid formed disseminated Cu-Pt-Pd rich grains in the oxide-rich gabbro layer [11]. In the Partridge River and South Kawishiwi intrusions, disseminated sulfide mineralizations of Cu and PGE are found within sheet like units of gabbro and troctolite, which were formed by fractional crystallizations of high-Al olivine tholeiitic (HOAT) parental magmas [5,31,49].

This mineralization-type is in striking contrast with the high-grade Ni, Cu-rich but PGE-poor, sulfide-rich deposits of the Eagle and Tamarack intrusions [13,16]. These deposits have been classified as conduit-type sulfide deposits, where the mineralization occurred during continuous upward movements of magmatic pulses from mantle source-rocks. Supply of external sulfur was critical to the formation of the high-grade, massive, semi-massive, and disseminated mineralizations observed in these intrusions [5]. During magmatic uplift, mineral fractionation was accompanied by crustal contamination and the sulfide liquid saturation of magma was caused by selective assimilation of sulfur-rich Paleoproterozoic meta-sedimentary country rocks the of Michigamme Formation and Upper Thomson Formation for the Eagle deposit and Tamarack deposit, respectively [32,33]. Sulfur-contribution from deeper Neoarchean granite-greenstone rocks has also been proposed by the above authors. The overall funnel-shaped cross-sections, as determined by drill-core studies of both intrusions, provide additional evidence for the proposed dynamic conduit-system models. Metallic upgrading, caused by the interaction of accumulated sulfide liquid in the magma conduit, with continuous upheaval of large volumes of metal-enriched magma [50,51], might also have contributed to the high ore-grades of these deposits. Crystallization modeling on high-FeO picritic basalt parental magmas for the Eagle [32] and Tamarack [33] deposits indicate that the magmas attained sulfide liquid saturation at about 17% crystallization [31]. Separation of sulfide liquid from magma early in the crystallization sequence caused the sulfide liquid to be Ni-rich. In contrast, in the case of late sulfide liquid saturation, the magma is considerably depleted in Ni by the fractional separation of olivine [52].

4.3. Implications from Echo Lake Intrusion

The Echo Lake intrusion is primarily composed of a layered sequence of troctolite, olivine gabbro, and peridotite (Figure 5). The occurrence of a Cu-PGE rich disseminated sulfide horizon in close association with a zone rich in Fe-Ti oxide hosted within olivine gabbro (Figures 5–7), poses a remarkable similarity not only with the Sonju Lake intrusion [8,11] in the MCR, but also with several other layered igneous complexes around the world, such as the Skaergaard Complex [39], southeastern Greenland, the Stella layered intrusion, South Africa [40], the Muskox intrusion, Nunavut, Canada [53], and the Kivakka layered intrusion, Russia [54]. PGE-sulfide deposits in these group of intrusions have been classified as "Skaergaard-type" deposits [53].

The variations of chemical proxies such as MgO, TiO_2, Cr_2O_3, and FeO (total) are shown in Figure 8. Strong positive correlations between Pt, Pd, and Cu enrichment in the layered intervals with high TiO_2- and Cr_2O_3-contents can clearly be observed. These are indicators of the peak concentrations of ilmenite and Ti- and Cr-rich magnetite (Table 4) in the mineralized intervals. Since the layered succession is primarily composed of olivine-rich rocks, there is no specific connection between the MgO content and the contents of PGE and Cu. The same is true for FeO (total), and postcumulus titaniferous magnetite is common in several horizons in the layered succession (Figures 3 and 4).

The lithological succession of the Echo Lake intrusion is similar to tholeiitic layered intrusions such as the Skaergaard Complex. The mechanism of PGE-enrichment in the Echo Lake intrusion is consistent with the attainment of sulfide-melt saturation in a high-Al olivine tholeiite (HAOT) magma by fractional crystallization in a closed system [5,8]. The HAOT magmatic composition has been reported from the South Kawishiwi intrusion of the Duluth Complex [55]. This magma is characterized by approximately 20 wt. % Al_2O_3, 7 wt. % MgO and 11 wt. % FeO. Ripley et al. [49] calculated that closed system crystallization of a HAOT magma can produce 17 wt. % olivine (Fo_{56}), 63 wt. % plagioclase (An_{60}), and 14 wt. % clinopyroxene (Mg# 57). This explains the origin of troctolites and olivine gabbros. This model of fractional crystallization in a closed system is in stark contrast to conduit-type, dynamic, open-systems argued for the Eagle and Tamarack intrusions [32,33].

As proposed by Ripley and Li [31], the HAOT magmas could form by fractional crystallization of a more primitive magma, such as a picritic basalt. This implies that the parental magma of the Echo Lake intrusion could potentially be a differentiated liquid from a more primitive magma. This clearly establishes a connection between the open system conduit-type magmatism and the closed system staging-chamber type magmatism associated with the Midcontinent Rift. This is consistent with the proposition that the predominant magma during early magmatism of the Midcontinent Rift was a low-Al, high-FeO picrite, while in the later stages, a well differentiated HAOT magma became prevalent [56]. Peridotites and olivine gabbros reported from the Eagle and Tamarack intrusions show more Mg-rich olivine compositions in the range of Fo_{75-85} [13] and Fo_{82-89} [16], respectively. However, at the Echo Lake intrusion, in troctolite, olivine gabbro, and magnetite-bearing olivine gabbro, the compositional range of olivine is Fo_{57-62} (Table 2). For plagioclase, the compositional range is An_{55-75} (Table 4). These compositions indicate a well-differentiated nature of the parental magma for the Echo Lake intrusion, similar to that reported for the Sonju Lake intrusion [8,11,56]. It is possible that at deeper undiscovered levels of the Echo Lake intrusion, there are peridotite layers with more primitive olivine and plagioclase compositions representative of early crystallization from an undifferentiated magma. An alternating layered succession of this type with dunite, peridotite, and pyroxenite have been reported from intrusions such as the Pados-Tundra ultramafic complex in Kola Peninsula, Russia [57].

Owing to the highly incompatible nature of PGE in silicate minerals, the concentrations of these metals in the magma increases by progressive fractional crystallization in a closed system. However, in an open system, the periodic influx of new magma could reduce the concentration of PGE in the mixed magma. Moreover, in an open system, crustal assimilation and mixing with siliceous magmas could potentially induce sulfide-melt saturation. This could lead to the separation of small batches of immiscible sulfide liquid, which could severely deplete the PGE content of the magma.

However, if sulfide-melt saturation is eventually reached at an advanced stage of magmatic crystallization by the consequent lowering of SCSS [31], PGE can accumulate within the immiscible sulfide liquid. Owing to the increased concentration of PGE in the evolved magma, the separated sulfide liquid, although small in volume, would be extremely enriched in PGE.

Combining all the available evidence from the layered lithological succession, variation trends of chemical constituents, and the localized abundances of chalcophile elements at the Echo Lake intrusion, and comparing these observations with the known reef-type PGE mineralizations in layered intrusions of MCR and around the world, the following mechanism of mineralization can be argued for the Echo Lake deposit. A HAOT-type magma [58] began crystallizing olivine, plagioclase, and relatively small quantities of pyroxene. The early crystallized olivine (Fo_{62}) and plagioclase (An_{74}) formed a cumulus mosaic texture with small grains of clinopyroxene in fine interstitial spaces. This caused the large observed volumetric abundances of troctolite and anorthosite in the layered sequence. Early crystallization of plagioclase enriched the magma in FeO. Crystallizations of olivine (Fo_{57-62}) and small proportions of augite lowered the FeO content slightly. Overall, the FeO-content of the magma increased steadily with fractional crystallization. This was accompanied by a progressive rise of H_2O and fO_2 in the magma. At an advanced stage in magmatic differentiation, a threshold point was reached, when the magma became super-saturated in FeO and this led to the crystallization of titaniferous magnetite, accompanied by ilmenite and chromite. Rapid precipitation of large quantities of oxide minerals created local zones of sulfide-melt saturation and the formation of disseminated Cu, PGE-enriched sulfide minerals. The highest concentrations of Cu and PGE are hosted within a Fe-Ti oxide rich layer of gabbro with disseminated sulfide minerals, between depths of 990 and 1010 m. It is possible that this mineralized interval was emplaced within the stratigraphic sequence by an H_2O-rich magmatic pulse mobilized from a separate part of the magma chamber.

The heightened abundances of oxide minerals are obvious indicators of high fO_2, which separated Fe as Fe^{3+} and thereby suddenly lowered the effective concentration of FeO in the melt. This caused localized zones of sulfide liquid saturation and the formation of tiny droplets of immiscible sulfide liquid. These immiscible droplets formed in isolated pockets, sequestered the chalcophile elements such as Pd, Pt, and Cu from the surrounding melt, and eventually crystallized to form the disseminated sulfide mineralized horizon. Isolated and small occurrences of delicate titaniferous magnetite-sulfide intergrowths and thin sulfide rims around larger titaniferous magnetite grain-boundaries (Figure 4) imply close relationships between the crystallizations of titaniferous magnetite and sulfide minerals like chalcopyrite. Local occurrences of biotite- titaniferous magnetite intergrowths (Figure 4) provide evidence of pockets of intercumulus melt enriched in H_2O.

In the complete layered interval, the concentration peaks for Pt and Pd overlap with each other, but the peak for Cu does not (Figures 5–7).This stratigraphic offset of concentration peaks has been reported from other layered-type sulfide deposits, such as the Sonju Lake intrusion [12], Skaergaard Complex [39], and the Bushveld Complex [59], and has been explained by differences in liquid/silicate partitioning coefficients for chalcophile metals, or by the fractionation of sulfide melt during compaction of layered successions of cumulates [40]. The low abundance of Ni in this mineralized sulfide interval can be explained by the depletion of Ni in the magma, caused by the large fractionation of olivine early in the crystallization history.

Discovery of the PGE-Cu enriched mineralized zone in the Echo Lake intrusion is significant for many reasons. This deposit shares several important characteristics with the layered PGE-rich units of the Duluth and Beaver Bay complexes, as inferred in this study. Although the reported observations in this study are based on a limited dataset generated from preliminary exploratory studies, the implications of these findings are profound and are indicative of potential reef-type PGE-mineralizations in layered mafic-ultramafic sequences in other parts of the Midcontinent Rift region, apart from the well-known layered intrusions in northern Minnesota.

Author Contributions: A.J.K. collected samples, generated data and wrote an initial draft of manuscript. J.T. supervised the project and wrote the final manuscript.

Funding: This research was funded by Altius Resources Inc. grant number 9622.

Acknowledgments: This study was performed with the support and assistance of Altius Minerals Inc. and Bitterroot Resources. We especially thank Michael Carr of Bitterroot Resources and Roderick Smith of Altius Minerals for giving us access to the drill core samples, core data, and for their useful comments during all phases of this work. Several people, including Melanie Humphrey from the Department of Environmental Quality, Michigan, and Nick Moleski from Western Michigan University, helped us to collect drill core samples. Comments from two anonymous reviewers and useful suggestions by the editor helped us to improve the quality of this manuscript. Peter Voice, Robb Gillespie, and R.V. Krishnamurthy provided feedback in early stages of this work.

Conflicts of Interest: The authors declare no conflict of interest. The funders provided access to samples and core data. The funders had no role in the design of the study; in the interpretation of data; in the writing of the manuscript, or in the decision to publish the results.

References

1. Nicholson, S.W.; Shirey, S.B.; Schulz, K.J.; Green, J.C. Rift-wide correlation of 1.1 Ga Midcontinent rift system basalts: Implications for multiple mantle sources during rift development. *Can. J. Earth Sci.* **1997**, *34*, 504–520. [CrossRef]
2. Hauck, S.; Severson, M.J.; Zanko, L.M.; Barnes, S.-J.; Morton, P.; Alminas, H.V.; Foord, E.E.; Dahlberg, E.H. An overview of the geology and oxide, sulfide, and platinum-group element mineralization along the western and northern contacts of the Duluth Complex. *Spec. Pap. Geol. Soc. Am.* **1997**, *312*, 137–185. [CrossRef]
3. Severson, M.J.; Miller, J.D., Jr.; Peterson, D.M.; Green, J.C.; Hauck, S.A. Mineral Potential of the Duluth Complex and related intrusions. In *RI-58 Geology and Mineral Potential of the Duluth Complex and Related Rocks of Northeastern Minnesota*; Minnesota Geological Survey Report of Investigations; Miller, J.D., Jr., Green, J.C., Severson, M.J., Chandler, V.W., Hauck, S.A., Peterson, D.M., Wahl, T.E., Eds.; Minnesota Geological Survey: Saint Paul, MN, USA, 2002; pp. 164–200.
4. Ripley, E.M.; Li, C. Ni-Cu-PGE mineralization associated with the Proterozoic Midcontinent rift system, USA. In *New Developments in Magmatic Ni-Cu and PGE Deposits*; Li, C., Ripley, E.M., Eds.; Geological Publishing House: Beijing, China, 2009; pp. 180–191.
5. Ripley, E.M. Ni-Cu-PGE Mineralization in the Partridge River, South Kawishiwi, and Eagle Intrusions: A Review of Contrasting Styles of Sulfide-Rich Occurrences in the Midcontinent Rift System. *Econ. Geol.* **2014**, *109*, 309–324. [CrossRef]
6. Miller, J.; Nicholson, S. Geology and mineral deposits of the 1.1 Ga midcontinent rift in the Lake Superior region—An overview. Field guide to copper-nickel-platinum group element deposits of the Lake Superior region. *Precambrian Res. Center Guideb.* **2013**, *13–01*, 1–49.
7. Altius Minerals. *Voyageur Lands—Michigan: Project Summary. 2015 Technical Report for Summary of Project Area*; Altius Minerals: St. John's, NL, Canada, 2015; pp. 1–9.
8. Miller, J.D., Jr.; Ripley, E.M. Layered Intrusion of the Duluth Complex, Minnesota, USA. In *Layered Intrusions*; Cawthorn, R.G., Ed.; Elsevier Scientific: Amsterdam, The Netherlands, 1996; pp. 257–301.
9. Miller, J.D., Jr.; Green, J.C.; Severson, M.J.; Chandler, V.W.; Hauck, S.A.; Peterson, D.M.; Wahl, T.E. *RI-58 Geology and Mineral Potential of the Duluth Complex and Related Rocks of Northeastern Minnesota*; Minnesota Geological Survey Report of Investigations; Minnesota Geological Survey: Saint Paul, MN, USA, 2002; 207p.
10. Paces, J.B.; Miller, J.D. Precise U-Pb ages of Duluth Complex and related mafic intrusions, northeastern Minnesota: Geochronological insights into physical, petrogenetic, paleomagnetic, and tectonomagnetic processes associated with the 1.1 Ga Midcontinent rift system. *J. Geophys. Res.* **1993**, *98*, 13997–14013. [CrossRef]
11. Miller, J.D., Jr. *Information Circular 44. Geochemical Evaluation of Platinum Group Element (PGE) mineralization in the Sonju Lake Intrusion, Finland, Minnesota*; Minnesota Geological Survey: Saint Paul, MN, USA, 1999; 32p.

12. Joslin, G.J. Stratiform Palladium-Platinum-Gold Mineralization in the Sonju Lake Intrusion, Lake County, Minnesota. Master's Thesis, University of Minnesota Duluth, Duluth, MN, USA, 2004.

13. Ding, X.; Li, C.; Ripley, E.M.; Rossell, D.; Kamo, S. The Eagle and East Eagle sulfide ore-bearing mafic-ultramafic intrusions in the Midcontinent Rift System, upper Michigan: Geochronology and petrologic evolution. *Geochem. Geophys. Geosyst.* **2010**, *11*, Q03003. [CrossRef]

14. Goldner, B.D. Igneous Petrology of the Ni-Cu-PGE Mineralized Tamarack Intrusion, Aitkin and Carlton Counties, Minnesota. Master's Thesis, University of Minnesota Duluth, Duluth, MN, USA, 2011.

15. Keays, R.R.; Lightfoot, P.C. Geochemical Stratigraphy of the Keweenawan Midcontinent Rift Volcanic Rocks with Regional Implications for the Genesis of Associated Ni, Cu, Co, and Platinum Group Element Sulfide Mineralization. *Econ. Geol.* **2015**, *110*, 1235–1267. [CrossRef]

16. Taranovic, V.; Ripley, E.M.; Li, C.; Rossell, D. Petrogenesis of the Ni-Cu-PGE sulfide bearing Tamarack Intrusive Complex, Midcontinent Rift System, Minnesota. *Lithos* **2015**, *212–215*, 16–31. [CrossRef]

17. Sims, P.K. *Geologic Map of Precambrian Rocks, Southern Lake Superior Region, Wisconsin and Northern Michigan*; U.S. Geological Survey Miscellaneous Investigations Map I-2185; USGS: Reston, VA, USA, 1992.

18. Bitterroot Resources Ltd. *Echo Lake Ni/Cu/PGE Project: Upper Peninsula of Michigan*; 1999 Technical Report on Summary of Project Area; Bitterroot Resources Ltd.: West Vancouver, BC, Canada, 1999; pp. 1–26.

19. Peterman, Z.E.; Zartman, R.E.; Sims, P.K. A protracted Archean history in the Watersmeet gneiss dome, northern Michigan. In *Shorter Contributions to Isotope Research*; Peterman, Z.E., Schnabel, D.C., Eds.; U.S. Geological Survey Bulletin: Reston, VA, USA, 1986; Volume 1622, pp. 51–64.

20. Van Schmus, W.R.; Bickford, M.E.L.; Zietz, I. Early and middle Proterozoic provinces in the central United States. In *Proterozoic Lithosphere Evolution*; Geodynamics Series; Kroner, A., Ed.; AGU: Washington, DC, USA, 1987; Volume 17, pp. 43–68.

21. Malone, D.H.; Stein, C.A.; Craddock, J.P.; Kley, J.; Stein, S.; Malone, J.E. Maximum depositional age of the Neoproterozoic Jacobsville Sandstone, Michigan: Implications for the evolution of the Midcontinent Rift. *Geosphere* **2016**, *12*, 1271–1282. [CrossRef]

22. Koerber, A. Geochemical and Petrological Investigation of the Prospective Ni-Cu-PGE Mineralization at the Echo Lake Intrusion in the Upper Peninsula of Michigan, USA. Master's Thesis, Western Michigan University, Kalamazoo, MI, USA, 2018.

23. Cannon, W.F.; Nicholson, S.W. *Geologic Map of the Keweenaw Peninsula and adjacent Area*; Michigan U.S. Geological Survey Miscellaneous Investigations Map I-2696; USGS: Reston, VA, USA, 2001.

24. Barnes, S.-J.; Maier, W.D. Platinum group elements and microstructures of normal Merensky Reef from Impala platinum mines, Bushveld Complex. *J. Petrol.* **2002**, *43*, 103–128. [CrossRef]

25. Campbell, I.H.; Naldrett, A.J.; Barnes, S.J. A model for the origin of the platinum-rich sulfide horizons in the Bushveld and Stillwater Complexes. *J. Petrol.* **1983**, *24*, 133–165. [CrossRef]

26. Boudreau, A.E.; Mathez, E.A.; McCallum, I.S. Halogen geochemistry of the Stillwater and Bushveld complexes: Evidence for transport of the platinum-group elements by Cl-rich fluids. *J. Petrol.* **1986**, *27*, 967–986. [CrossRef]

27. Donovan, J.J.; Tingle, T.N. An improved mean atomic number correction for quantitative microanalysis. *Microsc. Microanal.* **1996**, *2*, 1–7. [CrossRef]

28. Ripley, E.M.; Li, C. Sulfur isotopic exchange in the formation of magmatic Cu-Ni-(PGE) deposits. *Econ. Geol.* **2003**, *98*, 635–641. [CrossRef]

29. Seat, Z.; Beresford, S.W.; Grguric, B.A.; Mary Gee, M.A.; Grassineau, N.V. Reevaluation of the Role of External Sulfur Addition in the Genesis of Ni-Cu-PGE Deposits: Evidence from the Nebo-Babel Ni-Cu-PGE Deposit, West Musgrave, Western Australia. *Econ. Geol.* **2009**, *104*, 521–538. [CrossRef]

30. Keays, R.R.; Lightfoot, P.C. Crustal sulfur is required to form magmatic Ni–Cu sulfide deposits: Evidence from chalcophile element signatures of Siberian and Deccan Trap basalts. *Miner. Deposita* **2010**, *45*, 241–257. [CrossRef]

31. Ripley, E.M.; Li, C. Sulfide saturation in mafic magmas: Is external sulfur required for magmatic Ni-Cu-(PGE) ore genesis? *Econ. Geol.* **2013**, *108*, 45–58. [CrossRef]

32. Ding, X.; Ripley, E.M.; Shirey, S.B.; Li, C. Os, Nd, O and S isotope constraints on country rock contamination in the conduit-related Eagle Cu-Ni-(PGE) deposit, Midcontinent rift system, upper Michigan. *Geochim. Cosmochim. Acta* **2012**, *89*, 10–30. [CrossRef]

33. Taranovic, V.; Ripley, E.M.; Li, C.; Shirey, S.B. S, O, and Re-Os Isotope Studies of the Tamarack Igneous Complex: Melt-Rock Interaction During the Early Stage of Midcontinent Rift Development. *Econ. Geol.* **2018**, *113*, 1161–1179. [CrossRef]

34. Lehmann, J.; Arndt, N.; Windley, B.; Zhou, M.-F.; Wang, C.Y.; Harris, C. Field relationships and geochemical constraints on the emplacement of the Jinchuan intrusion and its Ni-Cu-PGE sulfide deposit, Gansu, China. *Econ. Geol.* **2007**, *102*, 75–94. [CrossRef]

35. Holwell, D.A.; Boyce, A.J.; McDonald, I. Sulfur isotope variations within the Platreef Ni-Cu-PGE deposit: Genetic implications for the origin of sulfide mineralization. *Econ. Geol.* **2007**, *102*, 1091–1110. [CrossRef]

36. Howarth, G.H.; Prevec, S.A. Trace element, PGE, and Sr–Nd isotope geochemistry of the Panzhihua mafic layered intrusion, SW China: Constraints on ore-forming processes and evolution of parent magma at depth in a plumbing-system. *Geochim. Cosmochim. Acta* **2013**, *120*, 459–478. [CrossRef]

37. Carmichael, I.S.E. The redox states of basic and silicic magmas: A reflection of their source regions? *Contrib. Mineral. Petrol.* **1991**, *106*, 129–141. [CrossRef]

38. Liu, Y.; Samaha, N.T.; Baker, D.R. Sulfur concentration at sulfide saturation (SCSS) in magmatic silicate melts. *Geochim. Cosmochim. Acta* **2007**, *71*, 1783–1799. [CrossRef]

39. Andersen, J.C.O.; Rasmussen, H.; Neilsen, T.F.D.; Ronsbo, J.G. The Triple Group and the Platinova gold and palladium reefs in the Skaergaard intrusion: Stratigraphic and petrographic relations. *Econ. Geol.* **1998**, *93*, 488–509. [CrossRef]

40. Maier, W.D.; Barnes, S.-J.; Gartz, V.; Andrews, G. Pt-Pd reefs in magnetites of the Stella layered intrusion, South Africa: A world of new exploration opportunities for platinum group elements. *Geology* **2003**, *31*, 885–888. [CrossRef]

41. Kelley, K.A.; Cottrell, E. The influence of magmatic differentiation on the oxidation state of Fe in a basaltic arc magma. *Earth Planet. Sci. Lett.* **2012**, *329–330*, 109–121. [CrossRef]

42. Holwell, D.A.; Keays, R.R. The Formation of Low-Volume, High-Tenor Magmatic PGE-Au Sulfide Mineralization in Closed Systems: Evidence from Precious and Base Metal Geochemistry of the Platinova Reef, Skaergaard Intrusion, East Greenland. *Econ. Geol.* **2014**, *109*, 387–406. [CrossRef]

43. Li, C.; Ripley, E.M. Sulfur contents at sulfide-liquid or anhydrite saturation in silicate melts: Empirical equations and example applications. *Econ. Geol.* **2009**, *104*, 405–412. [CrossRef]

44. Gaetani, G.A.; Grove, T.L. Partitioning of moderately siderophile elements among olivine, silicate melt, and sulfide melt: Constraints on core formation in the Earth and Mars. *Geochim. Cosmochim. Acta* **1997**, *61*, 1829–1846. [CrossRef]

45. Carroll, M.R.; Rutherford, M.J. Sulfur speciation in hydrous experimental glasses of varying oxidation state: Results from measured wavelength shifts of sulfur X-rays. *Am. Miner.* **1988**, *73*, 845–849.

46. O'Neill, H.St.C.; Mavrogenes, J.A. The sulfide capacity and the sulfur content at sulfide saturation of silicate melts at 1400°C and 1 bar. *J. Petrol.* **2002**, *43*, 1048–1087.

47. Jugo, P.J.; Luth, R.W.; Richards, J.P. Experimental data on the speciation of sulfur as a function of oxygen fugacity in basaltic melts. *Geochim. Cosmochim. Acta* **2005**, *69*, 497–503. [CrossRef]

48. Theriault, R.D.; Barnes, S.-J.; Severson, M.J. Origin of Cu-Ni-PGE sulfide mineralization in the Partridge River intrusion, Duluth Complex, Minnesota. *Econ. Geol.* **2000**, *95*, 929–944. [CrossRef]

49. Ripley, E.M.; Taib, N.I.; Li, C.; Moore, C.H. Chemical and mineralogical heterogeneity in the basal zone of the Partridge River Intrusion: Implications for the origin of Cu-Ni mineralization in the Duluth Complex, Midcontinent rift system. *Contrib. Mineral. Petrol.* **2007**, *154*, 35–54. [CrossRef]

50. Naldrett, A.J. World-class Ni-Cu-PGE deposits; key factors in their genesis. *Miner. Deposita* **1999**, *34*, 227–240. [CrossRef]

51. Li, C.; Ripley, E.M.; Naldrett, A.J. Compositional variation of olivine and sulfur isotopes in the Noril'sk and Talnakh intrusions, Siberia: Implications for ore-forming processes in dynamic magma conduits. *Econ. Geol.* **2003**, *98*, 69–86. [CrossRef]

52. Barnes, S.-J.; Francis, D. The distribution of platinum-group elements, nickel, copper, and gold in the Muskox layered intrusion, Northwest Territories, Canada. *Econ. Geol.* **1995**, *90*, 135–154. [CrossRef]

53. Miller, J.D., Jr.; Andersen, J.C.O. Attributes of Skaergaard-type PGE deposits [abs.]. In Proceedings of the 9th International Platinum Symposium, Billings, MT, USA, 21–25 July 2002.

54. Barkov, A.Y.; Nikiforov, A.A. Compositional variations of apatite, fractionation trends, and a PGE-bearing zone in the Kivakka layered intrusion, northern Karelia, Russia. *Can. Mineral.* **2016**, *54*, 475–490. [CrossRef]

55. Lee, I.; Ripley, E.M. Mineralogic and oxygen isotopic studies of open system magmatic processes in the South Kawishiwi intrusion, Spruce Road Area, Duluth Complex, Minnesota. *J. Petrol.* **1996**, *37*, 1437–1461. [CrossRef]

56. Park, Y.-R.; Ripley, E.M.; Miller, J.D.; Li, C.; Mariga, J.; Shafer, P. Stable isotopic constraints on fluid-rock interaction and Cu-PGE-S redistribution in the Sonju Lake intrusion. *Econ. Geol.* **2004**, *99*, 325–338. [CrossRef]

57. Barkov, A.Y.; Nikiforov, A.A.; Martin, R.F. The structure and cryptic layering of the Pados-Tundra ultramafic complex, Serpentinite belt, Kola Peninsula, Russia. *Bull. Geol. Soc. Finl.* **2017**, *89*, 35–56. [CrossRef]

58. Li, C.; Ripley, E.M. Empirical equations to predict the sulfur content of mafic magmas at sulfide saturation and applications to magmatic sulfide deposits. *Miner. Deposita* **2005**, *40*, 218–230. [CrossRef]

59. Maier, W.D.; Barnes, S.-J. Platinum-Group Elements in Silicate Rocks of the Lower, Critical and Main Zones at Union Section, Western Bushveld Complex. *J. Petrol.* **1999**, *40*, 1647–1671. [CrossRef]

Article

PGE–(REE–Ti)-Rich Micrometer-Sized Inclusions, Mineral Associations, Compositional Variations, and a Potential Lode Source of Platinum-Group Minerals in the Sisim Placer Zone, Eastern Sayans, Russia

Andrei Y. Barkov [1,*], Gennadiy I. Shvedov [2] and Robert F. Martin [3]

[1]   Research Laboratory of Industrial and Ore Mineralogy, Cherepovets State University, 5 Lunacharsky Avenue, 162600 Cherepovets, Russia

[2]   Institute of Mining, Geology and Geotechnology, Siberian Federal University, 95 Avenue Prospekt im. gazety "Krasnoyarskiy Rabochiy", 660025 Krasnoyarsk, Russia; g.shvedov@mail.ru

[3]   Department of Earth and Planetary Sciences, McGill University, 3450 University Street, Montreal, QC H3A 0E8, Canada; robert.martin@mcgill.ca

*   Correspondence: ore-minerals@mail.ru; Tel.: +7-8202-51-78-27

Received: 31 March 2018; Accepted: 19 April 2018; Published: 27 April 2018

Abstract: We report the results of a mineralogical investigation of placer samples from the upper reaches of the Sisim watershed, near Krasnoyarsk, in Eastern Sayans, Russia. The placer grains are predominantly Os–Ir–(Ru) alloys (80%) that host various inclusions (i.e., platinum-group elements (PGE)-rich monosulfide, PGE-rich pentlandite, Ni–Fe–(As)-rich laurite, etc.) and subordinate amounts of Pt–Fe alloys. Analytical data (wavelength- and energy-dispersive X-ray spectroscopy) are presented for all the alloy minerals and the suite of micrometer-sized inclusions that they contain, as well as associated grains of chromian spinel. The assemblage was likely derived from chromitite units of the Lysanskiy mafic–ultramafic complex, noted for its Ti–(V) mineralization. In the Os–Ir–(Ru) alloys, the ratio Ru/Ir is ≤ 1, Ir largely substitutes for Os, and compositional variations indicate the scheme $[Ir + Ru] \rightarrow 2Os$. In contrast, in the laurite–erlichmanite series, Ir and Os are strongly and positively correlated, whereas Ir and Ru are negatively correlated; Ru and Os are inversely correlated. These compositions point to the scheme $[Os^{2+} + 2Ir^{3+} + \square] \rightarrow 4Ru^{2+}$ or alternatively, to $Os^{2+} + Ir^{2+} \rightarrow 2Ru^{2+}$. We deduce a potential sequence of crystallization in the parental rock and address the effects of decreasing temperature and increasing fugacity of sulfur and arsenic on the assemblage. Inclusions of Ti-rich minerals in the alloy grains are consistent with the Lysanskiy setting; the complete spectrum of chromite–magnesiochromite compositions indicates that an important part of that complex was eroded. A localized fluid-dominated micro-environment produced the unique association of laurite with monazite-(Ce), again considered a reflection of the special attributes of the Lysanskiy complex.

Keywords: platinum-group elements; platinum-group minerals; PGE alloys; chromian spinel; schemes of substitution; Ti- and REE-rich inclusions; Sisim placer zone; Lysanskiy complex; Eastern Sayans; Russia

1. Introduction

The Sisim placer zone contains various deposits of minerals and unusual phases rich in the platinum-group elements (PGE). The deposits are found in an 80×30 km^2 area in the southern portion of the huge Krasnoyarskiy kray territory in central Siberia, not far from Krasnoyarsk in the southwestern part of the Eastern Sayans, Russia (Figure 1).

Figure 1. Map showing the location of the Sisim placer zone in the Krasnoyarskiy kray of the Russian Federation.

The Au–PGE-bearing placers are developed alluvially along the River Sisim and its tributaries, among which are the rivers Ko, Levyi Ko, Seyba, Malaya Alga, Kotel', and Koza. The placers were first mined for gold at the beginning of the 19th century, possibly even earlier. Vysotskiy [1] described the first occurrences of platinum-group minerals (PGM) from the Sisim placer zone, namely grains composed dominantly of "osmian iridium" along the Shirokiy brook in the middle reaches of the River Sisim. In 1985, D.I. Baykalov recognized abundant grains of PGM in a suite of placers that extend up to ~19 km along the River Ko, one of longest tributaries. Krivenko et al. [2] and Tolstykh & Krivenko [3] described the associations of PGM in those placers.

Here, we report new wavelength- and energy-dispersive X-ray spectroscopy results obtained on samples from placers of the Sisim watershed, originating from the upper reach of the River Sisim, including Ivanovka Creek. The PGM grains recovered are relatively small (generally <0.25–0.5 mm and, occasionally, up to 1.8 mm in the longest dimension). There is a predominance of Os–Ir–(Ru) alloy minerals, but these differ in composition from those known in other areas of the Eastern Sayans. The PGM grains at Sisim host inclusions of various types, which provide insight into the ore-forming environment and provenance. Some of these inclusions are rich in Ti or in the rare-earth elements (REE); the latter are especially unusual and are documented here for the first time as the observed intergrowth of PGE and REE-based minerals.

Our aims are the following: (1) to characterize the PGE mineralogy of the placer deposits of the entire River Sisim system, incorporating results obtained on associated placers along the River Ko [2,3]; (2) to examine mineral associations, compositional ranges, extents of solid solutions, and mechanisms of element substitutions in the PGM; (3) to discuss types and implications of various micrometer-sized inclusions hosted by placer grains of PGE alloy minerals; and (4) to suggest a potential lode source for the placer occurrences of PGM in the Sisim zone, on the basis of all of these findings and results obtained on a large set of associated grains of PGM and chromian spinel.

2. Materials and Methods

The compositions of placer grains of chromian spinel, PGE alloys, and various inclusions of amphiboles, PGM, and PGE-rich phases (Figures 2a–h and 3a–f), were investigated by wavelength-dispersive spectroscopy (WDS) using a Camebax-micro electron microprobe at the Sobolev Institute of Geology and Mineralogy, Russian Academy of Sciences, Novosibirsk, Russia.

The analytical conditions used for PGE-rich minerals were the following: 20 kV and 60 nA; the $L\alpha$ line was used for Ir, Rh, Ru, Pt, Pd, and As; the $M\alpha$ line was used for Os; and the $K\alpha$ line was used for S, Fe, Ni, Cu, and Co. We used as standards pure metals (for the PGE), $CuFeS_2$ (for Fe, Cu, and

S), synthetic FeNiCo (for Ni and Co), and arsenopyrite (for As). The minimum detection limit was ≤ 0.06 wt % in the results of the WDS analyses.

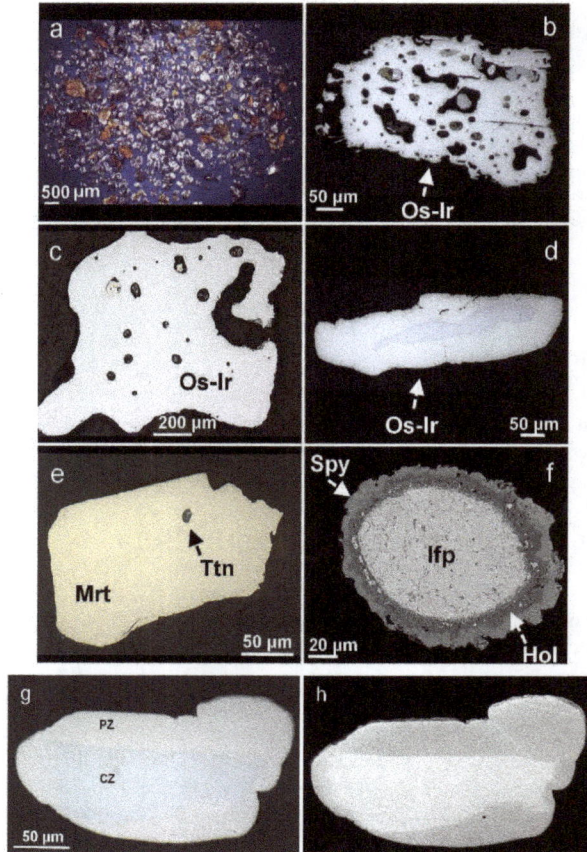

Figure 2. (**a**) Microscopic image showing grains of placer platinum-group mineral (PGM) concentrate, the size of which is generally ~0.1–0.5 mm, collected from the Sisim placer zone and analyzed in the present study. The bulk of these grains (up to ~80%) are Os–Ir–(Ru) alloy minerals (i.e., the minerals osmium and iridium). Grains of Pt–Fe alloy are subordinate (<20%); (**b,c**). Grains of Os–Ir alloy (Os–Ir), observed in a polished mount, contain various inclusions; (**d**) A grain of sharply zoned Os–Ir alloys, consisting of Os-rich core and Ir-rich rim; (**e**) Subhedral grain of mertieite-II (Mrt), hosting a tiny inclusion of titanite (Ttn); (**f**) Back-scattered electron (BSE) image displaying a rounded grain of isoferroplatinum (Ifp; Rh–Ir–Cu–bearing) $[(Pt_{2.57}Rh_{0.28}Ir_{0.17}Os_{0.09}Ru_{0.05})_{\Sigma 3.16}(Fe_{0.66}Cu_{0.17}Ni_{<0.01})_{\Sigma 0.84}]$, which is mantled by a composite rim that consists of a S-deficient phase of hollingworthite (Hol, dark grey: the internal rim) $[(Rh_{0.94}Pt_{0.18}Fe_{0.03}Ir_{0.01})_{\Sigma 1.84}As_{1.15}S_{0.69}]$, and an outer rim composed of a metal-deficient sperrylite (Spy) $[(Pt_{0.71}Rh_{0.17}Fe_{0.04}Ni_{<0.01})_{\Sigma 0.92}(As_{1.82}S_{0.26})_{\Sigma 2.08}]$; (**g**) A reflected light microphotograph and (**h**) BSE image showing a sharply zoned grain of Os–Ir–(Ru) alloy from the Sisim placer zone; the core zone (CZ, blue in Figure 2g) has the composition $Os_{92.6}Ir_{6.5}Ru_{0.7}Fe_{0.10}Ni_{0.10}$. The wavelength-dispersive spectroscopy (WDS) composition of the periphery zone (PZ) is $Os_{48.0}Ir_{29.1}Ru_{18.9}Pt_{2.5}Rh_{1.0}Fe_{0.47}Ni_{0.15}$.

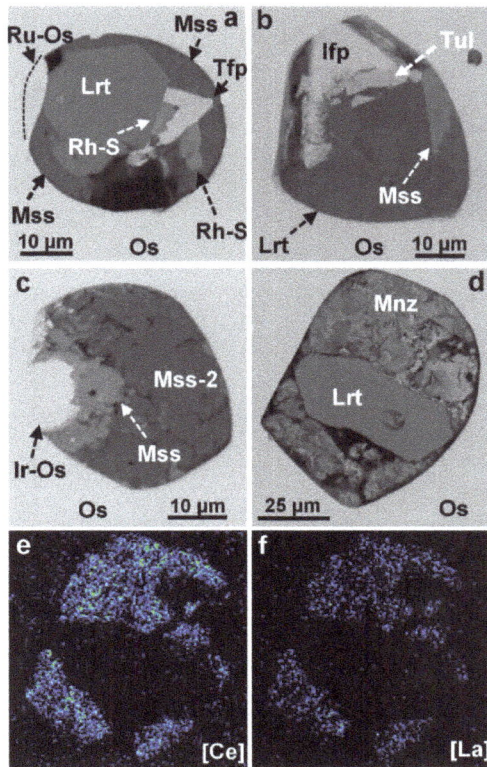

Figure 3. (**a**) BSE image displaying a multiphase inclusion enclosed by a placer grain of an Os-dominant alloy (Os) [$Os_{35.5}Ru_{32.7}Ir_{31.8}$] from the Sisim placer zone. It consists of a subhedral crystal of nearly end-member laurite (Lrt) [$Ru_{0.99}S_{2.01}$], a phase of tetraferroplatinum (Tfp) [$(Pt_{0.93}Pd_{0.06})_{\Sigma0.99}(Fe_{0.58}Cu_{0.24}Ni_{0.19})_{\Sigma1.01}$], a monosulfide-type phase (Mss) having the composition [$(Ni_{0.55}Fe_{0.33}Rh_{0.10})_{\Sigma0.99}S_{1.01}$], and a pentlandite-type phase (Rh–S); ideally $(Rh,Ni,Cu)_9S_8$ [$(Rh_{6.67-6.72}Ni_{1.70-1.76}Cu_{0.54-0.58}Fe_{0.15})_{\Sigma9.1-9.2}S_{7.8-7.9}$]. A Ru-dominant alloy (Ru–Os) [$Ru_{41.0}Ir_{32.4}Os_{19.7}Rh_{4.1}Fe_{1.6}Pt_{1.2}$] occurs as a narrow rim, forming part of this inclusion. (**b**) BSE image displaying a composite inclusion, hosted by an Os-dominant alloy [$Os_{53.0}Ir_{30.6}Ru_{16.4}$], which is associated with isoferroplatinum (Ifp) [$Pt_{2.95}(Fe_{0.98}Ni_{0.07})$], rimmed partly by a tulameenite-type phase (Tul; Pd-bearing) [$(Pt_{1.60}Pd_{0.56})Fe_{0.95}(Cu_{0.83}Ni_{0.06})$]. Also present are a monosulfide phase [$Ni_{0.23}Fe_{0.23}Ir_{0.16}Rh_{0.13}Cu_{0.12}Ru_{0.03})_{\Sigma0.90}S_{1.10}$] and a laurite-type phase (Lrt) [$(Ru,Fe,Ni,Rh)S_{2-x}$], which likely represents a solid solution toward pyrite and vaesite [$(Ru_{0.49}Fe_{0.26}Ni_{0.24}Rh_{0.11}Cu_{0.07}Ir_{0.04})_{\Sigma1.21}S_{1.80}$]. (**c**) BSE image of a core-like zone of Ir-dominant alloy (Ir–Os) [$Ir_{62.3}Os_{26.4}Ru_{11.2}$], which is surrounded by a monosulfide-type phase (Mss) [$(Ni_{0.28}Fe_{0.21}Ir_{0.17}Rh_{0.12}Cu_{0.11}Pt_{0.03})_{\Sigma0.92}S_{1.08}$]. The second phase of the latter type (Mss-2) has the following composition: [$(Ni_{0.52}Fe_{0.39}Rh_{0.10})_{\Sigma1.01}S_{1.00}$]. This inclusion is enclosed within a grain of Os-dominant alloy [$Os_{46.5}Ir_{41.0}Ru_{12.6}$]. (**d**) BSE image of a subhedral crystal of zoned laurite (Lrt) occurring at the center of inclusion with the composition $(Ru_{0.89}Ir_{0.05}Os_{0.04})_{\Sigma0.98}S_{2.02}$ in the central phase rimmed by a narrow zone rich in Ir [$(Ru_{0.74}Ir_{0.16}Rh_{0.07})_{\Sigma0.97}S_{2.03}$]. An aggregate of microcrystalline grains of a colloform rare-earth element (REE) phosphate that precipitated around the laurite core corresponds to monazite-(Ce) (Mnz). Inclusion hosted by a grain of Os-dominant alloy (Os) [$Os_{43.3}Ir_{42.2}Ru_{14.6}$]. (**e**,**f**) X-ray maps show the distribution of Ce (Figure 3e) and La (Figure 3f) in the grain of the REE phase of monazite-(Ce) shown in Figure 3d.

We analyzed (WDS) more than 50 detrital grains of chromian spinel, mostly subhedral and about 0.2–1 mm across, associated with grains of PGM in placer samples from the Sisim zone. The WDS analyses of chromite, amphiboles, and serpentine were acquired at 20 kV and 40 nA, using the $K\alpha$ line, and the following standards: chromite (for Fe, Mg, Al, and Cr), ilmenite (Ti), manganiferous garnet (for Mn), and synthetic V_2O_5 (for V). The amphibole analyses were done using the following standards: diopside (for Ca), albite (for Na), orthoclase (for K), pyrope (Mg, Fe, Al, and Si), a glass of diopside composition doped with 2 wt % TiO_2 for Ti, and manganiferous or chromiferous garnets (for Mn and Cr). For serpentine, we used an olivine standard (Mg, Si, Fe, and Ni), as well as diopside (Ca), and, as noted, garnets for Mn and Cr.

In addition, we employed scanning-electron microscopy (SEM) and energy-dispersive spectroscopy (EDS) in order to minimize interference from the host (or adjacent grains) during the analysis of phases 1–2 μm in size present in composite micro-inclusions (Figures 2a–f and 3a–d). These phases were analyzed at 20 kV and 1.2 nA using a Tescan Vega 3 SBH facility combined with an Oxford X-Act spectrometer at the Siberian Federal University, Krasnoyarsk, Russia. Pure elements (for the PGE, Fe, Cu, Nd, Pr, and Sm), as well as FeS_2 (for S), InAs (for As), CeO_2 (for Ce), LaB_6 (for La), GaP (for P), and CaF_2 (for Ca) were used as standards. The $L\alpha$ line was used for As, the REE, and the PGE, except for Pt ($M\alpha$ line); the $K\alpha$ line was used for Fe, Cu, Ni, S, P, and Ca. The results of WDS and EDS analyses (Tables 1–8) are in good agreement.

Table 1. Compositions of grains of magnesiochromite–chromite from the Sisim placer zone, Eastern Sayans, Russia.

#	TiO_2	Al_2O_3	Cr_2O_3	FeO (t)	FeO (calc.)	Fe_2O_3 (calc.)	MnO	MgO	NiO	Total
1	0.52	13.83	46.45	36.72	31.8	5.47	0.34	1.64	0.04	100.21
2	1.33	13.54	43.31	37.59	30.37	8.03	0.27	2.92	0.18	100.1
3	1.05	16.24	48.09	18.97	12.71	6.96	0.17	14.63	0.24	100.22
4	1.01	15.04	45.6	29.53	23.21	7.02	0.32	7.55	0.14	100.05
5	0.38	13.57	42.48	40.34	32.57	8.64	0.68	0.51	0.06	99.03
6	0.9	16.12	48.64	15.88	9.33	7.28	0.09	16.59	0.26	99.33
7	1.16	14.68	45.02	27.42	18.13	10.32	0.2	11.12	0.21	100.95
8	0.89	16.46	48.82	16.78	11.01	6.41	0.12	15.59	0.24	99.64
9	1.34	15.07	45.61	23.1	15.62	8.31	0.16	12.55	0.18	99.1
10	1.15	14.54	46.13	27.58	20.28	8.11	0.22	9.6	0.16	100.31

				Atomic Proportions (O = 4)							
#	Cr	Al	Fe^{3+}	Ti	Mg	Fe^{2+}	Mn	Ni	Mg#	Cr#	Fe^{3+}#
1	1.27	0.56	0.14	0.01	0.08	0.92	0.01	0.001	8	69	7
2	1.18	0.55	0.21	0.03	0.15	0.87	0.01	0.005	15	68	11
3	1.19	0.6	0.16	0.02	0.68	0.33	<0.01	0.006	67	67	8
4	1.19	0.59	0.17	0.03	0.37	0.64	0.01	0.004	36	67	9
5	1.19	0.56	0.23	0.01	0.03	0.96	0.02	0.002	3	68	12
6	1.2	0.59	0.17	0.02	0.77	0.24	<0.01	0.006	76	67	9
7	1.14	0.55	0.25	0.03	0.53	0.49	0.01	0.005	52	67	13
8	1.2	0.6	0.15	0.02	0.72	0.29	<0.01	0.006	71	67	8
9	1.16	0.57	0.2	0.03	0.6	0.42	<0.01	0.005	59	67	10
10	1.19	0.56	0.2	0.03	0.47	0.55	0.01	0.004	46	68	10

Note: Results of WDS analyses acquired with a Camebax-micro microprobe are quoted in wt %. These grains of magnesiochromite–chromite (subhedral; 0.1–0.4 mm in size) were collected in the Sisim placer area and at Ivanovka Creek associated. FeO (t) is all Fe as FeO. FeO (calc.) and Fe_2O_3 (calc.) are values calculated on the basis of stoichiometry and charge balance per four oxygen atoms (O = 4). The index Mg# is 100 Mg/(Mg + Fe^{2+} + Mn); Cr# is 100 Cr/(Cr + Al); and Fe^{3+}# is 100 Fe^{3+}/(Fe^{3+} + Cr + Al).

3. Results

3.1. Placer and Lode Occurrences of Chromian Spinel

In placer samples of chromian spinel from the Sisim zone, we find a broad range of compositions along the magnesiochromite–chromite solid solution, with Mg# [100 Mg/(Mg + Fe^{2+} + Mn)] in the range 76–3 and a MgO content up to 16.6 wt % (Table 1). In contrast to the variations in Mg#, this series displays a fairly uniform Cr# [100 Cr/(Cr + Al)] (Figure 4, Table 1). In addition to placer grains, we included several grains of chromite from serpentinite exposed along the River Levyi Ko. We believe that these bodies of serpentinite belong to the Lysanskiy ultramafic–mafic layered complex, which hosts significant Ti–(V) ores. The chromite grains analyzed in the lode serpentinite are fairly similar in composition to the detrital grains recovered from the Sisim placer zone (Figure 4).

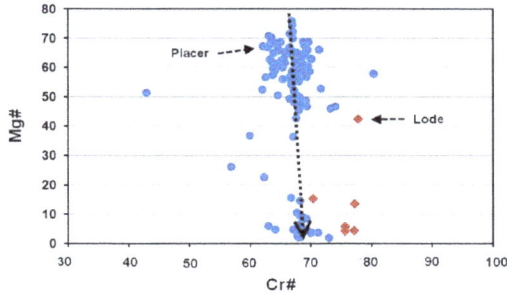

Figure 4. Plot of values of the index Mg# [100 Mg/(Mg + Fe^{2+} + Mn)] vs. Cr# [100 Cr/(Cr + Al)] in compositions of placer grains of chromian spinel (chromite–magnesiochromite series) from the Sisim River system; a total of 106 data-points are plotted. For comparison, WDS data are presented for several grains of chromite collected from lode outcrops of serpentinite exposed in the area at the River Levyi Ko (this study).

3.2. Placer Grains of Os–Ir–(Ru) and Pt–Fe Alloys

Os- and Ir-rich alloys (Table 2) predominate in the placer concentrates at Sisim, with grain sizes generally ≤0.5 mm and a maximum of 1–1.5 mm. Zonation in the grains of Os–Ir–(Ru) alloy is recognized optically and in back-scattered electron (BSE) images (Figure 2d,g). Both discrete and cryptic zonation is characterized by a relative increase in Ir and Ru toward the margin.

The observed compositions of the matrix and inclusions of the Os–Ir–(Ru) alloys define coherent pairs (Table 2); these phases coexist in fields in the Os–Ru–Ir diagram (Figure 5), indicating that they achieved a mutual equilibrium during crystallization. In contrast to alloys of ophiolite origin (e.g., [4,5]), the Os–Ir–(Ru) alloys at Sisim contain rather subordinate amounts of Ru. Few compositions correspond to the mineral ruthenium. The bulk of the alloy species are classified as either osmium and iridium; rutheniridosmine is less abundant (Figure 5). The overall field of compositions observed at Sisim does not differ essentially from the field recognized [2] at the River Ko. The similarity implies a common provenance for these placer occurrences.

In the Os–Ir–Ru compositional field (Figure 5), the solid solutions display a general decrease in Ru with increasing Os content. Interestingly, the inferred boundary of the compositional field is nearly linear and extends along the line Ru/Ir = 1. Note that the trendline observed for the sharply zoned crystal of Os–Ir–(Ru) alloy (Figure 2g,h) is close and subparallel to the line Ru/Ir = 1 (Figure 5). The core zone (CZ) in this grain is [Os$_{92.6}$Ir$_{6.5}$Ru$_{0.7}$Fe$_{0.10}$Ni$_{0.10}$]; its periphery zone (PZ, Figure 2g) corresponds to [Os$_{48.0}$Ir$_{29.1}$Ru$_{18.9}$Pt$_{2.5}$Rh$_{1.0}$Fe$_{0.47}$Ni$_{0.15}$]. The evolutionary trend thus reveals a strong increase in the content of Ir and Ru, with a minor buildup in Pt, Rh, and Fe, and a correspondingly strong decrease in Os. The compositions based on a total of 202 data-points (Figure 5) yield a fairly strong negative correlation of Os vs Ir, with a correlation coefficient R of −0.86.

Table 2. Compositions of Os–Ir–Ru alloy minerals from the Sisim placer zone, Eastern Sayans.

#			Ru	Os	Ir	Rh	Pt	Fe	Ni	Total
1	Os-dominant	Matrix	6.72	55.45	37	0.06	0.23	0.13	0.03	99.62
2			11.58	45.6	38.48	1.13	2.86	0.16	0.02	99.83
3			5.82	74.44	20	0.05	0.04	0.03	bdl	100.38
4			5.42	65.97	27.4	0.24	0.43	0.03	0.05	99.54
5			11.94	42.79	41.28	1.1	1.61	0.26	0.05	99.03
6		Inclusion	11.76	53.15	34.38	0.24	0.39	0.11	0.09	100.12
7			0.25	77.34	21.83	bdl	0.09	0.06	0.03	99.6
8			0.89	88.05	10.5	bdl	bdl	0.06	bdl	99.5
9			3.08	71.01	25.65	bdl	bdl	0.06	0.05	99.85
10			4.54	54.2	39.53	0.26	1.58	0.18	bdl	100.3
11	Ir-dominant	Matrix	6.36	14.58	75.29	0.52	1.81	0.58	0.09	99.23
12			3.55	12.63	82.53	0.14	0.79	0.8	0.09	100.53
13			1.11	27.12	70.78	bdl	bdl	0.35	0.06	99.41
14			5	7.21	84.83	0.15	0.68	1.07	0.15	99.08
15		Inclusion	13.17	37.08	48.38	bdl	0.55	0.23	0.04	99.45
Atomic Proportions (per a Total of 100 at. %)										
#			Ru	Os	Ir	Rh	Pt	Fe	Ni	
1			12	52.5	34.7	0.1	0.21	0.42	0.1	
2			19.6	41.1	34.3	1.88	2.51	0.49	0.06	
3			10.4	70.6	18.8	0.09	0.03	0.1	0	
4			9.8	63.2	26	0.42	0.4	0.11	0.14	
5			20.3	38.6	36.9	1.84	1.42	0.81	0.14	
6			20	48	30.7	0.39	0.35	0.34	0.26	
7			0.5	77.5	21.6	0	0.08	0.2	0.11	
8			1.7	87.8	10.4	0	0	0.19	0.01	
9			5.7	69.2	24.7	0	0	0.21	0.15	
10			8.2	51.9	37.4	0.47	1.47	0.59	0.01	
11			11.3	13.7	70.3	0.91	1.66	1.85	0.28	
12			6.4	12	77.8	0.25	0.73	2.59	0.28	
13			2.1	27	69.6	0	0	1.18	0.18	
14			8.9	6.8	79.5	0.26	0.63	3.45	0.44	
15			22.3	33.3	43.1	0	0.48	0.72	0.1	

Note: Results of electron-microprobe spectroscopy (EDS) are quoted in weight percent. Pd is close to the detection limit; bdl is below detection limit.

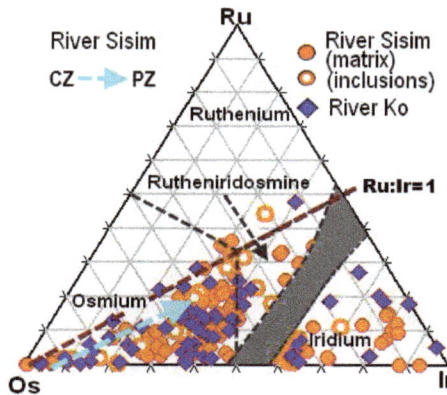

Figure 5. Compositional variations of grains of Os–Ir–(Ru) alloys from placer deposits associated with the River Sisim (this study) and River Ko [2] in terms of the Ru–Os–Ir diagram (atomic %). The blue arrow shows the variation, from the CZ toward the PZ, which exists in the zoned placer grain of Os–Ir–(Ru) alloy from Sisim (Figure 2h). The nomenclature and miscibility gap are in [6].

Placer grains of Pt–Fe alloy at Sisim are ≤0.5 mm in size, and they typically contain elevated levels of Cu (up to 1.9 wt %, Table 3). These grains exhibit variable ΣPGE/(Fe + Cu + Ni) ratios, typically close to 3, thus corresponding to isoferroplatinum (or ferroan platinum).

Some of the Pt–Fe alloy grains are rimmed by Pt-rich phases of arsenide or sulfarsenide compositions, or both, as in the case of the grain shown in Figure 2f. It has the composition of isoferroplatinum [$(Pt_{2.57}Rh_{0.28}Ir_{0.17}Os_{0.09}Ru_{0.05})_{\Sigma 3.16}(Fe_{0.66}Cu_{0.17}Ni_{<0.01})_{\Sigma 0.84}$] and is rimmed by successive rims, the inner one composed of a S-deficient phase of hollingworthite-type [$(Rh_{0.94}Pt_{0.18}Fe_{0.03}Ir_{0.01})_{\Sigma 1.84}As_{1.15}S_{0.69}$] and the outer one consisting of metal-deficient sperrylite [$(Pt_{0.71}Rh_{0.17}Fe_{0.04}Ni_{<0.01})_{\Sigma 0.92}(As_{1.82}S_{0.26})_{\Sigma 2.08}$]. The observed deviations from stoichiometry are likely related to conditions of their deposition at low temperatures from late fluids, possibly involving the effects of metastable crystallization. Interestingly, a similar phase of nonstoichiometric sperrylite [$(Pt_{0.93}Rh_{0.13}Fe_{0.07})_{\Sigma 1.15}(As_{1.65}S_{0.20})_{\Sigma 1.85}$], also rich in Rh and S, was described in a similar textural association from British Columbia, Canada [7].

Table 3. Compositions of Pt–Fe–(Cu–Ni) alloy minerals from the Sisim placer zone, Eastern Sayans.

#		Ru	Os	Ir	Rh	Pt	Pd	Fe	Ni	Cu	Total
1	Matrix	0.41	0.71	4.96	bdl	84.62	bdl	7.15	0.03	0.42	98.3
2		1.12	0.3	5.44	5.6	79.63	bdl	6.8	0.15	1.07	100.1
3		1.2	0.31	5.81	4.56	80.01	bdl	7.38	0.09	0.65	100
4		0.9	2.19	4.55	4.38	80.02	bdl	5.65	0.01	1.93	99.6
5		0.79	2.67	5.24	4.55	79.28	bdl	5.82	0.01	1.75	100.1
6		2.52	2.35	6.82	1.03	79.82	bdl	6	0.12	0.81	99.5
7	Inclusion	bdl	0.15	9.72	0.27	77.55	0.25	8.72	0.62	0.11	97.4
8		bdl	0.06	8.33	0.28	78.98	0.3	9	0.7	0.2	97.9
9		bdl	0.16	10.84	0.22	76.4	0.18	8.69	0.59	0.11	97.2
10		0.15	0.08	9.57	0.38	78.92	bdl	8.7	0.68	0.22	98.7
11		0.15	0.31	8.82	8.47	69.51	bdl	9.19	0.85	0.4	97.7
12		bdl	bdl	8.9	4.2	71.6	bdl	12.7	3.7	0.9	102
13		bdl	bdl	20.9	3.5	64.4	bdl	10.2	2.6	bdl	101.6
14		bdl	bdl	10.6	7.9	66.9	bdl	10.5	2.6	bdl	98.5
15		bdl	bdl	4.8	2.9	69.8	1.2	13.9	5.6	1.5	99.7
16		bdl	bdl	10.8	6.7	73.7	bdl	10.1	1.9	bdl	103.2
17		bdl	bdl	bdl	bdl	71.7	2.5	12.8	4.4	6.1	97.5
18		bdl	bdl	bdl	bdl	80.2	6.9	10.8	1.6	0.9	100.4
19		bdl	bdl	bdl	bdl	92.9	bdl	8.8	0.7	bdl	102.4
20		bdl	bdl	bdl	bdl	92	bdl	8.6	0.7	bdl	101.3
21		bdl	bdl	bdl	bdl	66.7	12.8	11.3	0.7	11.3	102.8

Atomic Proportions (per a Total of 100 at. %)										
#	Ru	Os	Ir	Rh	Pt	Pd	Fe	Ni	Cu	ΣPGE/(Fe + Ni + Cu)
1	0.7	0.6	4.3	0	72	0	21.2	0.1	1.1	3.46
2	1.7	0.2	4.4	8.4	63.3	0	18.9	0.4	2.6	3.57
3	1.8	0.3	4.7	6.9	63.9	0	20.6	0.2	1.6	3.46
4	1.4	1.8	3.8	6.8	65.3	0	16.1	0	4.8	3.77
5	1.2	2.2	4.3	7	64.3	0	16.5	0	4.4	3.79
6	4.1	2	5.8	1.6	66.6	0	17.5	0.3	2.1	4.03
7	0	0.1	8.1	0.4	63.9	0.4	25.1	1.7	0.3	2.69
8	0	0.1	6.9	0.4	64.2	0.4	25.6	1.9	0.5	2.58
9	0	0.1	9.1	0.3	63.2	0.3	25.1	1.6	0.3	2.7
10	0.2	0.1	7.9	0.6	64.1	0	24.7	1.8	0.6	2.69
11	0.2	0.2	6.8	12.2	53	0	24.5	2.1	0.9	2.63
12	0	0	6.1	5.4	48.4	0	30	8.3	1.9	1.49
13	0	0	15.5	4.9	47.2	0	26.1	6.3	0	2.08
14	0	0	7.8	10.9	48.5	0	26.6	6.3	0	2.04
15	0	0	3.2	3.6	45.3	1.4	31.5	12.1	3	1.15
16	0	0	7.9	9.1	53	0	25.4	4.5	0	2.34
17	0	0	0	0	46.5	3	29	9.5	12.1	0.98
18	0	0	0	0	57.8	9.1	27.2	3.8	2	2.03
19	0	0	0	0	73.8	0	24.4	1.8	0	2.81
20	0	0	0	0	74	0	24.2	1.9	0	2.84
21	0	0	0	0	40	14.1	23.7	1.4	20.8	1.18

Note: Results of WDS (# 1–11) and scanning-electron microscopy (SEM)/EDS (# 12–21) analyses are quoted in weight percent; bdl is below detection limit, and PGE is platinum-group elements. The atomic proportions are based on a total of 100%. Numbers 1–11, 19, and 20 represent isoferroplatinum or ferroan platinum; # 12–14, 16, and 18 are Pt–Fe alloy ("Pt₂Fe"-type); # 15 refers to a member of the tetraferroplatinum–ferronickelplatinum series; # 17 pertains to tetraferroplatinum; and # 21 corresponds to tulameenite (Pd-bearing).

In addition, the Pt–Fe alloy grains at Sisim are substantially enriched in Rh (up to 5.6 wt %) and Ir (up to 6.8 wt %, Table 3); elevated contents of the components chengdeite (Ir$_3$Fe) and "Rh$_3$Fe" are implied in these variants of isoferroplatinum or ferroan platinum.

3.3. Inclusions of Hydrous Silicates and High-Ti Micrometer-Sized Mixtures

Serpentine and various amphiboles occur as inclusions in placer grains of PGM at Sisim. The serpentine inclusions appear to contain other minerals as micro-impurities, which result in deviations from stoichiometry. However, the analyzed level of MgO is unusually high (45.6 wt %, WDS), thus indicating a high-Fo composition of the olivine precursor coexisting with the Os–Ir alloy. Calcic and sodic–calcic amphiboles, identified as magnesio-hornblende, barroisite, and edenite, were recognized in inclusions enclosed within the PGM grains (Table 4). They are moderately high in magnesium (Mg# up to 83.6); some compositions display elevated contents of Na (up to 3.6% Na$_2$O), K (1.1% K$_2$O), and Ti (up to 1.7% TiO$_2$).

Interestingly, heterogeneous micro-inclusions, apparently composed of mixtures of different phases, also were analyzed; these are hosted by grains of PGM and invariably contain high levels of Ti (54.5–68.2% TiO$_2$), elevated Al (9.0–10.3% Al$_2$O$_3$), Fe (2.9–8.4% FeO$_{total}$), Si (5.8–11.2% SiO$_2$), and substantial Cr (0.4–1.0% Cr$_2$O$_3$). The main material is likely a Ti-rich oxide or a mixture of oxides, which fill these inclusions along with a subordinate silicate.

Table 4. Compositions of amphibole inclusions hosted by grains of platinum-group minerals in the Sisim placer zone, Eastern Sayans.

#	SiO$_2$	TiO$_2$	Al$_2$O$_3$	Cr$_2$O$_3$	FeO	MnO	MgO	CaO	Na$_2$O	K$_2$O	Total
1	45.92	1.2	10.28	0.16	6.59	0.14	18.15	10.56	2.03	0.15	95.18
2	45.83	0.84	8.27	0.18	7.95	0.2	19.04	8.38	1.53	0.11	92.33
3	44.31	1.63	10.26	0.27	11.78	0.17	14.97	10.2	0.61	0.22	94.42
4	43.61	1.62	9.91	0.2	12.05	0.22	14.7	10.49	0.7	0.21	93.71
5	47.04	1	10.34	0.17	10.63	0.19	14.31	8.16	1.04	0.02	92.9
6	48.63	1.02	8.2	0.21	11.32	0.34	13.48	11.93	0.73	0.84	96.7
7	48.37	1.5	13.32	0.28	9.9	0.18	10.28	9.02	2.17	1.06	96.08
8	44.39	1.75	10.1	0.9	12.06	0.16	13.2	5.68	3.64	0.25	92.13
9	48.67	1.72	13.6	0.12	6.78	0.18	7.81	13.81	2.38	0.38	95.45
10	51.11	1.3	14.4	0.13	6.65	0.17	8.45	12.18	2.4	0.38	97.17

					Atomic Proportions (O = 23)							
#	Si	$^{[4]}$Al	Al	Fe^{3+}	Ti	Cr	Fe^{2+}	Mn	Mg	Ca	Na	K	Mg#
1	6.68	1.32	0.44	0	0.13	0.02	0.8	0.02	3.94	1.65	0.57	0.03	82.8
2	6.84	1.16	0.29	0.2	0.09	0.02	0.8	0.03	4.24	1.34	0.44	0.02	83.6
3	6.58	1.42	0.37	0.44	0.18	0.03	1.02	0.02	3.31	1.62	0.18	0.04	76.1
4	6.54	1.46	0.29	0.55	0.18	0.02	0.97	0.03	3.29	1.68	0.2	0.04	76.7
5	7.02	0.98	0.83	0	0.11	0.02	1.33	0.02	3.18	1.3	0.3	<0.01	70.2
6	7.1	0.9	0.52	0	0.11	0.02	1.38	0.04	2.94	1.87	0.21	0.16	67.4
7	6.99	1.01	1.26	0	0.16	0.03	1.2	0.02	2.22	1.4	0.61	0.2	64.5
8	6.81	1.19	0.63	0	0.2	0.11	1.55	0.02	3.02	0.93	1.08	0.05	65.8
9	7.04	0.96	1.36	0	0.19	0.01	0.82	0.02	1.69	2.14	0.67	0.07	66.8
10	7.18	0.82	1.57	0	0.14	0.01	0.78	0.02	1.77	1.83	0.65	0.07	68.9

Note: Results of WDS analyses are quoted in wt %; bdl is "below detection limit". Numbers 1, 3, 4, and 6 pertain to magnesio-hornblende; # 2, 5, 7, and 8–10 correspond to edenite. The atomic proportions are based on 23 oxygen atoms per formula unit, a.p.f.u. (O = 23).

3.4. Inclusions of REE- and Ti-Rich Minerals Coexisting with PGM

A colloform REE-rich phase seems to fill a cavity in the host Os–Ir alloy around a central inclusion of laurite (Figure 3d). Low analytical totals for the phosphate material (# 1, 2, Table 5) are ascribed to its porosity. Nevertheless, the SEM/EDS compositions led to a stoichiometric formula (Ce,La,Nd,Ca,Pr)PO$_4$ of monazite-(Ce) (Table 5). The X-ray maps for Ce and La show that they are distributed uniformly over the entire grain (Figure 3e,f).

Furthermore, titanite is documented (# 3, Table 5) in a single micro-inclusion hosted by a placer grain of mertieite-II [$Pd_8Sb_{2.3}As_{0.6}$] (Figure 2e, # 1, 2, Table 6); the latter is close in composition to the ideal composition, $Pd_8Sb_{2.5}As_{0.5}$ [8].

Table 5. Compositions of inclusions of REE- and Ti-rich minerals hosted by grains of platinum-group minerals from the Sisim placer zone, Eastern Sayans.

#	P_2O_5	SiO_2	TiO_2	Al_2O_3	Cr_2O_3	Ce_2O_3	La_2O_3	Sm_2O_3	Nd_2O_3	Pr_2O_3	FeO	CaO	Na_2O	Total
1	28.52	bdl	bdl	bdl	bdl	24.91	12.42	bdl	8.38	2.22	bdl	2.83	bdl	79.3
2	30.93	bdl	bdl	bdl	bdl	24	14.84	1.09	9.91	2.71	1.22	3.13	bdl	87.8
3	0	29.77	37.22	0.46	0.12	bdl	bdl	bdl	bdl	bdl	0.65	26.81	0.06	95.1

						Atomic Proportions							
#	P	Si	Ti	Al	Cr	Ce	La	Sm	Nd	Pr	Fe	Ca	Na	Σ(REE + Ca)
1	1.08	0	0	0	0	0.41	0.2	0	0.13	0.04	0	0.14	0	0.92
2	1.06	0	0	0	0	0.36	0.22	0.02	0.14	0.04	0.04	0.14	<0.01	0.94
3	0	1.02	0.96	0.02	<0.01	0	0	0	0	0	0.02	0.99	0	1.01

Note: Results of SEM/EDS analyses are quoted in wt %; bdl is below detection limit. Numbers 1 and 2 pertain to monazite-(Ce); analysis # 3 corresponds to titanite. Atomic proportions are calculated per 4 oxygen a.p.f.u. for monazite-(Ce) and per 5 oxygen a.p.f.u. for titanite. Zero stands for "not detected".

Table 6. Compositions of various platinum-group minerals from the Sisim placer zone, Eastern Sayans.

#	Mineral	Ru	Os	Ir	Rh	Pt	Pd	Fe	Ni	Co	Cu	S	As	Sb	Total
1	Mertieite-II	bdl	bdl	bdl	bdl	bdl	72	bdl	bdl	bdl	bdl	bdl	3.48	23.5	99
2		bdl	bdl	bdl	bdl	bdl	71.86	bdl	bdl	bdl	bdl	bdl	3.5	23.28	98.6
3	Hollingworthite	0.95	0.29	5.89	31.14	13.2	bdl	0.85	0.03	bdl	bdl	9.82	34.58	bdl	96.8
4		3.03	0.73	18.24	17.5	15.95	bdl	0.1	0.06	bdl	bdl	10.78	31.3	bdl	97.7
5	Cherepanovite	3.51	bdl	2.1	47.11	5	bdl	0.13	0.02	bdl	bdl	3.13	39.62	bdl	100.6
6		2.73	bdl	2.6	45.52	6.27	bdl	0.06	0.02	bdl	bdl	3.43	39.41	bdl	100
7	Laurite (Fe-Ni-rich)	31.5	bdl	4.9	6.9	bdl	bdl	9.1	8.9	bdl	2.7	36.4	bdl	bdl	100.4
8	Laurite (As-rich)	46.8	3.7	3	bdl	bdl	bdl	1.1	0.6	1.8	bdl	22.1	23.4	bdl	102.5
9	Kashinite	bdl	bdl	77.8	0.7	bdl	bdl	bdl	bdl	bdl	bdl	22.5	bdl	bdl	101
10	Kashinite (Cu-rich)	bdl	bdl	64.5	bdl	bdl	bdl	3.7	bdl	bdl	6.4	24.6	bdl	bdl	99.2

					Atomic Proportions								
#	Ru	Os	Ir	Rh	Pt	Pd	Fe	Ni	Co	Cu	S	As	Sb
1	0	0	0	0	0	8.12	0	0	0	0	0	0.56	2.32
2	0	0	0	0	0	8.13	0	0	0	0	0	0.56	2.3
3	0.02	<0.01	0.08	0.76	0.17	0	0.04	<0.01	0	0	0.77	1.16	0
4	0.08	0.01	0.25	0.45	0.22	0	<0.01	<0.01	0	0	0.89	1.1	0
5	0.06	0	0.02	0.79	0.04	0	<0.01	<0.01	0	0	0.17	0.91	0
6	0.05	0	0.02	0.77	0.06	0	<0.01	<0.01	0	0	0.19	0.92	0
7	0.49	0	0.04	0.11	0	0	0.26	0.24	0	0.07	1.89	0	0
8	0.89	0.04	0.03	0	0	0	0	0.02	0.06	0	1.33	0.6	0
9	0	0	1.82	0.03	0	0	0	0	0	0	3.15	0	0
10	0	0	1.32	0	0	0	0.26	0	0	0.4	3.02	0	0

Note: Results of WDS (# 1–6) and SEM/EDS (# 7–10) analyses are quoted in weight %; bdl is below detection limit. Numbers 1 and 2 pertain to placer grain. Numbers 3–10 pertain to inclusions hosted by grains of PGE alloy minerals. The atomic proportions are based on a total of 11 a.p.f.u. for mertieite-II, 3 a.p.f.u. for hollingworthite and laurite, 2 a.p.f.u. for cherepanovite, and 5 a.p.f.u. for kashinite.

3.5. Variations and Element Correlations in the Laurite–Erlichmanite Series at Sisim

Extensive variations and broad ranges of compositions (Figure 6a–c, Table 7) are documented for members of the laurite–erlichmanite series, which occur as inclusions (<50 μm across) in placer grains of PGE alloy minerals. Element correlations are based on the analyzed specimens at Sisim (this study) and the River Ko [3]. Iridium correlates positively and strongly with Os; the value of R is 0.84, based on a total of 55 data-points (Figure 6a). In contrast, the Ir–Ru correlation is negative (R = −0.88, Figure 6b). The Os–Ru correlation is inverse (R = −0.97), as expected (Figure 6c).

In addition, we encountered an unusual laurite-type phase that is anomalously enriched in Ir (0.40 a.p.f.u., # 9, Table 7). A phase of similar composition (0.35 Ir a.p.f.u., # 10, Table 7) was reported from the Miass placer zone, southern Urals, Russia [4]. Note that both phases are represented by erlichmanite (i.e., the Os-rich end-member) and contain minor Ru (~0.1 a.p.f.u.); this feature agrees well with the sympathetic Os–Ir covariations observed in the series at Sisim (Figure 6c). More studies are required to explain the anomalous Ir-enrichment, which could be a result of an unusual environment of crystallization. On the other hand, it cannot be excluded that the Ir anomaly may reflect the presence of "invisible" exsolution-induced domains of orthorhombic IrS_2.

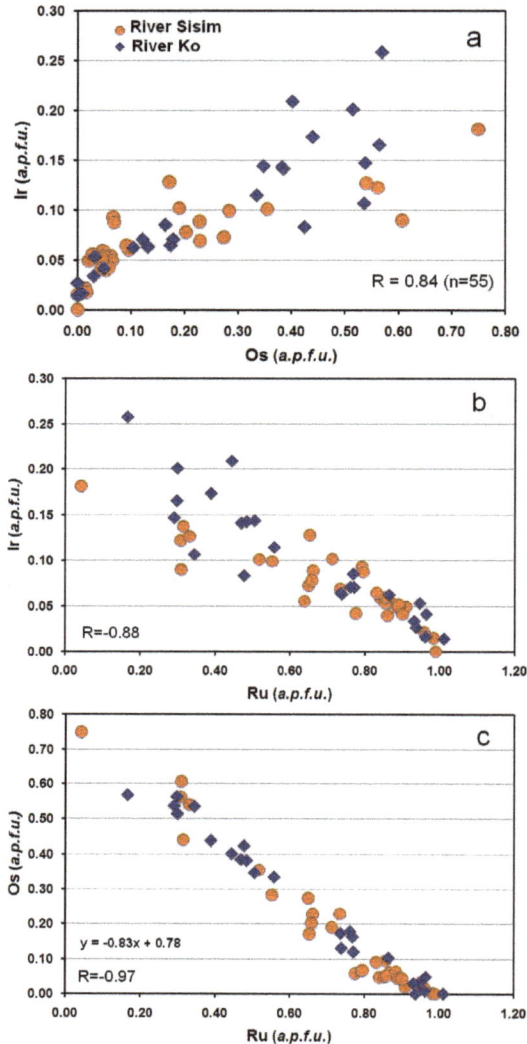

Figure 6. Correlations of Ir and Os (**a**); Ir and Ru (**b**); and Os and Ru (**c**) observed in compositions of members of the laurite–erlichmanite series, which occur as inclusions in placer grains of Os–Ir–(Ru) alloys from the Sisim River system (this study), compared with laurite–erlichmanite grains from the River Ko placer [3]. The equation of linear regression and values of the correlation coefficient (R) are based on a total of 55 data-points (this study).

Table 7. Compositions of laurite–erlichmanite inclusions hosted by grains of PGE alloy minerals in the Sisim placer zone, Eastern Sayans.

#	Ru	Os	Ir	Rh	Pt	Fe	Ni	S	As	Total
1	14.91	45.58	10.79	bdl	bdl	bdl	0.02	28.49	bdl	99.79
2	49.16	6.67	5.79	0.82	bdl	0.01	0.01	36.01	bdl	98.47
3	1.7	57.75	14.09	bdl	bdl	0.24	0.2	26.02	0.17	100.17
4	34.24	22.2	8.71	0.66	bdl	bdl	bdl	32.94	0.08	98.83
5	46.17	9.55	6.8	0.34	bdl	bdl	0.04	35.31	0.08	98.29
6	28.8	27.8	9.8	bdl	bdl	bdl	bdl	34.2	bdl	100.6
7	35.6	20.6	8	bdl	bdl	bdl	bdl	35.3	bdl	99.5
8	57.3	1.9	2	bdl	bdl	bdl	bdl	37.9	bdl	99.1
9	4.44	34.39	31.63	0.33	0.09	0.28	bdl	26.43	bdl	97.6
10	3.46	44.46	26.6	bdl	bdl	bdl	bdl	24.58	0.86	99.96

Atomic Proportions (per a Total of 3 a.p.f.u.)									
#	Ru	Os	Ir	Rh	Pt	Fe	Ni	S	As
1	0.33	0.54	0.13	0	0	0	0.001	2	0
2	0.87	0.06	0.05	0.01	0	0	0	2	0
3	0.04	0.75	0.18	0	0	0.01	0.008	2	0.006
4	0.66	0.23	0.09	0.01	0	0	0	2.01	0.002
5	0.83	0.09	0.06	0.01	0	0	0.001	2	0.002
6	0.55	0.28	0.1	0	0	0	0	2.07	0
7	0.66	0.2	0.08	0	0	0	0	2.06	0
8	0.96	0.02	0.02	0	0	0	0	2	0
9	0.11	0.44	0.4	0.01	<0.01	0.01	0	2.02	0
10	0.09	0.59	0.35	0	0	0	0	1.94	0.03

Note: Results of WDS (# 1–5 and 9) and SEM/EDS (# 6–8) analyses are quoted in weight %.; bdl is below detection limit. Analyses # 1–8 pertain to inclusions of members of the laurite–erlichmanite series, which are hosted by grains of PGE alloy minerals from the Sisim placer zone (this study). Numbers 9 and 10 represent unusual phases of erlichmanite that are anomalously enriched in Ir, which were collected at Sisim (# 9, this study) and at Rudnaya, Eastern Sayans (# 10) [4].

3.6. Monosulfide and Pentlandite-Type Inclusions in Grains of PGE Alloys

Compositionally, there are two types of Ni–Fe–PGE sulfide inclusions associated with laurite, Pt–Fe–(Cu) and Ir–Os–(Ru) alloys (Figure 3a–c). Grains of the first type have their $\Sigma Me/S$ value, in which ΣMe is the total content of metals, close to 1:1 (or lower, with a minimum of ~0.7), and correspond to monosulfide (Mss). In the phases representative of the second type, values of $\Sigma Me/S$ approach 9:8 (ideally 1.125), being characteristic of pentlandite. These phases are invariably enriched in Rh, and to a lesser extent, in Ir (Table 8).

Figure 3c shows genetically informative textural relations; they involve a core zone of Ir–Os alloy and two phases of the Mss-type (Mss and Mss-2), which are developed as overgrowths in a composite inclusion hosted by a placer grain of Os-dominant alloy [$Os_{46.5}Ir_{41.0}Ru_{12.6}$]. The micrometer-sized core of Ir-dominant alloy has the composition [$Ir_{62.3}Os_{26.4}Ru_{11.2}$]; it is rimmed by the first Mss-type phase [$(Ni_{0.28}Fe_{0.21}Ir_{0.17}Rh_{0.12}Cu_{0.11}Pt_{0.03})_{\Sigma 0.92}S_{1.08}$]. The second phase (Mss-2) is $(Ni_{0.52}Fe_{0.39}Rh_{0.10})_{\Sigma 1.01}S_{1.00}$. The first phase appears to have formed early as a result of buildup in sulfur fugacity (fS_2) after crystallization of the Ir–Os alloy nucleus. As a result of crystallization at a high temperature, this phase is relatively enriched in Ir and ΣPGE. In contrast, the second phase is essentially devoid of Ir, poorer in Rh, and rich in Ni.

The overall variations (Figure 7) observed for the Mss- and pentlandite-type phases at Sisim compare well to those recorded from placers of the River Ko [3]. These phases all define a single linear trend that extends toward the compositions that are rich in (Ni + Fe) and relatively poorer in S, with a corresponding decrease in ΣPGE. Vacancies could likely exist at the metal sites of the pentlandite-type phases rich in the PGE from these placers, as suggested for ferhodsite [$(Fe,Rh,Ir,Ni,Cu,Co,Pt)_{9-x}S_8$] discovered in the Nizhniy Tagil complex, Russia [9]. Ferhodsite is tetragonal and has a pentlandite-derivative structure. Owing to the possible presence of vacancies, compositions of $(Ni,Fe,PGE)_{9-x}S_8$ and $(Ni,Fe,PGE)_{1-x}S$ could almost coincide for some data-points in the plot $(Ni + Fe)–\Sigma PGE–S$ (Figure 7).

Table 8. Compositions of monosulfide- and pentlandite-type inclusions hosted by grains of PGE alloy minerals in the Sisim placer zone, Eastern Sayans.

#	Ru	Os	Ir	Rh	Pt	Pd	Fe	Ni	Co	Cu	S	Total
1	bdl	bdl	0.95	12.69	0.25	0.02	23.49	29.26	0.26	0.41	30.65	98
2	bdl	bdl	0.57	12.25	0.13	0.15	24.23	30.11	0.2	0.23	30.02	97.9
3	6.98	bdl	0.88	10.34	bdl	bdl	20.54	25.03	bdl	0.81	31.14	95.7
4	bdl	bdl	bdl	11.5	bdl	bdl	18.7	33.1	bdl	bdl	32.1	95.4
5	bdl	bdl	bdl	11.7	bdl	bdl	19.3	34	bdl	bdl	33.8	98.8
6	bdl	bdl	bdl	11.3	bdl	bdl	19.5	33.7	bdl	bdl	34	98.5
7	bdl	bdl	bdl	10.5	bdl	bdl	24.7	31	bdl	bdl	34.6	100.8
8	bdl	bdl	3.1	10.2	bdl	1	22.3	25.6	bdl	bdl	34.4	96.6
9	bdl	bdl	bdl	10.8	bdl	bdl	23	32.1	bdl	bdl	33.9	99.8
10	bdl	bdl	bdl	12.1	bdl	bdl	18.9	35.3	bdl	bdl	34.9	101.2
11	31.5	bdl	4.9	6.9	bdl	bdl	9.1	8.9	bdl	2.7	36.4	100.4
12	bdl	bdl	bdl	11.4	bdl	bdl	24	28.7	bdl	bdl	33.5	97.6
13	bdl	bdl	2.7	29.6	bdl	bdl	11.6	20.1	bdl	1.9	31.9	97.8
14	bdl	bdl	2.6	29.8	bdl	bdl	10.9	20.4	bdl	2.3	32.5	98.5
15	bdl	bdl	27.4	10.2	4.5	bdl	9.6	13.7	bdl	5.9	28.8	100.1
16	1.5	13.4	26	7.6	5.5	bdl	7.3	10.1	bdl	5.2	23.5	100.1
17	2.7	bdl	27.1	11.8	bdl	bdl	10.9	11.5	bdl	6.3	30.5	100.8
18	bdl	bdl	64.5	bdl	bdl	bdl	3.7	bdl	bdl	6.4	24.6	99.2
Atomic Proportions (per a Total of 100 at. %)												
#	Ru	Os	Ir	Rh	Pt	Pd	Fe	Ni	Co	Cu	S	Me/S
1	0	0	0.2	6.1	0.06	0.01	20.9	24.7	0.2	0.3	47.4	1.11
2	0	0	0.1	5.9	0.03	0.07	21.5	25.5	0.2	0.2	46.5	1.15
3	3.5	0	0.2	5.1	0	0	18.8	21.8	0	0.7	49.8	1.01
4	0	0	0	5.6	0	0	16.6	28	0	0	49.8	1.01
5	0	0	0	5.4	0	0	16.5	27.7	0	0	50.4	0.99
6	0	0	0	5.2	0	0	16.7	27.4	0	0	50.7	0.97
7	0	0	0	4.7	0	0	20.6	24.5	0	0	50.2	0.99
8	0	0	0.8	4.9	0	0.46	19.6	21.5	0	0	52.8	0.89
9	0	0	0	4.9	0	0	19.4	25.8	0	0	49.9	1.01
10	0	0	0	5.5	0	0	15.8	28	0	0	50.7	0.97
11	16.4	0	1.3	3.5	0	0	8.6	8	0	2.2	59.9	0.67
12	0	0	0	5.3	0	0	20.7	23.6	0	0	50.4	0.99
13	0	0	0.7	15.3	0	0	11.1	18.2	0	1.6	53	0.89
14	0	0	0.7	15.3	0	0	10.3	18.3	0	1.9	53.5	0.87
15	0	0	8.6	6	1.39	0	10.3	14	0	5.6	54.1	0.85
16	1	4.9	9.4	5.1	1.96	0	9.1	11.9	0	5.7	50.9	0.96
17	1.5	0	8.2	6.7	0	0	11.3	11.4	0	5.8	55.2	0.81
18	0	0	26.4	0	0	0	5.2	0	0	7.9	60.4	0.65

Note: Results of WDS (# 1–3) and SEM/EDS (# 4–18) analyses are quoted in wt %; bdl is below detection limit.

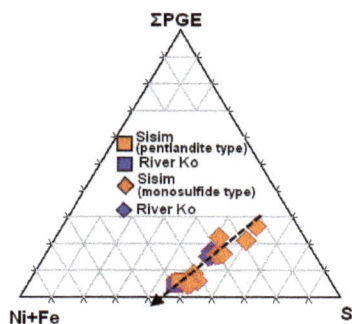

Figure 7. Variation of compositions of sulfide phases of the monosulfide and pentlandite types, which occur as inclusions hosted by placer grains of Os–Ir–(Ru) alloys from the River Sisim (this study) and the River Ko [3] in terms of ternary plot $\Sigma PGE–Ni + Fe$ (+Co)–S (atomic %).

3.7. Other Unusual Phases in Micrometric Inclusions at Sisim

Some phases of unusual compositions rich in Pd, Cu, and Ni were found in inclusions hosted by PGE alloy minerals. A Pd-bearing tetraferroplatinum (# 15, 17, Table 3) [$(Pt_{0.91-0.93}Rh_{0-0.07}Ir_{0-0.06}Pd_{0.03-0.06})_{\Sigma0.99-1.07}(Fe_{0.58-0.63}Ni_{0.19-0.24}Cu_{0.06-0.24})_{\Sigma0.93-1.01}$] represents a solid solution with ferronickelplatinum. A phase of Pd-rich tulameenite corresponds to $(Pt_{1.60}Pd_{0.56})_{\Sigma2.16}Fe_{0.95}(Cu_{0.83}Ni_{0.06})_{\Sigma0.89}$ (# 21, Table 3).

The other phases, analyzed in micrometer-sized inclusions or as a rim around Pt–Fe alloy grains, are hollingworthite, cherepanovite, and kashinite (Cu-free and Cu-rich varieties, Table 6). Especially interesting are Fe-, Ni- and As-rich phases (# 7, 8, Table 6) related to laurite. The first of them appears to be a new and hitherto unreported example of a solid solution involving pyrite-type components: laurite, pyrite, and vaesite. Previously, Barkov et al. reported on a member of the pyrite–laurite series from the Imandra complex, Russia [10]. The second phase of laurite is anomalously enriched in As (23.4% or 0.60 As a.p.f.u., # 8, Table 6); it likely indicates the existence of a solid solution with anduoite (orthorhombic $RuAs_2$).

4. Discussion

4.1. Crystallization History of Associations of PGM at Sisim

We presume that the Os-dominant alloy phase was first to crystallize in the ultramafic lode source in inferred association with chromian spinel and cumulus olivine, proposed on the basis of inclusions of high-Mg serpentine. The core of the zoned crystals (Figure 2d,g) of the Os–Ir–(Ru) alloys nucleated first; their periphery formed at a later stage, at lower temperature. The evolutionary trend is expressed by the change of compositions from $Os_{92.6}Ir_{6.5}Ru_{0.7}Fe_{0.10}Ni_{0.10}$ (core) to $Os_{48.0}Ir_{29.1}Ru_{18.9}Pt_{2.5}Rh_{1.0}Fe_{0.47}Ni_{0.15}$ (periphery). This trend (Figure 5) is consistent with the following, descending order of melting temperatures known from the literature [11]: Os (3030 °C) → Ir (2447 °C) → Ru (2310 °C) → Rh (1963 °C) → Pt 1772 °C → Pd (1554 °C). Thus, Os, as the high-temperature component, was preferentially incorporated into the core; the rim is enriched in Ir and Ru, which are less refractory, during fractional crystallization. This example provides evidence of a simple and effective mechanism of fractionation of Os from Ru and Ir in natural systems. Minor Pt and Rh, along with Fe, are also relatively enriched in the periphery. Palladium is not important here, as is typically observed in Os–Ir–(Ru) alloys associated with chromitite. Nevertheless, we note a relative Pd-enrichment that is characteristic of the late phases of alloys associated with inclusions or rims.

In contrast with other PGE deposits, the overall field of compositions of the alloys at Sisim and River Ko is limited to the Ru-poor portion of the Os–Ir–Ru system by the line Ru/Ir = 1 (Figure 5). Furthermore, the trend of zonation observed in the Os–Ir–(Ru) alloy extends subparallel to this boundary. This feature likely implies the existence of a geochemical relationship involving Ir and Ru, so that the alloy phases with a Ru/Ir ratio exceeding 1 were not stable under the local conditions of crystallization.

Grains of Pt–Fe alloys likely crystallized after the grains of Os–Ir alloy phases. The observed variants of isoferroplatinum, enriched in the chengdeite [Ir_3Fe] component, appear to have formed first among the Pt–Fe alloys. At a late stage and at subsolidus temperatures, the other Pt–Fe alloys rich in Pd, Cu, and Ni, such as Pd-bearing tetraferroplatinum and tulameenite (also Pd-rich), formed as components of inclusions hosted by grains of PGE alloys.

Levels of fS_2 and fAs_2 likely increased during the advanced stages of ore formation, leading to the deposition of various species of sulfide, sulfarsenide, and arsenide rich in PGE. These occur in two textural forms, as follows: as late phases crystallized from microvolumes associated with the inclusions, or as a rim that consists of hollingworthite, a sperrylite-like phase, cherepanovite, kashinite, and its Cu-rich variant. The development of the composite rims of hollingworthite and sperrylite (Figure 2f) provides a clear indication of the S–As enrichment in a late fluid.

As noted, laurite is fairly common in the form of multicomponent or single inclusions in association with the Mss-type phases (Figure 3a,b). Laurite-type phases that are atypically rich in Fe and Ni imply the existence of a solid solution involving various pyrite-type components: RuS_2 (laurite), FeS_2 (pyrite), and NiS_2 (vaesite). One of the laurite phases is anomalously rich in as (23.4 wt %); it is indicative of considerable, though likely limited, solid solution with anduoite (orthorhombic $RuAs_2$). These occurrences of laurite–erlichmanite phases at Sisim likely formed at low temperatures at a subsolidus stage. They are unlikely to have a high-temperature magmatic origin, which laurite reveals in other environments (e.g., [12,13]). In fluid-saturated systems rich in volatiles, laurite or ruthenoan pyrite belong to a hydrothermal paragenesis described from the Imandra complex, Russia [10,14]. Unusual laurite–clinochlore intergrowths also crystallized relatively late, from microvolumes of H_2O-bearing fluid enriched in Ru, S, and lithophile elements in the Pados-Tundra complex, Russia [15]. In addition, laurite can form at a late stage as a result of a solution-and-redeposition reaction involving the original grains of Os-rich alloy in environments of increasing fS_2, as observed at Miass, Russia [4].

At Sisim, a late generation of Ir–Os alloy precipitated as a core-like phase surrounded by the Mss-type phases, which all are enclosed within a grain of Os–Ir alloy (Figure 3c). As noted, the first of the Mss phases formed after the alloy core, as a result of buildup in fS_2 in the micro-environment. Contents of Ir and ΣPGE decreased sharply, followed by a corresponding increase in Ni in the Mss-2 phase (Figure 3c), likely owing to a normal drop in temperature during crystallization.

We observe that Rh is a main constituent of the PGE documented in the Mss- and pentlandite-type inclusions. This feature is not unusual, as rhodian pentlandite is known in various deposits (e.g., [16–19]). Presumably, Rh is better accommodated by the structure of these sulfides, at least under the given conditions of crystallization. High-temperature conditions would presumably promote the incorporation of greater amounts of Rh and ΣPGE into these phases.

This suggestion is corroborated by experimental results. In the system Fe–Rh–S, at 900 °C, the pyrrhotite phase ($Fe_{1-x}S$) dissolves up to 25.7 at. % Rh; with decreasing temperature, the maximum solubility strongly decreases to 2.8 at. % Rh at 500 °C [20]. The solubility of Ir, also determined in $Fe_{1-x}S$, is much lower than that of Rh; even at a higher temperature of synthesis in the system Fe–Ir–S, pyrrhotite dissolves 5.8 at. % Ir at 1100 °C, 3.4 at. % Ir at 1000 °C, and only 1.0 at. % Ir at 800 °C [21]. The maximum levels of solubility of Pt and Os are much lower: 1.1 at. % Pt at 1100 °C, as observed in the system Fe–Pt–S [22], and 0.7 at. % Os at 1180 °C (or 0.3 at. % at 900 °C) in the system Fe–Os–S [23].

The PGE-bearing phases of Mss- and pentlandite-type included in grains of PGE alloys at Sisim and River Ko show a linear trend of crystallization (Figure 7). They evolved in the direction of a decrease in ΣPGE and a corresponding increase in Ni and Fe, presumably as temperature dropped. The overall content of S is relatively high in the ΣPGE-rich sulfide phases; note that the general level of S decreased during crystallization (Figure 7). The S-excess compositions of the phases richest in the PGE presumably compensate an excess in positive charges resulting from the incorporation of Rh^{3+} (and Ir^{3+}) in the place of Ni^{2+} and Fe^{2+} in crystal structures.

4.2. Contrasting Behavior of Ir and Mechanisms of Element Substitutions

The behavior of Ir in Os–Ir–(Ru) alloys differs from that in Os–Ru–(Ir) disulfides at Sisim. In the alloys, the compositional field of which is restricted by the line Ru/Ir = 1 (Figure 5), the Ir content correlates negatively and strongly with Os (R = −0.86). Thus, Ir largely substitutes for Os in these structures; an idealized scheme, $[Ir + Ru]^0 \rightarrow 2Os^0$, seems relevant and operates along the boundary. In contrast, the laurite–erlichmanite series displays the well-recognized sympathetic covariation of Ir and Os, with R = 0.84 (Figure 6a). The Ir–Ru correlation is antipathetic, with a value of R = −0.88 (Figure 6b). These variations are consistent with the substitution mechanism $[Os^{2+} + 2 Ir^{3+} + \square] \rightarrow 4Ru^{2+}$. In addition, we cannot exclude that the $Ir^{2+}(S_2)^{2-}$ component is involved, so that the alternative scheme of substitution is $Os^{2+} + Ir^{2+} \rightarrow 2Ru^{2+}$.

The extent of $[Ir_x(Ru,Os)_{1-x}]S_2$ solid solution in the laurite–erlichmanite series is limited to ~20 mol % "IrS_2", whereas a continuous solid-solution exists between the RuS_2 and OsS_2 end-members, cf. [24]. This pattern may well result from the existence of vacancy-type defects and related complications arising from the incorporation of Ir [25]. The proposed mechanism is 0.667 Ir^{3+} + 0.333 $^{Me}\square$ = $(Ru + Os)^{2+}$; the Ir is assumed to be trivalent in $Ir_{0.67}S_2$ (i.e., the ideal phase of pyrite structure "IrS_3"). A second coupled substitution is $[(Ir + Rh)^{3+} + (AsS)^{3-} = (Ru + Os)^{2+} + (S_2)^{2-}]$, in which Ir (and Rh) are incorporated as the irarsite–hollingworthite component [25].

We suggest that the contrasting behavior of Ir at Sisim is controlled principally by levels of fS_2 in the system. At very low levels, Ir substitutes preferentially for Os in the alloy phases. The positive covariation of Ir and Os, documented in the laurite–erlichmanite series, is consistent with a gradual increase in fS_2. It is known that the phase OsS_2 crystallizes at a substantially higher level of fS_2 than RuS_2; the position of the $Ir–Ir_2S_3$ buffer is relatively close to the $Os–OsS_2$ buffer, and they both are well above the $Ru–RuS_2$ buffer [26]. Therefore, the incorporation of the "$Ir_{1-x}S_2$" component into the laurite–erlichmanite series would clearly require elevated levels of fS_2, which likely led to the positive covariation of Ir with Os, not Ru. The proposed mechanism $[Os^{2+} + 2 Ir^{3+} + \square] \rightarrow 4Ru^{2+}$ may well account for other occurrences of laurite–erlichmanite (e.g., at Pados-Tundra, Russia), in which there is a positive correlation of Ir and Os, and members of the $RuS_2–RuSe_2$ series occur [15].

4.3. Potential Provenance of PGM in the Sisim Placer Zone

The observed terrane affinities strongly suggest that the Lysanskiy complex is the common source of placer associations of detrital grains of PGM in the area of the rivers Sisim and Ko. This complex is layered; it represents a suite of ultramafic–mafic bodies ~0.5 × 30 km in extent [27], which have tectonic contacts with the host rocks of the Bakhtinskaya suite of the Upper Proterozoic age. The complex includes several massifs: Lysanskiy, Podlysanskiy, and Kedranskiy, along with many small and fragmented bodies. These are composed of sequences of serpentinite, wehrlite, lherzolite, harzburgite, clinopyroxenite, websterite, gabbronorite, troctolite, gabbro, and anorthosite. Some massifs are dominantly ultramafic, whereas others contain mostly gabbroic rocks. The Lysanskiy complex has a high potential for Ti–(V) mineralization; it hosts podiform ilmenite—titanian magnetite orebodies up to 1–2 km across, namely the Rossyp', Piramida, Bezymyannyi, and Malyi Lysan deposits.

A wide spectrum of compositions is here documented for detrital grains of chromian spinel (magnesiochromite–chromite series) from the Sisim zone. Compositions of lode grains of chromite from outcrops of serpentinite, attributed to the Lysanskiy complex, are fairly similar (Figure 4). Some of the detrital grains of magnesiochromite display very high levels of Mg, attaining 16.6 wt % MgO (with 48.6% Cr_2O_3, Table 1). This level of magnesium enrichment is unusual and even exceeds the highest contents reported from the Lower Zone in the Bushveld complex, South Africa (up to ~14% MgO and 57% Cr_2O_3 [28] or from the Monchepluton layered complex, Russia (~14% MgO and 56% Cr_2O_3) [29]. On the other hand, these grains are substantially enriched in Ti (up to 1.3% TiO_2; Table 1), which could reflect the presence of exsolved, submicrometric lamellae rich in the ulvöspinel component. The relative enrichment in Ti is unusual for such a high-magnesium phase (cf. 0.3 wt % TiO_2 in the chromian spinel at Monchepluton quoted above). The presence of Ti-bearing amphiboles and, especially, high-Ti micromixture inclusions (54.5–68.2% TiO_2) within the grains of PGE alloys is noteworthy. The pattern of Ti enrichment strongly points to the Lysanskiy complex as the lode source of placer grains of PGM and chromian spinel. The placer grain of mertieite-II that hosts the inclusion of titanite (Figure 2e, Table 5) is the only example of a Pd placer mineral found at Sisim; it likely was derived from a more evolved PGE-bearing zone of the Lysanskiy complex.

Interestingly, at a final stage of ore genesis, the unique association of laurite and monazite-(Ce) appeared as consequences of the accumulation of the incompatible elements S, P, and the REE in a microvolume of residual aqueous fluid (Figure 3d–f). The laurite core could have formed at a subsolidus stage, with microparticles of monazite-(Ce) (Table 5) precipitated from a colloidal solution around this core. Colloidal monazite-type nanoparticles have been produced experimentally (e.g., [30]).

5. Conclusions

(1) We attribute the PGM-bearing placer deposits in the Sisim watershed to the Lysanskiy ultramafic–mafic layered complex, Eastern Sayans. The PGE mineralization is strongly dominated by Os–Ir alloy minerals poor in Ru and is thus distinct from deposits in an ophiolite setting.

(2) The Os–Ir–(Ru) alloy minerals and associated Pt–Fe alloys were likely derived from chromitite units of the complex, whose unusual degree of Mg enrichment suggest a picritic parental melt, unusual, however, for its level of titanium. The completeness of the magnesiochromite–chromite series in the placer grains suggests that large volumes of the source rocks were completely eroded.

(3) The limitation of the Os–Ir–(Ru) alloys at Sisim to the Ru-poor portion of the Os–Ir–Ru system by the line Ru/Ir = 1 implies a close geochemical relationship of Ir and Ru, manifested by the scheme Ir + Ru $\rightarrow$ 2Os. A drop in temperature, leading to a decrease in Os, is recorded in zoned grains. This zonation indicates the existence of a simple and effective mechanism of fractionation of Os from Ru and Ir in natural systems.

(4) In contrast, we document a strong positive covariation of Ir and Os along with a negative Ir–Ru correlation in the laurite–erlichmanite series, likely promoted by locally high levels of fS_2. This relationship points to the scheme $[Os^{2+} + 2Ir^{3+} + \square] \rightarrow 4Ru^{2+}$. Alternatively, the IrS_2 component $[Ir^{2+}(S_2)^{2-}]$ is involved; if so, the incorporation of essential, though limited amounts of Ir is governed by the scheme $Os^{2+} + Ir^{2+} \rightarrow 2Ru^{2+}$.

(5) The inferred sequence of crystallization of PGE alloys at Sisim is as follows: (1) grains (Os-rich) of Os–Ir–(Ru) alloy; the core (Os-rich) of the zoned grains of Os–Ir–(Ru) alloy $\rightarrow$ (2) grains (Ir-rich) and periphery zones enriched relatively in Ir–Ru of the zoned grains of Os–Ir–(Ru) alloy $\rightarrow$ (3) isoferroplatinum or ferroan platinum (rich in Ir $\rightarrow$ poor in Ir) $\rightarrow$ (4) various Pt–(Pd)–Fe–Cu–Ni alloys, all likely formed under subsolidus conditions $\rightarrow$ (5) various S–As-rich phases deposited to form inclusions (or a late rim) as a result of buildup in levels of fS_2 and fAs_2 in the micro-environments.

(6) Inclusions of the PGE-bearing phases of monosulfide and pentlandite types, hosted by grains of PGE alloys, follow a linear trend of crystallization, which reflects a decrease in temperature. The decrease in ΣPGE and overall S was accompanied by an increase in Ni and Fe. The observed S-excess in the ΣPGE-rich sulfide phases likely compensates the excess in positive charges owing to the Rh^{3+} $(+Ir^{3+})$-for-$(Ni + Fe)^{2+}$ substitution.

(7) A unique association of laurite with micrometer-sized particles of monazite-(Ce) is documented in a composite inclusion. The juxtaposition reflects an increase in levels of incompatible elements (S, P, and the REE) in a residual microvolume of aqueous fluid. This could be another expression of the unusual character of the parental magma of the Lysanskiy complex.

Author Contributions: The authors (A.Y.B., R.F.M., and G.I.S.) discussed the results and wrote the article together, and G.I.S. conducted the field work in the Sisim placer area, collected the samples analyzed, and studied jointly during the present project.

Acknowledgments: We thank Liana Pospelova (Novosibirsk) and Sergey Silyanov (Krasnoyarsk, Russia) for their expert assistance with the analytical data. We thank the three anonymous reviewers and members of the editorial board whose comments and suggestions helped improve this manuscript. A.Y.B. gratefully acknowledges a partial support of this investigation by the Russian Foundation for Basic Research (project # RFBR 16-05-00884).

Conflicts of Interest: The authors declare no conflict of interest.

References

1. Vysotskiy, N.K. Platinum and areas of its mining. In *The Natural Producing Forces of Russia. Part 5—A Review of Platinum Deposits Outside The Urals*; The USSR Academy of Sciences: Leningrad, Russia, 1933; p. 240. (In Russian)

2. Krivenko, A.P.; Tolstykh, N.D.; Nesterenko, G.V.; Lazareva, E.V. Types of mineral assemblages of platinum metals in auriferous placers of the Altai-Sayan region. *Russian Geol. Geophys.* **1994**, *35*, 58–65.

3. Tolstykh, N.D.; Krivenko, A.P. On the composition of sulfides containing the platinum-group elements. *Zap. Vseross Miner. Obshch* **1994**, *123*, 41–49. (In Russian)

4. Barkov, A.Y.; Tolstykh, N.D.; Shvedov, G.I.; Martin, R.F. Ophiolite-related associations of platinum-group minerals at Rudnaya, western Sayans, and Miass, southern Urals, Russia. *Mineral. Mag.* **2018**, in press. [CrossRef]

5. Melcher, F.; Grum, W.; Simon, G.; Thalhammer, T.V.; Stumpfl, E.F. Petrogenesis of the ophiolitic giant chromite deposits of Kempirsai, Kazakhstan: A study of solid and fluid inclusions in chromite. *J. Petrol.* **1997**, *38*, 1419–1458. [CrossRef]

6. Harris, D.C.; Cabri, L.J. Nomenclature of platinum-group-element alloys: Review and revision. *Can. Mineral.* **1991**, *29*, 231–237.

7. Barkov, A.Y.; Fleet, M.E.; Nixon, G.T.; Levson, V.M. Platinum-group minerals from five placer deposits in British Columbia, Canada. *Can. Mineral.* **2005**, *43*, 1687–1710. [CrossRef]

8. Karimova, O.; Zoloratev, A.; Evstigneeva, T.L.; Johanson, B. Mertieite-II, $Pd_8Sb_{2.5}As_{0.5}$, crystal structure refinement and formula revision. *Mineral. Mag.* **2018**. [CrossRef]

9. Begizov, V.D.; Zavyalov, E.N. Ferhodsite (Fe,Rh,Ir,Ni,Cu,Co,Pt)$_{9-x}$S$_8$—A new mineral from the Nizhniy Tagil ultramafic complex. *New Data Miner.* **2016**, *51*, 8–11. (In Russian)

10. Barkov, A.Y.; Halkoaho, T.A.A.; Laajoki, K.V.O.; Alapieti, T.T.; Peura, R.A. Ruthenian pyrite and nickeloan malanite from the Imandra layered complex, northwestern Russia. *Can. Mineral.* **1997**, *35*, 887–897.

11. Roberts, P.M. *Introduction to Brazing Technology*; CRC Press: Boca Raton, FL, USA, 2016; p. 340. [CrossRef]

12. Andrews, D.R.A.; Brenan, J.M. Phase-equilibrium constraints on the magmatic origin of laurite and Os–Ir alloy. *Can. Mineral.* **2002**, *40*, 1705–1716. [CrossRef]

13. Bockrath, C.; Ballhaus, C.; Holzheid, A. Stabilities of laurite RuS$_2$ and monosulfide liquid solution at magmatic temperature. *Chem. Geol.* **2004**, *208*, 265–271. [CrossRef]

14. Barkov, A.Y.; Fleet, M.E. An unusual association of hydrothermal platinum-group minerals from the Imandra layered complex, Kola Peninsula, northwestern Russia. *Can. Mineral.* **2004**, *42*, 455–467. [CrossRef]

15. Barkov, A.Y.; Nikiforov, A.A.; Tolstykh, N.D.; Shvedov, G.I.; Korolyuk, V.N. Compounds of Ru–Se–S, alloys of Os–Ir, framboidal Ru nanophases and laurite–clinochlore intergrowths in the Pados-Tundra complex, Kola Peninsula, Russia. *Eur. J. Mineral.* **2017**, *29*, 613–622. [CrossRef]

16. Cabri, L.J. (Ed.) *The Geology, Geochemistry, Mineralogy, Mineral Beneficiation of the Platinum-Group Elements*; Canadian Institute of Mining, Metallurgy and Petroleum: Westmount, QC, Canada, 2002; Volume 54, p. 852.

17. Junge, M.; Oberthür, T.; Melcher, F. Cryptic variation of chromite chemistry, platinum group element and platinum group mineral distribution in the UG-2 chromitite: An example from the Karee mine, western Bushveld complex, South Africa. *Econ. Geol.* **2014**, *109*, 795–810. [CrossRef]

18. Junge, M.; Wirth, R.; Oberthür, T.; Melcher, F.; Schreiber, A. Mineralogical siting of platinum-group elements in pentlandite from the Bushveld Complex, South Africa. *Miner. Depos.* **2014**, *50*, 41–54. [CrossRef]

19. Makovicky, E.; Karup-Møller, S. The Pd–Ni–Fe–S phase system at 550 and 400 °C. *Can. Mineral.* **2016**, *54*, 377–400. [CrossRef]

20. Makovicky, E.; Makovicky, M.; Rose-Hansen, J. The system Fe–Rh–S at 900° and 500 °C. *Can. Mineral.* **2002**, *40*, 519–526. [CrossRef]

21. Makovicky, E.; Karup-Møller, S. The phase system Fe–Ir–S at 1100, 1000 and 800 °C. *Mineral. Mag.* **1999**, *63*, 379–385. [CrossRef]

22. Majzlan, J.; Makovicky, M.; Makovicky, E.; Rose-Hansen, J. The system Fe–Pt–S at 1100 °C. *Can. Mineral.* **2002**, *40*, 509–517. [CrossRef]

23. Karup-Møller, S.; Makovicky, E. The system Fe–Os–S at 1180°, 1100° and 900 °C. *Can. Mineral.* **2002**, *40*, 499–507. [CrossRef]

24. Cabri, L.J.; Harris, D.C.; Weiser, T.W. Mineralogy and distribution of platinum-group mineral (PGM) placer deposits of the world. *Explor. Min. Geol.* **1996**, *5*, 73–167.

25. Barkov, A.Y.; Fleet, M.E.; Martin, R.F.; Alapieti, T.T. Zoned sulfides and sulfarsenides of the platinum-group elements from the Penikat layered complex, Finland. *Can. Mineral.* **2004**, *42*, 515–537. [CrossRef]

26. Stockman, H.W.; Hlava, P. Platinum-group minerals in Alpine chromitites from Southwestern Oregon. *Econ. Geol.* **1984**, *79*, 491–508. [CrossRef]

27. Glazunov, O.M. *The Geochemistry and Petrology of the Gabbro-Pyroxenite Formation of the Eastern Sayans*; Nauka: Novosibirsk, Russia, 1975; p. 188. (In Russian)

28. Hulbert, L.J.; von Gruenewaldt, G. Textural and compositional features of chromite in the Lower and Critical zones of the Bushveld complex south of Potgietersrus. *Econ. Geol.* **1985**, *80*, 872–895. [CrossRef]
29. Barkov, A.Y.; Korolyuk, V.N.; Martin, R.F. The maximum extent of Mg-enrichment in olivine of layered intrusions: Evidence from Monchepluton (Fo$_{96}$) and Pados-Tundra (Fo$_{93}$), Kola Peninsula, Russia. *Eur. J. Mineral.* **2018**, under review.
30. Hickmann, K.; John, V.; Oertel, A.; Koempe, K.; Haase, M. Investigation of the early stages of growth of monazite-type lanthanide phosphate nanoparticles. *J. Phys. Chem. C* **2009**, *113*, 4763–4767. [CrossRef]

Article

Platinum-Group Minerals of Pt-Placer Deposits Associated with the Svetloborsky Ural-Alaskan Type Massif, Middle Urals, Russia

Sergey Yu. Stepanov [1], Roman S. Palamarchuk [2,*], Aleksandr V. Kozlov [2], Dmitry A. Khanin [3,4], Dmitry A. Varlamov [3] and Daria V. Kiseleva [1]

[1]   The Zavaritsky Institute of Geology and Geochemistry UB RAS, 15 Akademika Vonsovskogo str., 620016 Ekaterinburg, Russia; Stepanov-1@yandex.ru (S.Y.S.); podarenka@mail.ru (D.V.K.)
[2]   Department of Geology and Exploration Deposit, St. Petersburg Mining University, 2 21st Line, 199106 St. Petersburg, Russia; akozlov@spmi.ru
[3]   Institute of Experimental Mineralogy RAS, 4 Academika Osypyana str., 142432 Chernogolovka, Russia; mamontenok49@yandex.ru (D.A.K.); dima@iem.ac.ru (D.A.V.)
[4]   Department of Mineralogy, Lomonosov Moscow State University, GSP-1, Leninskie Gory, 119991 Moscow, Russia
*   Correspondence: palamarchuk22@yandex.ru; Tel.: +7-981-979-93-95

Received: 24 December 2018; Accepted: 22 January 2019; Published: 28 January 2019

check for updates

Abstract: The alteration of platinum group minerals (PGM) of eluval, proximal, and distal placers associated with the Ural-Alaskan type clinopyroxenite-dunite massifs were studied. The Isovsko-Turinskaya placer system is unique regarding its size, and was chosen as research object as it is PGM-bearing for more than 70 km from its lode source, the Ural-Alaskan type Svetloborsky massif, Middle Urals. Lode chromite-platinum ore zones located in the Southern part of the dunite "core" of the Svetloborsky massif are considered as the PGM lode source. For the studies, PGM concentrates were prepared from the heavy concentrates which were sampled at different distances from the lode source. Eluvial placers are situated directly above the ore zones, and the PGM transport distance does not exceed 10 m. Travyanistyi proximal placer is considered as an example of alluvial ravine placer with the PGM transport distance from 0.5 to 2.5 km. The Glubokinskoe distal placer located in the vicinity of the Is settlement are chosen as the object with the longest PGM transport distance (30–35 km from the lode source). Pt-Fe alloys, and in particular, isoferroplatinum prevail in the lode ores and placers with different PGM transport distance. In some cases, isoferroplatinum is substituted by tetraferroplatinum and tulameenite in the grain marginal parts. Os-Ir-(Ru) alloys, erlichmanite, laurite, kashinite, bowieite, and Ir-Rh thiospinels are found as inclusions in Pt-Fe minerals. As a result of the study, it was found that the greatest contribution to the formation of the placer objects is made by the erosion of chromite-platinum mineralized zones in dunites. At a distance of more than 10 km, the degree of PGM mechanical attrition becomes significant, and the morphological features, characteristic of lode platinum, are practically not preserved. One of the signs of the significant PGM transport distance in the placers is the absence of rims composed of the tetraferroplatinum group minerals around primary Pt-Fez alloys. The sie of the nuggets decreases with the increasing transport distance. The composition of isoferroplatinum from the placers and lode chromite-platinum ore zones are geochemically similar.

Keywords: Ural Platinum belt; Ural-Alaskan massif; Svetloborsky massif; placer system; platinum group elements; platinum group minerals

1. Introduction

At the beginning of the 20th century, clinopyroxenite-dunite massifs, also called Ural-Alaskan type massifs, were found to be the lode sources of the globally unique Ural placer deposits [1–3]. However, for almost 200 years of industrial history of the Ural placer deposit development, only the Solovyeva Mountain lode deposit of the Nizhnetagilsky massif was discovered [3]. At the same time, despite the obvious industrial significance of placers, clinopyroxenite-dunite massifs were studied to a greater extent [4–14]. The last comprehensive work on the geology of placers and the conditions of their formation, as well as platinum-group mineral (PGM) assemblages is the work of N. K. Vysotsky [2], published at the beginning of the last century. Regardless of a number of separate contemporary studies, including those characterizing the PGM assemblages from the placers associated with the Nizhnetagilsky massif [15–17], the platinum-group minerals from other placer systems (e.g., Isovsko-Turinskaya, Nyasminskaya, etc.) have not been studied using modern analytical techniques.

Furthermore, the changes in the platinum-group minerals occurring during their transport from a lode source to a placer were not studied in detail. Despite the significant amount of work devoted to the Koryak-Kamchatka region, as well as the Bushveld complex and the Great Dyke (South Africa), where the formation of the platinum placers [18–24] was investigated in detail, similar studies of the Ural Platinum Belt placers have not been conducted yet.

The aim of this work is to establish the changes in the platinum-group minerals during their transport from the lode source to the alluvial placer by the example of the placer system associated with the Svetloborsky clinopyroxenite-dunite massif.

The present work determined the morphological feature obtained during alteration accompanying the transport of detrital material from the lode source to the placers, and the relationships of mineral assemblages from the alluvial placers of the Isovsko-Turinskaya placer system and the chromite-platinum ore zones of the Svetloborsky massif.

2. Materials and Methods

2.1. Geological Setting

Like other clinopyroxenite-dunite massifs of the Urals, the Svetloborsky massif is located in the western part of the Tagilo-Magnitogorsk megazone (Figure 1a), 15 km east of the Main Uralian Fault [5]. The massif is composed of rocks of the Late Ordovician Kachkanar dunite-clinopyroxenite-gabbro complex and is a tectonic detachment occurring in Silurian metabasalts. The geological structure of the massif is typical of zonal clinopyroxenite-dunite massifs (Figure 1b). Its major part comprises a dunite core composed of fine and medium-grained dunites surrounded by a clinopyroxenite rim of variable thickness. The valleys of transient or weak watercourses forming numerous ravines are well-represented across the massif's area. The Svetloborsky massif is drained by the river Kosya valley in the Southern part, and by the river Is valley in its Northern part. All ravines and river valleys are platiniferous to different degrees.

Two types of lode platinum mineralization are distinguished within the Svetloborsky massif, (1) Platinum-bearing dunites and (2) chromite-platinum mineralization. The first type of mineralization was described in detail by N. D. Tolstykh et al. [9]. The parameters of chromite-platinum mineralization are given in Reference [14].

The manifestation of chromite-platinum mineralization is found for the area of the Vershinniy exploration site, which is located in the South-Western part of the massif. In main trenches dug during geological exploration, vein-disseminated and massive chromitite zones were identified. These zones are located in the fine to medium-grained dunite transition zone with platinum in the concentration range from two to 50 g/t. The majority of individual grains and aggregates of platinum-group minerals in such zones are spatially associated with chromitite segregations [14]. To the West of the Vershinniy exploration site, in the vicinity of the contact between the dunite core and pyroxenites, the Vysotsky site is located, where platinum-bearing dunites were observed. The zone is a linear stockwork with an intensely serpentinized dunite substrate with numerous dykes, lenticular bodies, and veins of pyroxenite, hornblendite and isite (local name, fine-grained melanocratic vein variety of hornblendite named after the river Is in the Urals, Russia). Here, the chromites do not form significant concentrations, and the platinum-group minerals are found directly in the olivine-serpentine matrix without any spatial connection to the chromites [9].

The mineralized zones were subjected to erosion followed by the formation of eluvial placers. The transport of material from eluvium and its subsequent redeposition led to the formation of alluvial placers.

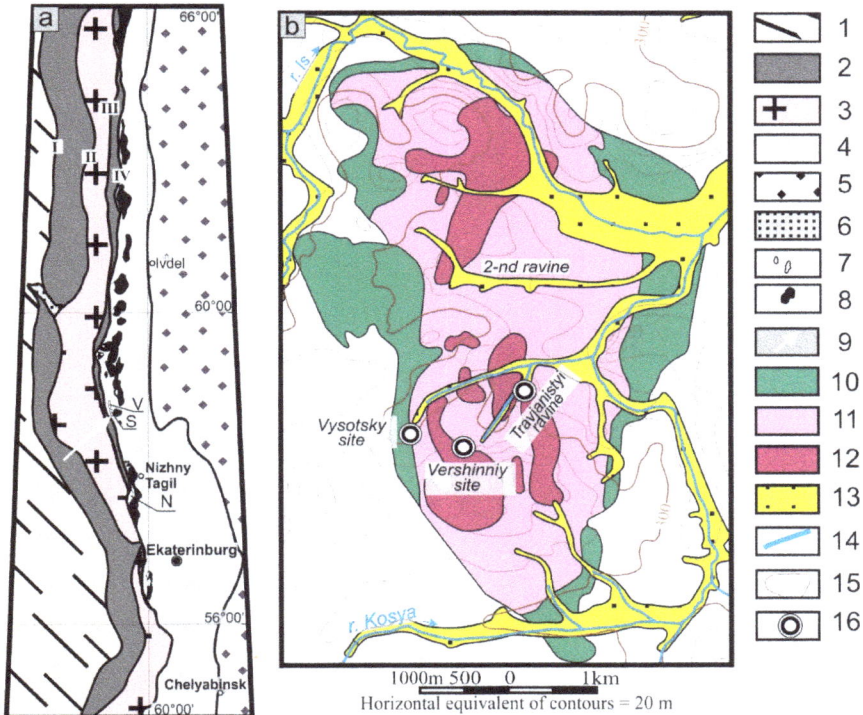

Figure 1. (**a**) The Ural-Alaskan type massifs in the structure of the Urals (compiled from state geological maps of 1:1,000,000 scale): 1—Paleozoic of the East European Platform; 2—Western Ural fold-thrust zone; 3—Central Ural uplift; 4—Tagilo-Magnitogorskaya megazone; 5—sedimentary cover of the West Siberian platform; 6—Polyudovsk uplift; 7, 8—the massifs of the Ural Platinum Belt: 7—dunite bodies, 8—pyroxenites, gabbros, volcanites; 9—the location of Svetloborsky clinopyroxenite-dunite massif. Roman numerals indicate the main faults (thrusts): I—Main Western Ural; II—Osevoy; III—Prisalatimsky; IV—Main Uralian Fault. Letters denote the Ural-Alaskan type massifs: N—Nizhnetagilsky, S—Svetloborsky; and V—Veresovoborsky. (**b**) The Svetloborsky massif's geological structure, compiled from [5] with additions: 10—pyroxenites; 11–12—dunites: 11—fine-, small-grained, 12—medium-grained; 13—alluvial sediment; 14—rivers, streams; 15—contour lines; and 16—sites of detailed sampling.

All studied placers belong to a single connected drainage system. The section of modern sediments of the Vershinniy exploration site (Figure 1b) overlapping the allocated chromite-platinum zone is given as an example of eluvial placer [14]. The structure, characteristics and composition of the PGM assemblages from this placer were described in detail earlier [25]. Travyanistyi ravine is taken as an example of a ravine, proximal or alluvial placer with a small transport distance of the detrital material, whose sources are located on the Vershinniy site. The transport distance of the detrital material from the lode source to the sampling site is slightly more than 1 km. The sampling site of the alluvial placer with a transport distance of more than 30 km (distal placer) is represented by the Glubokinskoe site belonging to the Isovsko-Turinskaya placer system (Figure 2).

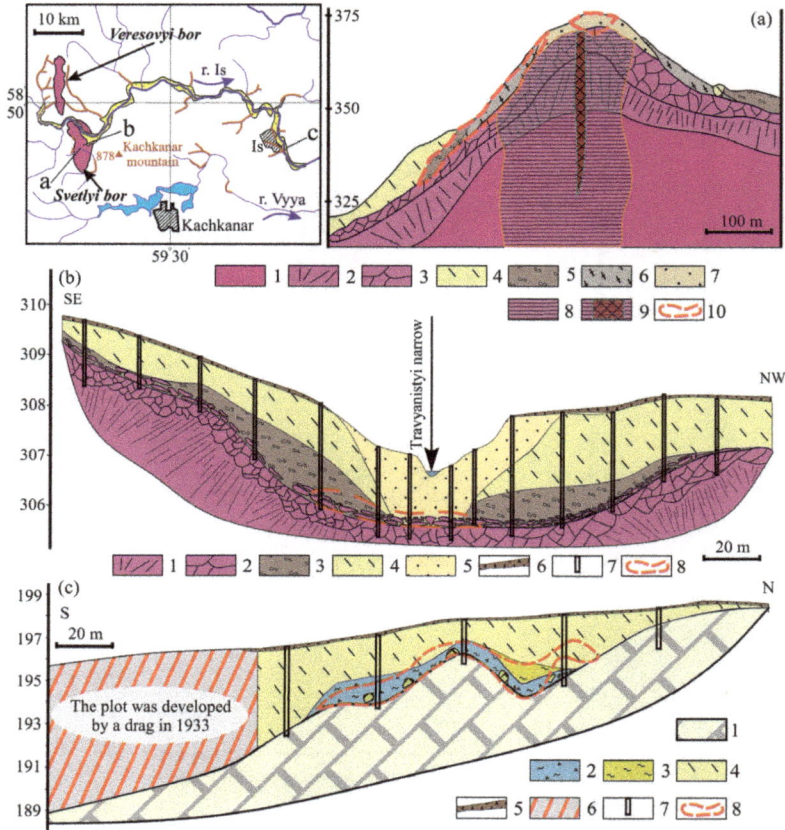

Figure 2. On the map are the rivers Is and Viya, Is and Kachkanar villages, as well as the sampling sites: Vershinniy eluvial (**a**), Travyanistyi proximal (**b**) and Glubokinskoe distal (**c**) placers. Platinum placers are designated by brown (**a**): 1—dunites, 2—serpentinized fractured dunites, 3—weathered serpentinized dunites, 4—sediments of temporary watercourses, 5—clay-eluvial sediments, 6—eluvial with low clay amount, 7—eluvial, 8—vein-disseminated chromitites, 9—massive chromitites, 10—the contour with the platinum content of more than 200 mg/m^3. (**b**): 1—serpentinized fractured dunites, 2—weathered serpentinized dunites, 3—clay-eluvial sediments, 4—sediments of temporary watercourses, 5—alluvial sediments, 6—soil-turf layer, 7—dug holes, 8—the contour with the platinum content of more than 200 mg/m^3. (**c**): 1—Silurian limestone, 2—Jurassic alluvial, 3—Neogenic sediments of temporary watercourses, 4—sediments of temporary watercourses, 5—soil-turf layer, 6—anthropogenic sediments, 7—dug holes, and 8—the contour with the platinum content of more than 200 mg/m^3.

During geological exploration, a heavy concentrate survey was carried on a 40 × 20 m grid at the Vershinniy site. As a result, a platinum anomaly was found in eluvial sediments, and its contours were drawn up according to the platinum contents of 0.2 g/m³. The eluvial deposits enriched with platinum minerals are located at the top of the hill (Figure 2a), directly above the development zone of massive chromitites. These deposits are represented by unsorted gravel-sand mixtures having a small fraction of clay component and an uneven PGM distribution. The contour of the platinum anomaly generally follows the contour of the lode chromite-platinum ore zone.

The eluvial deposits occurring on the slope of the hill are characterized by a larger amount of clay and a smaller fraction of the dunite fragments compared to eluvial deposits from the top of the hill. PGM in such sediments are slightly concentrated in the lower parts of the section, but generally, the sorting of the material remains poor.

The modern sediments, largely sorted during the further transport of detrital material by temporary watercourses, cover the previously formed eluvial-clay sediments in many parts of the placer (Figure 2b). In the valley's central part, directly in the river bed, the alluvial deposits are formed with a section typical for such placers (from top to bottom): soils (up to 1 m thick), clay deposits (up to 1 m), sometimes a small layer of gravel, and a layer of sand (about 0.2 m). The rock names in such sections are given after the predominant fraction in the composition of clay-sand-gravel mixtures. Most PGM are concentrated in the sand layer.

The tested alluvial placer deposits having a long transport distance belong to the site of the first terrace of the Is river valley. The Mesozoic alluvial (according to state geological maps of 1:200,000 scale) deposits occurring on Silurian limestone (Figure 2c) are platinum-bearing. In the Paleogene, these deposits were largely washed up, resulting in a platinum-bearing Paleogene sediment formation. The Jurassic and Paleogene sediments are covered by modern sediments of up to 4 m thick, with platinum-bearing areas being present in some of its parts, represented by washed-up ancient sediments. The large thicknesses of clay and gravel deposits are noted for the alluvial placer with a long transport distance as compared to the section of the proximal placer sediments. The major amount of PGM is also concentrated in the sands.

2.2. Sample Collection and Preparation

In order to extract the platinum-group minerals from the chromitites of the Svetloborsky massif (Figure 3), bulk samples of chromitite rocks weighing 60–80 kg were collected from the lode outcrops and main trenches, which were studied during the geological exploration survey for the lode platinum mineralization by ZAO Ural-MPG (closed joint stock company Ural-MPG). The samples were crushed to a fraction of −1 mm and enriched with a centrifugal concentrator.

The eluvial placer of the Vershinniy site was tested by a heavy concentrate survey with the sampling of modern sediments with a volume of 20 L. These heavy concentrate samples were kindly given to us for further research by A. V. Korneev, ZAO Ural-MPG chief geologist.

The proximal placer of the Travyanistyi ravine and the distal placer of the Isovsko-Turinskaya placer system were tested during independent expeditionary work. Samples with a volume of 50 L each were taken from the already processed sites. The samples were enriched using a centrifugal concentrator.

The concentrates and heavy mineral concentrates consist mainly of chromit and PGM grains. The PGM were extracted from the concentrates using the "blow-off", one of the varieties of the air separation method that can be used on-site during the fieldwork. It is based on the difference in the density of chromit and PGM. Under the action of air flow from human lungs, less dense chromites are removed from the concentrate, while denser PGM remain. PGM grains were studied under a binocular microscope, followed by their mounting on carbon conductive adhesive tape, and studied by scanning electron microscopy. PGM compositions of the grains were analyzed by EPMA (electron probe microanalyzer) after the grains were placed in polished sections made of epoxy resin.

Figure 3. (**a**) Photo of chromitite in bedrock of Svetloborsky massif, (**b**) dunite gutter with modern sediments at proximal placer, (**c**) heavy mineral concentrate with PGM, and (**d**) the last dredge at the Isovsko-Turinskaya placer system.

2.3. Analytical Methods

The morphological features of PGM were studied using a CamScan MX2500 scanning electron microscope (VSEGEI, Saint-Petersburg, Russia). The morphological features of PGM, as well as their internal structure and composition, were examined using a CamScan MV2300 SEM with the INCA Energy 350 detector at an accelerating voltage of 20 kV, working distance 25 mm, and spectral accumulation time of 70 s (IEM RAS, Chernogolovka, Russia). The following standards were used: Pure metals for platinum-group elements (using the Lα-line), Cu, Fe, Ni, Co, FeS_2 synt for S, InAs for As, pure element for Sb. The size of the electron beam spot on the surface of the sample varied from 115 to 140 nm, in a scanning mode to 60 nm, while the excitation zone can reach 4–5 μm (depending on the microrelief, structure and composition of samples). The SEM images were obtained in the backscattered electron mode with material contrast and $10\times$ to $2500\times$ magnification.

The chemical composition of the PGM was determined using a Camebax SX50 X-ray microanalyzer in WDS-mode at an accelerating voltage of 20 kV and a probe current of 30 nA (MSU, Moscow, Russia). The following reference materials were used for calibration: pure metals for Ru, Rh, Pd, Os, Ir and Pt; $CuSbS_2$ for Sb and Cu, CoAsS for Co; NiS for Ni; FeS for Fe and S. Detection limits were (wt. %): Os—0.08, Ir—0.1, Ru—0.05, Rh—0.05, Pd—0.05, Pt—0.05, Fe—0.03, Ni—0.03, Cu—0.03, S—0.05, As—0.05, Co—0.03, Pb—0.08. and Bi—0.1.

3. Results

3.1. Morphological Features

The platinum-group minerals from the bedrock are characterized by a variety of surfaces. Idiomorphic cubic crystals with sizes less than 50 μm, described for different types of lode mineralization, are distinguished [9,14], as well as relatively large aggregates cementing the chromit grains, rarely exceeding 1 mm in size. Xenomorphic grains of similar size with dissolved surfaces are found in the dunite type of mineralization. Quite often, individual grains with idiomorphic and xenomorphic surfaces are intergrown and form relatively large (about 300 um on average) aggregates with well-developed own growth and inherited impression surfaces. For most grains, numerous plane-faced surfaces are observed as well as well-pronounced edges and vertices of the crystals. Many surfaces are characterized by the presence of growth striations that formed both in the process of joint growth of Pt-Fe alloys with chromits (Figure 3b–d) and as a result of the simple form alternation of individual platinum grains during the crystal growth.

The individuals and aggregates of Pt-Fe alloys from the eluvial placer of the Vershinniy site are characterized by morphological features similar to the PGM from the lode sources. There are single individuals with a cubic faceting (Figure 4e). The grains with the plane-faced surfaces and growth striations are quite widespread. Their edges and vertices are not as clearly expressed as those of the Pt-Fe alloy grains from the chromitites. Some grains show signs of mechanical deformations such as poor rounding and irregular grooves (Figure 4f). Hexagonal osmium plates retaining their idiomorphic form were found in a single case.

PGM nuggets from the Travyanistyi proximal placer are characterized both by the abundance of plane-faced growth surfaces (Figure 5a,b) and by the prevalence of deformed fragments (Figure 5c). Growth striations are rare. About 60% of PGM nuggets have size of 0.5–1.5 mm. Hexagonal pinacoidal osmium inclusions are found in some of the individuals (Figure 5d).

The grains from the distal placer show both elongated (Figure 5e,f) and isometric forms (Figure 5g,h). Almost all edges and vertices of the individuals are smoothed. Single grooves formed by mechanical abrasion during the transport of the detrital material are rarely found on the surfaces of Pt-Fe alloy grains. In this placer, PGM nuggets have a prevailing size of 0.1–0.5 mm comprising about 75% of PGM in heavy concentrate.

Pt-Fe alloys retain the primary morphological features characteristic of lode PGM in the placers with a small distance of clastic material transport. At a long distance (more than 10 km), the degree of mechanical attrition of PGM becomes significant, and the morphological features characteristic of lode PGM assemblages (plane-faced surfaces and growth striations) are practically not preserved. During of transport the rim integrity around isoferroplatinum, commonly composed of tetraferroplatinum group minerals were destroyed. In the placer with longest transport distance they are completely destroyed. The size of the nuggets decreases with an increasing transport distance from lode chromite-platinum zone oin prximal and distal placers.

3.2. Chemical Composition of Platinum-Group Minerals

PGM assemblages of the Ural-Alaskan type massifs has been studied to a great extent. All PGM nuggets are Pt-Fe alloys with rare inclusions of accessory minerals (Os-Ir-Ru alloys and PGE sulfides). The content of accessory do not exceed 3% of grain volume. Iridium nuggets occure extremely seldom. Pt-Fe alloys (native platinum, isoferroplatinum and ferroan platinum according to classification [26]) and minerals of the tetraferroplatinum group (tetraferroplatinum-tulameenite-ferronickelplatinum solid solution), are predominating. The mineral inclusions of Os-Ir-(Ru) alloys are abundant in Pt-Fe alloys. Sulfides of the laurite-erlichmanite isomorphous series are also common as inclusions in Pt-Fe minerals. Sulfides of the kashinite-bowieite isomorphous series, as well as the minerals of the Pt-Ir-Rh thiospinel group are relatively rare, and by analogy with other sulfides and Os-Ir-(Ru) alloys, form inclusions in Pt-Fe alloys.

Some Pt-Fe alloys that can be defined only by their crystal structure, and therefore, the mineral with Pt_3Fe composition or close to it is referred to as isoferroplatinum, and Pt_2Fe composition or close to it as ferroan platinum.

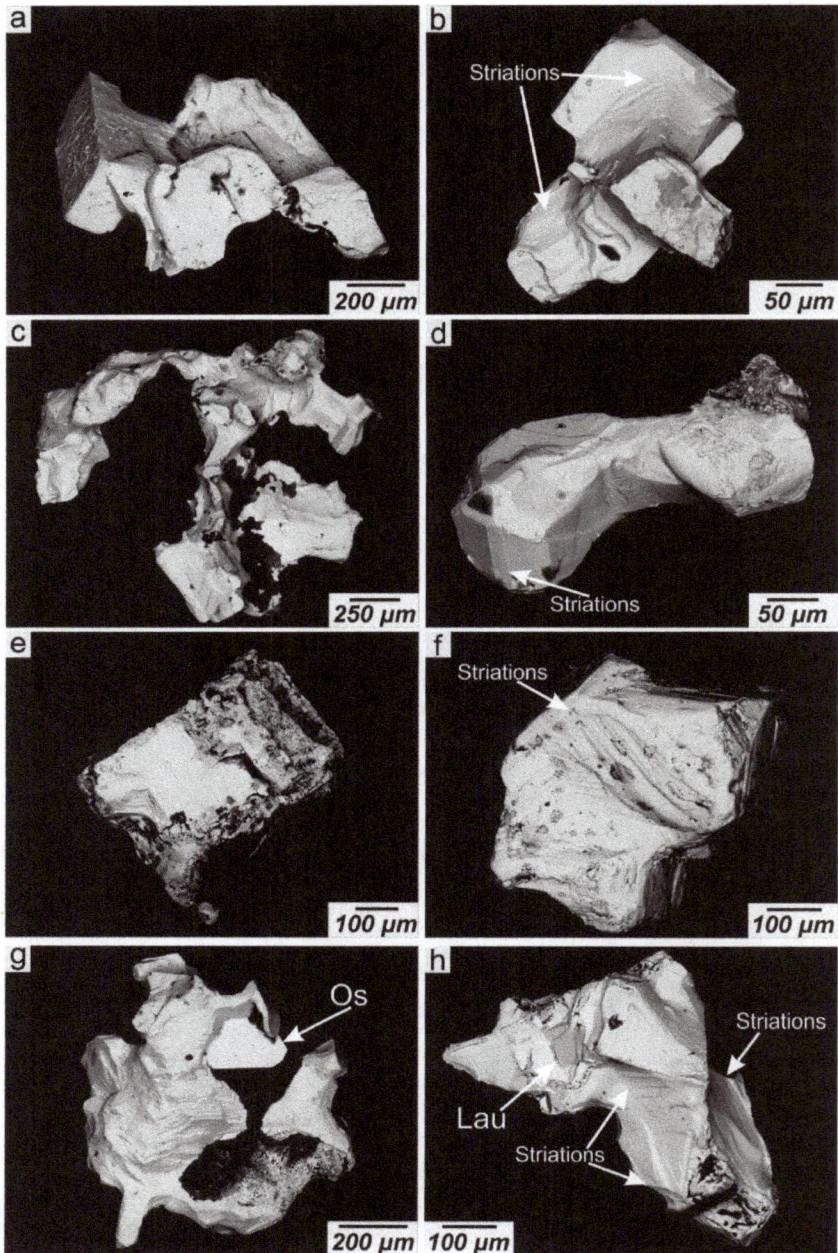

Figure 4. The morphological features of Pt-Fe alloy grains from lode chromite-platinum mineralization (**a–d**) and Vershinniy eluvial placer (**e–h**). Os—osmium, Lau—laurite. The **g** photo was published in Reference [25]. BSE images (SEM).

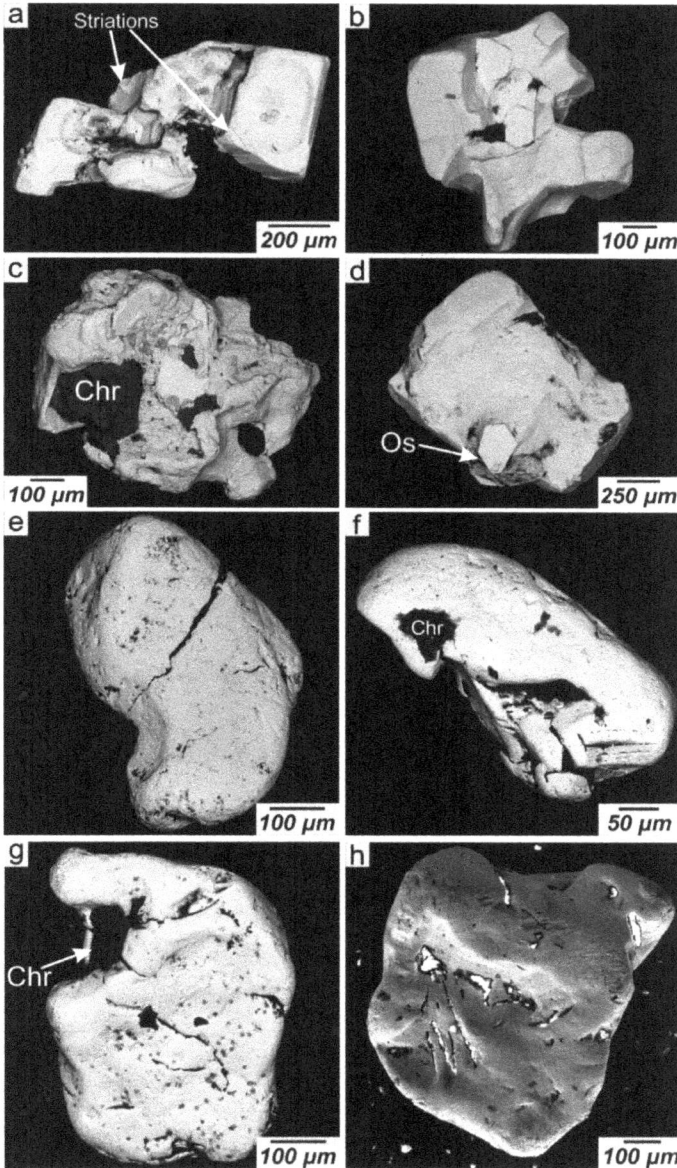

Figure 5. Morphological features of Pt-Fe alloy grains from the Travyanistyi proximal placer (**a–d**) and Glubokinskoe distal placer (**e–h**). Os—osmium, Chr—chromite. BSE images (SEM).

3.2.1. Pt-Fe Alloys

The assemblages of Pt-Fe alloys from the bedrock of the Svetloborsky massif and the placers is characterized by the predominance of isoferroplatinum (>95%). Native platinum is not present, and ferroan platinum is observed as single small inclusions in isoferroplatinum.

Isoferroplatinum mainly forms homogeneous grains both in primary ores and in the placer assemblages (Figure 6). However, the Pt content of isoferroplatinum varies considerably from 55.6 to 76.9 at. %, while the amount of total PGE lies within a smaller range and comprises 74 at. % on average, close to the isoferroplatinum theoretical formula (Table 1). Such a consistent average content of the PGE totals, with a significant fluctuation of Pt is due to the significant concentrations of impurity components, which can reach 12.8 at. %. Iridium is characterized by the highest concentrations (up to 7 at. %). Rh and Pd contents are relatively consistent, and on average do not exceed 1 at. %. Ru is characterized by very low concentrations, on average below 0.2 at. %. Os content can reach 1 at. %, however, this may be due to small inclusions of the Os-Ir-Ru alloys. Cu shows consistent concentrations ranging from 1.3 to 1.8 at. %. Ni does not exceed 0.2 at. %, however, its content is characterized by an inverse relationship with the amount of the PGE, reaching 1.6 at. %.

Table 1. Compositions of isoferroplatinum (1–20) and ferroan platinum (18–22) from the chromite-platinum ore zones of the Svetloborsky massif and associated placers.

No.	Fe	Ni	Cu	Ru	Rh	Pd	Ir	Pt	Total	No.	Fe	Ni	Cu	Ru	Rh	Pd	Ir	Pt
				wt. %										*at. %*				
1	7.85	0.13	0.86	0.12	1.03	0.26	0.51	88.67	99.43	1	22.43	0.35	2.16	0.19	1.60	0.39	0.42	72.46
2	8.72	0.13	0.84	bdl	0.73	0.54	bdl	88.76	99.72	2	24.46	0.35	2.07	bdl	1.11	0.79	bdl	71.22
3	8.73	0.02	0.51	bdl	0.67	0.77	1.03	87.22	98.95	3	24.79	0.05	1.27	bdl	1.03	1.15	0.85	70.86
4	7.89	0.07	0.87	bdl	1.24	0.18	3.44	85.90	99.59	4	22.51	0.19	2.18	bdl	1.92	0.27	2.85	70.08
5	7.83	0.09	0.73	0.02	0.83	1.03	1.18	89.07	100.78	5	22.13	0.24	1.81	0.03	1.27	1.53	0.97	72.02
6	7.41	bdl	1.03	bdl	0.17	0.47	1.95	89.58	100.61	6	21.27	bdl	2.60	bdl	0.26	0.71	1.62	73.54
7	7.70	0.06	0.70	bdl	0.54	0.58	4.60	86.42	100.60	7	21.99	0.17	1.76	bdl	0.83	0.87	3.81	70.57
8	7.47	0.02	0.75	bdl	0.59	0.57	7.52	83.45	100.37	8	21.46	0.05	1.89	bdl	0.92	0.86	6.27	68.55
9	7.84	0.09	0.59	0.24	0.85	bdl	1.60	89.51	100.72	9	22.33	0.24	1.48	0.38	1.31	bdl	1.32	72.94
10	7.84	0.04	0.63	0.56	0.88	0.41	1.85	88.50	100.71	10	22.22	0.11	1.57	0.88	1.35	0.61	1.52	71.74
11	8.56	0.37	0.84	bdl	0.73	0.11	bdl	90.08	100.69	11	23.87	0.98	2.06	bdl	1.10	0.16	bdl	71.83
12	8.19	0.05	0.10	0.20	1.02	0.56	0.15	90.64	100.91	12	23.23	0.13	0.25	0.31	1.57	0.83	0.12	73.56
13	8.80	bdl	0.44	0.21	0.49	0.47	5.07	84.01	99.49	13	24.94	bdl	bdl	1.14	bdl	0.06	4.67	69.19
14	8.55	0.25	0.62	0.31	1.13	0.84	3.93	83.48	99.11	14	24.03	0.67	1.53	0.48	1.72	1.24	3.21	67.12
15	8.64	0.66	1.16	0.24	0.87	0.38	4.06	83.88	99.89	15	23.83	1.73	2.81	0.37	1.30	0.55	3.25	66.16
16	8.62	0.19	0.69	0.11	0.39	0.29	5.15	83.56	99.00	16	24.47	0.51	1.72	0.17	0.60	0.43	4.24	67.86
17	8.27	bdl	0.41	bdl	0.50	0.80	bdl	90.58	100.56	17	23.48	bdl	1.02	bdl	0.77	1.19	bdl	73.54
18	13.59	bdl	1.03	bdl	bdl	bdl	bdl	84.16	98.78	18	35.24	bdl	2.35	bdl	bdl	bdl	bdl	62.41
19	11.54	bdl	1.26	bdl	bdl	bdl	bdl	88.13	100.93	19	30.49	bdl	2.91	bdl	bdl	bdl	bdl	66.60
20	11.63	bdl	1.09	0.05	0.92	bdl	1.26	85.82	100.77	20	30.58	bdl	2.52	0.07	1.32	bdl	0.96	64.55
21	14.60	0.51	0.82	bdl	0.80	0.44	6.00	76.25	99.42	21	36.48	1.21	1.80	bdl	1.08	0.58	4.35	54.50
22	13.36	0.56	0.27	bdl	0.37	0.47	6.10	78.25	99.38	22	34.50	1.37	0.61	bdl	0.52	0.64	4.57	57.79

Locations: the chromite-platinum ore zones of Svetloborsky massif (No. 1–4; 18–19), Vershinniy eluvial placer (No. 5–8; 20), Travyanistyi proximal placer (No. 9–12) and Glubokinskoe distal placer (No. 13–17; 21–22). bdl—below detection limits.

A single grain of ferroan platinum was found in the Glubokinskoe distal placer. Basically, ferroan platinum usually forms single small inclusions in isoferroplatinum. The composition of the ferroan platinum is variable and rarely corresponds to theoretical Pt_2Fe. The minerals of the tetraferroplatinum group occur in subordinate quantities. They form rims around isoferroplatinum with a thickness of up to 50 μm (Figure 6c). Tetraferroplatinum and tulameenite are identified, and the compositions of all studied grains are close to theoretical values (Table 2). The amount of the tetraferroplatinum group minerals regularly decreases from the lode source to the most distant placers, where the rims composed of these minerals are absent.

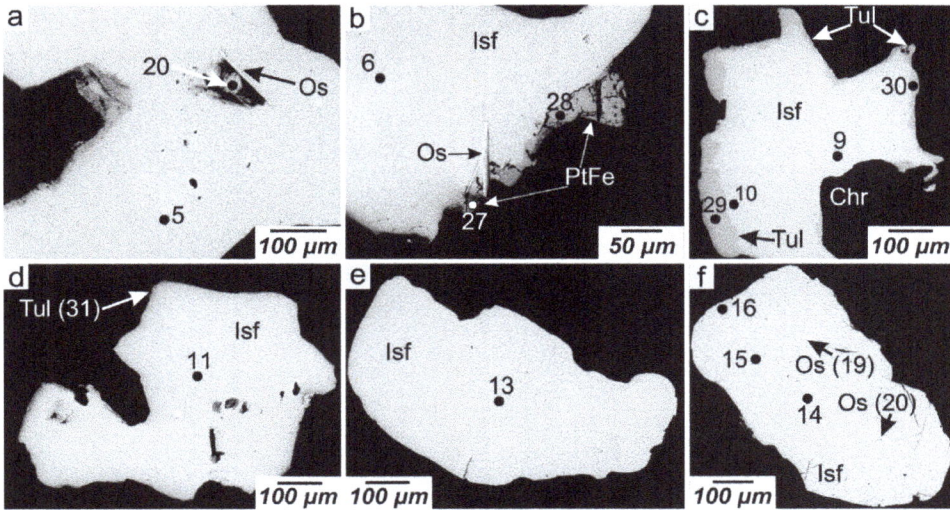

Figure 6. BSE images (SEM) of Pt-Fe aggregates from Vershinniy eluvial (**a,b**), Travyanistyi proximal (**c,d**) and Glubokinskoe distal (**e,f**) placers. The number of points corresponds to the data in Table 1 for isoferroplatinum (No. 5–6, 9–11, 13–16) and ferroan platinum (No. 20), in Table 2 for tetraferroplatinum–tulameenite (No. 27–31) and in Table 3 for osmium (No. 19, 20). Isf—isoferroplatinum, PtFe—tetraferroplatinum Tul—tulameenite, Os—osmium, Ir—iridium, and Chr—chromite. The **a** and **b** photo were published in [25].

Table 2. Compositions of tetraferroplatinum group minerals from the chromite-platinum ore zones of the Svetloborsky massif and associated placers.

No.	Fe	Ni	Cu	Ru	Rh	Pd	Ir	Pt	Total	No.	Fe	Ni	Cu	Ru	Rh	Pd	Ir	Pt
				wt. %										*at. %*				
23	10.66	bdl	13.23	bdl	bdl	bdl	bdl	76.11	100.00	23	24.20	bdl	26.38	bdl	bdl	bdl	bdl	49.42
24	18.90	bdl	3.26	bdl	bdl	bdl	bdl	77.41	99.57	24	43.04	bdl	6.52	bdl	bdl	bdl	bdl	50.44
25	16.28	bdl	2.72	bdl	bdl	bdl	bdl	80.35	99.35	25	39.09	bdl	5.73	bdl	bdl	bdl	bdl	55.18
26	14.93	bdl	1.67	bdl	bdl	bdl	bdl	82.26	98.86	26	37.40	bdl	3.67	bdl	bdl	bdl	bdl	58.93
27	17.44	0.09	1.18	bdl	0.02	0.31	0.39	79.53	98.96	27	41.92	0.21	2.49	bdl	0.03	0.39	0.27	54.69
28	16.78	0.10	2.30	bdl	0.34	1.71	0.06	77.78	99.07	28	39.72	0.23	4.78	bdl	0.44	2.12	bdl	52.67
29	8.75	0.23	13.28	bdl	0.38	0.09	2.86	75.08	100.67	29	20.26	0.51	27.00	bdl	0.48	0.11	1.92	49.72
30	9.31	0.22	12.44	0.49	1.05	0.14	2.52	73.71	99.88	30	21.57	0.48	25.31	0.63	1.32	0.17	1.69	48.83
31	10.61	0.27	11.80	0.15	0.41	0.36	bdl	76.84	100.44	31	24.28	0.59	23.71	0.19	0.51	0.43	bdl	50.29
32	18.17	0.30	1.19	0.41	bdl	0.06	bdl	78.96	99.09	32	42.91	0.67	2.47	0.53	bdl	0.07	bdl	53.35

Locations: the chromite-platinum ore zones of Svetloborsky massif (No. 1–4), Vershinniy eluvial placer (No. 5–6), Travyanistyi proximal placer (No. 7–10). bdl—below detection limits.

A comparative analysis of the Pt-Fe alloy assemblages from the various placers established a general coincidence of the mineral assemblages (Figure 7). The predominance of isoferroplatinum with a relatively consistent composition is characteristic of all the placers studied. Ferroan platinum occurs in the form of small inclusions in the Pt-Fe alloys from the eluvial placer and in the form of a relatively large individual from the Glubokinskoe distal placer. On the contrary, the secondary minerals of the tetraferroplatinum group are found in relatively large amounts in the Pt-Fe alloys from the Travyanistyi ravine; they are present as a single grain in the eluvial placer and are not detected in the Glubokinskoe distal placer. The latter can be explained by a significant transport distance, resulting in the mechanical destruction of the low-hardness tetraferroplatinum group minerals that compose thin peripheral rims.

Table 3. Composition of osmium from the chromite-platinum ore zones of Svetloborsky massif and associated placers.

No.	Ru	Rh	Pd	Os	Ir	Pt	Total	No.	Ru	Rh	Pd	Os	Ir	Pt
				wt. %								*at. %*		
1	6.20	bdl	bdl	50.59	43.21	bdl	100.0	1	11.11	bdl	bdl	48.18	40.71	bdl
2	0.95	bdl	bdl	68.10	30.93	bdl	99.98	2	1.78	bdl	bdl	67.77	30.45	bdl
3	1.65	bdl	bdl	72.02	26.28	bdl	99.95	3	3.07	bdl	bdl	71.22	25.71	bdl
4	1.05	bdl	bdl	73.68	25.27	bdl	100.0	4	1.96	bdl	bdl	73.20	24.84	bdl
5	1.34	bdl	bdl	74.87	23.78	bdl	99.99	5	2.50	bdl	bdl	74.18	23.32	bdl
6	0.14	0.34	bdl	62.54	34.73	1.10	98.85	6	0.26	0.63	bdl	63.27	34.76	1.08
7	0.17	0.68	bdl	67.41	30.86	0.55	99.67	7	0.33	1.26	bdl	67.36	30.51	0.54
8	0.55	0.21	0.16	67.84	29.79	0.19	98.74	8	1.04	0.38	0.29	68.39	29.72	0.18
9	2.43	0.16	0.20	68.36	28.90	1.14	101.19	9	4.44	0.28	0.34	66.17	27.69	1.08
10	0.83	0.46	bdl	89.91	6.57	2.79	100.56	10	1.54	0.83	bdl	88.55	6.40	2.68
11	0.71	0.32	bdl	93.29	5.23	bdl	99.55	11	1.33	0.59	bdl	92.93	5.15	bdl
12	0.68	0.48	bdl	93.1	4.66	bdl	98.92	12	1.28	0.89	bdl	93.21	4.62	bdl
13	0.71	0.34	bdl	92.73	4.87	bdl	98.65	13	1.34	0.63	bdl	93.19	4.84	bdl
14	0.87	0.34	bdl	92.74	5.23	bdl	99.18	14	1.63	0.63	bdl	92.57	5.17	bdl
15	1.01	0.41	bdl	92.72	4.88	bdl	99.02	15	1.90	0.76	bdl	92.52	4.82	bdl
16	2.04	0.89	bdl	76.05	21.34	bdl	100.32	16	3.74	1.61	bdl	74.08	20.57	bdl
17	2.18	0.65	bdl	75.47	22.07	bdl	100.37	17	4.00	1.17	bdl	73.55	21.28	bdl
18	1.05	0.69	bdl	93.26	4.29	bdl	99.29	18	1.95	1.26	bdl	92.58	4.21	bdl
19	6.97	0.30	0.60	60.54	30.40	0.39	99.19	19	12.40	0.52	1.01	57.26	28.45	0.36
20	8.58	0.55	bdl	59.10	28.96	2.82	100.02	20	15.00	0.95	bdl	54.88	26.62	2.55

Locations: the chromite-platinum ore zones of Svetloborsky massif (No. 1–5), the Vershinniy eluvial placer (No. 6–10), the Travyanistyi proximal placer (No. 11–17), and the Glubokinskoe distal placer (18–20). bdl—below detection limits.

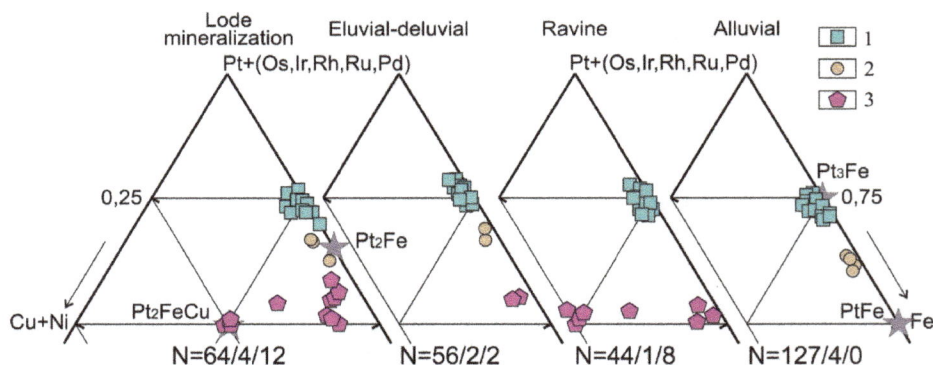

Figure 7. Composition (at. %) of isoferroplatinum (1), ferroan platinum (2) and tetraferroplatinum-tulameenite (3) from placers associated with the Svetloborsky massif and lode chromite-platinum ore zones. N—number of isoferroplatinum/ferroan platinum/tetraferroplatinum-tulameenite. The grey Pentagrams are stoichiometric formulae.

3.2.2. Os-Ir-Ru Alloys

Os-Ir-Ru alloys form four minerals: Osmium, iridium, ruthenium, and rutheniridosmine. However, like in other placers, associated with the Ural-Alaskan type massifs, only osmium and iridium are observed in the placers studied.

Osmium forms predominantly pinacoidal hexagonal inclusions in Pt-Fe alloy grains or regular accretions of lamellar subindividuals (Figure 8a,b). In some Pt-Fe aggregates, osmium is clearly tending towards the phase boundaries of the isoferroplatinum and chromit (Figure 8a,b); in others, it occurs in the form of inclusions in the central parts of the PGM grains (Figure 8c).

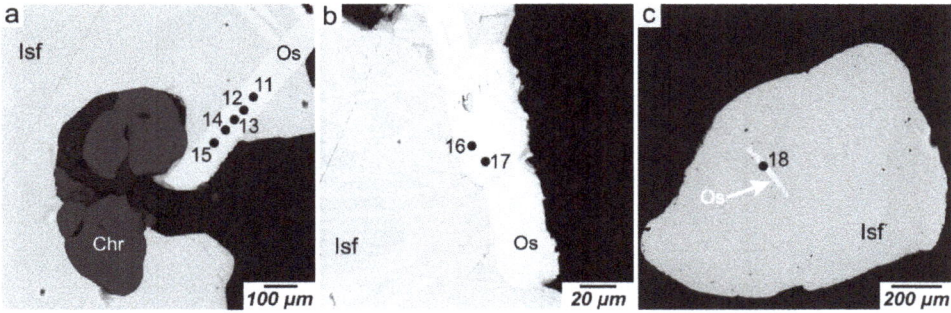

Figure 8. BSE images (SEM) of osmium inclusions in isoferroplatinum from the Travyanistyi proximal placer (**a**,**b**) and the Glubokinskoe distal placer (**c**). The number of points corresponds to the data in Table 3 (No. 11–18). Isf—isoferroplatinum, Os—osmium, and Chr—chromite.

Osmium composition varies considerably (Table 3). Like most Os-Ir-Ru inclusions in Pt-Fe alloy grains from placers associated with the Ural-Alaskan type massifs and their lode sources [6–14,18–24,26–33], Ru is low in the Os-Ir-Ru solid solutions (Figure 9), and its concentration does not exceed 15 at. %. Low Rh concentrations do not exceed 1.6 at. %. In some analyzes, Pd is present and not exceeding 1.0 at. %.

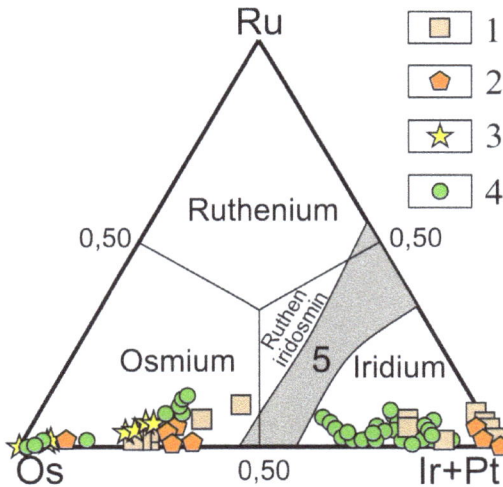

Figure 9. Composition (at. %) of Os-Ir-Ru inclusions in Pt-Fe alloys from the lode chromite-platinum mineralization (1), the Vershinniy eluvial placer (2), the Travyanistyi proximal placer (3), and Glubokinskoe distal placer (4). (5)—miscibility gap.

Iridium is most often encountered as small isometric inclusions (Figure 10a–e). Occasionally it forms larger roundish inclusions in isoferroplatinum (Figure 10c). In the eluvial placer, a nugget of iridium about 0.6 mm in size was found, overgrown by isoferroplatinum (Figure 10f).

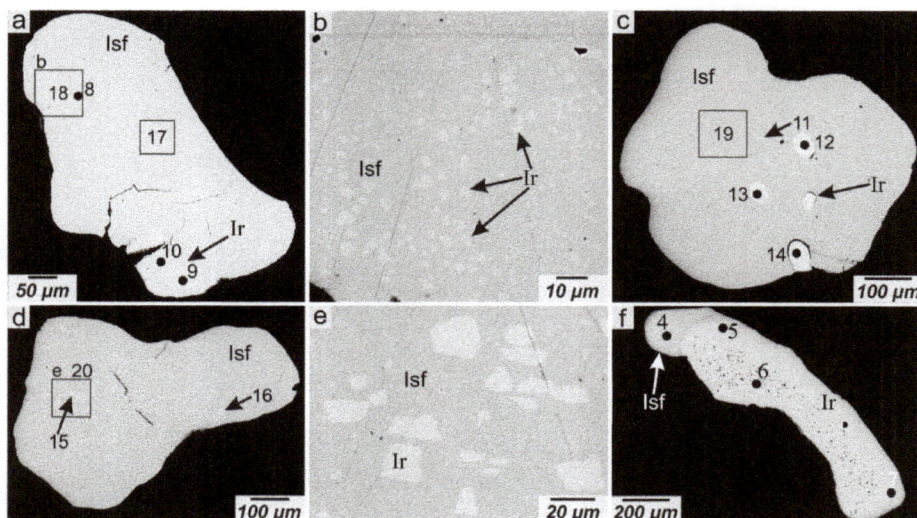

Figure 10. BSE images (SEM) of Ir exsolutions (**a–e**) in isoferroplatinum from the Glubokinskoe distal placer and (**f**) nugget of iridium from the Vershinniy eluvial placer. The number of points and fields corresponds to the analyses in Table 4 (No. 4–6, 8–20). The figures **b** and **e** are magnifications of figures **a** and **d**, respectively. Isf—isoferroplatinum, Os—osmium, Ir—iridium. The **f** photo was published in Reference [25].

The composition of iridium varies widely (see Figure 9). Ru impurities are low. However, unlike osmium, significant impurities of other PGE (Table 4) in iridium comprise Pt, with contents that may reach 16 at. %. Rh is noted in amounts up to 6.3 at. %. Pd concentrations do not exceed 1.4 at. %. In some analyzes, along with PGE, Fe concentrations (up to 30 at. %) are noted.

In summary, the inclusions of Os-Ir-Ru alloys show a wide variation of compositions. Osmium is found in relatively small quantities in all samples, however iridium exsolutions are characteristic and primarily of the Glubokinskoe distal placer, and iridium nuggets are found only in the eluvial placer.

3.2.3. PGE sulfides

PGE sulfides are represented by minerals of two solid solution series: laurite-erlichmanite (RuS_2–OsS_2) and kashinite-bowieite (Ir_2S_3–Rh_2S_3). In general, the minerals corresponding to erlichmanite composition are clearly predominant in Pt-Fe alloys in all types of placers (Figure 11a–d). Erlichmanite forms either large individuals with complex faceting and inhomogeneous zoned structure (Figure 11d) or small rounded inclusions (Figure 11d). Laurite is less common and is found in the form of small idiomorphic zonal inclusions in isoferroplatinum (Figure 11e,f). Zonality is caused by an increase in Os/Ru ratios towards the edge of the crystals.

Table 4. Composition of Os-rich iridium from the chromite-platinum ore zones of Svetloborsky massif and associated placers.

No.	Fe	Ni	Cu	Ru	Rh	Pd	Os	Ir	Pt	Total
					wt. %					
1	11.09	bdl	bdl	bdl	1.41	bdl	bdl	72.58	14.72	99.80
2	1.96	bdl	bdl	4.60	3.83	bdl	bdl	79.03	10.59	100.01
3	0.43	bdl	bdl	2.64	1.53	bdl	23.49	66.27	5.64	100.00
4	10.31	0.56	0.23	bdl	1.08	0.64	bdl	15.95	71.66	100.43
5	2.86	0.18	bdl	2.71	2.42	0.82	2.83	79.63	8.75	100.20
6	2.52	0.15	bdl	2.38	2.02	0.69	1.82	73.68	16.82	100.08
7	2.72	0.14	bdl	1.22	1.73	0.68	2.93	73.08	17.35	99.85
8	0.74	bdl	0.14	3.43	3.50	bdl	15.53	68.37	8.18	99.89
9	1.42	0.16	0.72	2.55	2.52	bdl	18.71	57.56	15.64	99.28
10	1.04	bdl	bdl	2.76	2.14	0.87	22.01	62.78	7.99	99.59
11	0.55	bdl	bdl	2.62	2.79	0.10	16.70	70.19	7.03	99.98
12	0.35	0.06	bdl	3.37	2.20	bdl	21.91	59.09	12.26	99.24
13	0.39	0.36	0.28	3.90	1.35	bdl	18.83	62.07	12.02	99.20
14	0.14	0.17	0.20	3.79	2.08	0.08	13.75	68.90	11.11	100.22
15	0.39	0.07	0.25	2.82	2.23	0.17	20.23	65.74	7.58	99.48
16	0.53	bdl	0.15	1.30	3.53	bdl	12.93	71.32	9.74	99.50
Field										
17	7.67	0.30	0.70	bdl	0.94	0.71	bdl	14.42	80.76	105.50
18	8.21	0.03	0.72	0.25	1.14	0.32	bdl	13.27	77.33	101.27
19	7.88	bdl	1.12	0.24	0.39	0.44	bdl	10.14	82.77	102.98
20	7.16	0.15	0.23	0.30	0.54	0.05	bdl	15.99	74.29	98.71

No.	Fe	Ni	Cu	Ru	Rh	Pd	Os	Ir	Pt
					at. %				
1	29.87	bdl	bdl	bdl	2.06	bdl	bdl	56.73	11.34
2	6.02	bdl	bdl	7.80	6.38	bdl	bdl	70.49	9.31
3	1.41	bdl	bdl	4.79	2.72	bdl	22.62	63.16	5.30
4	27.79	1.45	0.54	bdl	1.58	0.91	bdl	12.48	55.25
5	8.74	0.52	bdl	4.58	4.00	1.32	2.54	70.65	7.65
6	7.85	0.44	bdl	4.08	3.41	1.13	1.66	66.47	14.96
7	8.55	0.41	bdl	2.11	2.94	1.12	2.70	66.59	15.58
8	2.36	bdl	0.40	6.03	6.05	bdl	14.51	63.20	7.45
9	4.49	0.48	2.00	4.46	4.32	bdl	17.34	52.75	14.13
10	3.34	bdl	bdl	4.88	3.73	1.47	20.73	58.52	7.33
11	1.77	bdl	bdl	4.68	4.91	0.18	15.88	66.06	6.52
12	1.15	0.19	bdl	6.09	3.91	bdl	21.04	56.14	11.48
13	1.27	1.11	0.79	6.99	2.37	bdl	17.91	58.41	11.15
14	0.45	0.52	0.57	6.76	3.64	0.14	13.03	64.62	10.27
15	1.26	0.22	0.73	5.07	3.94	0.30	19.32	62.10	7.06
16	1.73	bdl	0.44	2.35	6.26	bdl	12.40	67.71	9.11
Field									
17	20.88	0.78	1.67	bdl	1.39	1.01	bdl	11.39	62.88
18	22.96	0.08	1.77	0.39	1.73	0.47	bdl	10.77	61.83
19	21.85	bdl	2.73	0.37	0.59	0.64	bdl	8.16	65.66
20	21.13	0.42	0.60	0.49	0.86	0.08	bdl	13.70	62.72

Note: Iridium exsolutions in isoferroplatinum from the chromite-platinum ore zones of Svetloborsky massif (No. 1–3), Glubokinskoe distal placer (No. 8–20), and the composition of iridium nugget from Vershinniy eluvial placer (No. 4–7). The analyses No 4, 17–20 are Ir-containing platinum. bdl—below detection limits.

Table 5. The composition of laurite–erlichmanite (1–20) and kashinite–bowieite (20–25).

No.	S	Ru	Rh	Pd	Os	Ir	Total	Calculated Mineral Formulae
								Laurite-Erlichmanite
1	25.61	8.86	bdl	bdl	59.60	6.30	100.37	$(Os_{0.77}Ru_{0.21}Ir_{0.08})_{1.06}S_{1.94}$
2	28.00	14.06	bdl	bdl	50.56	6.09	98.71	$(Os_{0.61}Ru_{0.32}Ir_{0.07})_{1.00}S_{2.00}$
3	31.38	22.79	2.09	bdl	36.64	6.26	99.16	$(Ru_{0.46}Os_{0.40}Ir_{0.07}Rh_{0.04})_{0.97}S_{2.03}$
4	33.15	28.65	1.71	bdl	30.05	5.54	99.10	$(Ru_{0.56}Os_{0.31}Ir_{0.06}Rh_{0.03})_{0.96}S_{2.04}$
5	25.94	0.79	2.85	bdl	64.95	4.60	99.13	$(Os_{0.84}Ru_{0.02}Rh_{0.07}Ir_{0.06})_{0.99}S_{2.01}$
6	27.03	0.85	2.66	bdl	66.11	2.36	99.01	$(Os_{0.85}Ru_{0.02}Rh_{0.06}Ir_{0.03})_{0.96}S_{2.04}$
7	31.37	18.88	1.32	bdl	42.15	6.15	99.87	$(Os_{0.46}Ru_{0.39}Ir_{0.07}Rh_{0.03})_{0.95}S_{2.05}$
8	29.05	13.63	1.90	bdl	49.38	5.93	99.89	$(Os_{0.58}Ru_{0.30}Ir_{0.07}Rh_{0.04})_{0.99}S_{2.01}$
9	29.26	15.25	2.31	bdl	50.16	2.14	99.12	$(Os_{0.59}Ru_{0.33}Rh_{0.05}Ir_{0.02})_{0.99}S_{2.01}$
10	29.40	12.95	1.91	0.47	52.48	1.97	99.18	$(Os_{0.62}Ru_{0.28}Rh_{0.04}Ir_{0.02}Pd_{0.01})_{0.97}S_{2.03}$
11	27.77	12.23	1.91	bdl	54.79	4.65	101.35	$(Os_{0.65}Ru_{0.28}Ir_{0.06}Rh_{0.04})_{1.03}S_{1.97}$
12	29.26	20.33	2.51	bdl	43.25	4.98	100.33	$(Os_{0.49}Ru_{0.43}Ir_{0.06}Rh_{0.05})_{1.03}S_{1.97}$
13	37.43	51.82	1.38	bdl	5.91	4.36	100.90	$(Ru_{0.88}Os_{0.06}Ir_{0.04}Rh_{0.02})_{1.00}S_{2.00}$
14	38.15	50.88	1.06	bdl	6.96	3.87	100.92	$(Ru_{0.86}Os_{0.06}Ir_{0.03}Rh_{0.02})_{0.97}S_{2.03}$
15	32.27	28.97	1.64	bdl	33.51	4.19	100.58	$(Ru_{0.58}Os_{0.35}Ir_{0.04}Rh_{0.03})_{1.00}S_{2.00}$
16	36.02	47.71	2.53	0.72	8.88	4.89	100.75	$(Ru_{0.84}Os_{0.09}Ir_{0.04}Rh_{0.04}Pd_{0.01})_{1.02}S_{1.98}$
17	31.57	33.72	2.46	0.54	24.13	6.60	99.02	$(Ru_{0.66}Os_{0.25}Ir_{0.07}Rh_{0.05}Pd_{0.01})_{1.04}S_{1.96}$
18	32.60	33.51	2.90	0.38	23.72	6.46	99.57	$(Ru_{0.65}Os_{0.24}Ir_{0.07}Rh_{0.05}Pd_{0.01})_{1.02}S_{1.98}$
19	29.36	17.63	1.95	bdl	45.78	5.80	100.52	$(Os_{0.52}Ru_{0.38}Ir_{0.07}Rh_{0.04})_{1.01}S_{1.99}$
								Kashinite-Bowieite
20	22.13	bdl	8.21	bdl	bdl	70.12	100.46	$(Ir_{1.61}Rh_{0.35})_{1.96}S_{3.04}$
21	23.53	bdl	15.34	bdl	bdl	61.34	100.21	$(Ir_{1.33}Rh_{0.62})_{1.95}S_{3.05}$
22	25.35	bdl	22.42	bdl	bdl	51.88	99.65	$(Ir_{1.06}Rh_{0.85})_{1.91}S_{3.09}$
23	26.49	0.14	30.95	bdl	bdl	41.46	99.04	$(Rh_{1.12}Ir_{0.80}Ru_{0.01})_{1.93}S_{3.07}$
24	23.70	bdl	21.04	bdl	bdl	55.10	99.84	$(Ir_{1.17}Rh_{0.83})_{2.00}S_{3.00}$
25	24.09	bdl	22.60	bdl	bdl	53.58	100.27	$(Ir_{1.12}Rh_{0.87})_{1.99}S_{3.01}$

Location: PGM from the chromite-platinum ore zones of the Svetloborsky massif (No. 1–4; 20–23), the Vershinniy eluvial placer (No. 5–8; 24–25) and the Glubokinskoe distal placer (No. 13–19). Formulae of analyses No 1–19 are calculated on the basis of 3 atom per formulae, those of analyses No. 20–25 are calculated on the basis of 5 apfu. bdl—below detection limits.

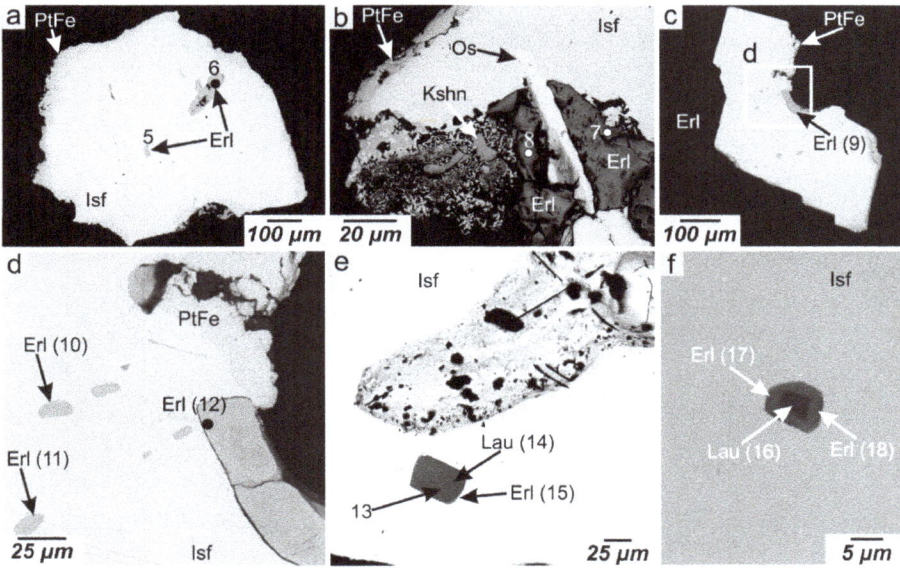

Figure 11. BSE images (SEM) of laurite–erlichmanite (Lau-Erl) in isoferroplatinum from (**a,b**) the Vershinniy eluvial placer, (**c,d**) the Travyanistyi proximal placer and (**e,f**) the Glubokinskoe distal placer. The number of points corresponds to the analyses in Table 5 (No. 5–18). Isf—isoferroplatinum, Pt-Fe—tetraferroplatinum, Erl—erlichmanite, Kshn—kashinite, and Lau—laurite. The **a** and **b** photo were published in Reference [25].

The compositional range of laurite-erlichmanite varies within wide limits (Figure 12). The concentration of Ir varies from 0.6 to 2.7 at. %, whereas Rh does not exceed 2.3 at. %. Pt and Pd contents are below detection limit.

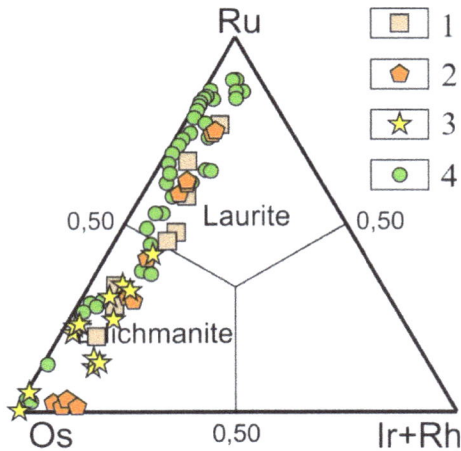

Figure 12. Compositional range (at. %) of laurite–erlichmanite from chromite-platinum ore zones of the Svetloborsky massif (1), the Vershinniy eluvial placer (2), the Travyanistyi proximal placer (3), and the Glubokinskoe distal placer (4).

Kashinite-bowieite grains were only detected in the Vershinniy eluvial placer, associated with other PGE sulfides and Pt-Ir-Rh thiospinels (Figures 11b and 13a). In the studied placers, only kashinite is found with a composition close to bowieite. In contrast to the minerals of the laurite-erlichmanite isomorphous series, the almost complete absence of impurity components is characteristic of all the studied kashinite samples (Table 5)

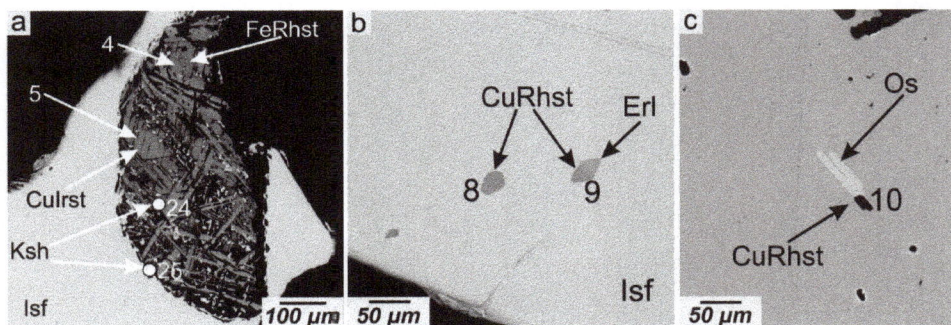

Figure 13. BSE images (SEM) of Pt-Ir-Rh thiospinels from the Vershinniy eluvial (**a**), the Travyanistyi proximal (**b**) and the Glubokinskoe distal placers (**c**). The number of points corresponds to the data in Table 6 for thiospinels (No. 4, 5, 8–10) and Table 5 for kashinite (No. 24, 25). Isf—isoferroplatinum, Erl—erlichmanite, Os—osmium, CuIrst—cuproiridsite, FeRhst—ferrorhodsite, and CuRhst—cuprorhodsite.

Table 6. Composition of Ir-Rh-Pt thiospinels from the chromite-platinum ore zones of the Svetloborsky massif (1–4), the Vershinniy eluvial placer (5–6), the Travyanistyi proximal placer (7–10), and the Glubokinskoe distal placer (11–13).

No.	S	Fe	Cu	Rh	Ir	Pt	Total	Calculated Mineral Formulae
1	24.86	bdl	11.77	6.53	41.73	14.68	99.57	$(Cu_{0.99}Fe_{0.00})_{0.99}(Ir_{1.15}Pt_{0.40}Rh_{0.34})_{1.89}S_{4.12}$
2	28.85	7.50	7.02	35.86	15.92	4.39	99.54	$(Fe_{0.59}Cu_{0.48})_{1.07}(Rh_{1.53}Ir_{0.36}Pt_{0.10})_{1.99}S_{3.94}$
3	30.32	6.63	6.99	36.64	15.30	3.15	99.03	$(Fe_{0.51}Cu_{0.47})_{0.98}(Rh_{1.53}Ir_{0.34}Pt_{0.07})_{1.94}S_{4.08}$
4	31.25	7.04	7.25	41.68	8.27	3.84	99.33	$(Fe_{0.53}Cu_{0.47})_{1.00}(Rh_{1.69}Ir_{0.18}Pt_{0.08})_{1.95}S_{4.05}$
5	25.34	5.18	6.00	13.50	50.22	bdl	100.24	$(Cu_{0.48}Fe_{0.47})_{0.95}(Ir_{1.33}Rh_{0.67})_{2.00}S_{4.05}$
6	25.48	5.50	5.13	30.16	34.51	bdl	100.78	$(Fe_{0.48}Cu_{0.39})_{0.87}(Rh_{1.42}Ir_{0.87})_{2.29}S_{3.86}$
7	26.08	bdl	11.91	19.23	21.20	21.32	99.74	$Cu_{0.93}(Rh_{0.93}Ir_{0.55}Pt_{0.54})_{2.02}S_{4.05}$
8	26.18	bdl	11.99	19.88	20.52	21.15	99.72	$Cu_{0.93}(Rh_{0.96}Ir_{0.53}Pt_{0.54})_{2.02}S_{4.05}$
9	26.48	bdl	12.31	20.07	20.71	20.05	99.62	$Cu_{0.94}(Rh_{0.96}Ir_{0.53}Pt_{0.50})_{1.99}S_{4.05}$
10	26.29	bdl	12.11	19.66	20.85	20.33	99.24	$Cu_{0.94}(Rh_{0.95}Ir_{0.54}Pt_{0.52})_{2.00}S_{4.06}$
11	25.87	bdl	11.81	18.77	32.47	10.22	99.14	$Cu_{0.94}(Rh_{0.91}Ir_{0.85}Pt_{0.26})_{2.02}S_{4.04}$
12	26.03	bdl	11.85	19.26	32.35	10.42	99.91	$Cu_{0.93}(Rh_{0.93}Ir_{0.84}Pt_{0.27})_{2.03}S_{4.04}$
13	26.02	bdl	12.18	19.53	32.57	9.80	100.10	$Cu_{0.95}(Rh_{0.94}Ir_{0.84}Pt_{0.25})_{2.03}S_{4.02}$
14	31.23	14.30	5.08	24.58	10.82	11.39 * 1.87 **	99.27	$S_{53.19}Fe_{13.99}Rh_{13.04}Ni_{10.60}Cu_{4.37}Ir_{3.07}Co_{1.73}$

Note: Analyses No. 1, 5—cuproiridsite, No. 2–4—ferrorhodsite, No. 6—Ir-bearing ferrorhodsite, No. 7–13—cuprorhodsite, the formulae are calculated on the basis of 7 apfu, No. 14—unnamed phase, the formula is calculated for 100 at. %. *—Ni containing, **—Co containing. bdl—below detection limits.

The distribution of PGE sulfides in different placers is significantly different. Kashinite was found only in the Vershinniy eluvial placer. Among Os and Ru sulfides, only erlichmanite was found in the Travyanistyi proximal placer, while both erlichmanite and laurite were found in the eluvial and distal placers. At the same time, all sulfides are characterized by very wide variations of chemical composition.

3.2.4. Pt-Ir-Rh Thiospinels

Minerals of the thiospinel group ((Fe,Cu)(Rh,Ir,Pt)$_2$S$_4$) are rare in the placers studied. Cuprorhodsite (CuRh$_2$S$_4$) prevails forming small idiomorphic individuals (Figure 13b,c). It is characterized by consistent copper concentrations, which are the largest possible for thiospinels, and varying Rh, Ir, and Pt contents (Table 6; Figure 14). Generally, cuprorhodsite occupies an intermediate position in the malanite (CuPt$_2$S$_4$)-cuprorhodsite (CuRh$_2$S$_4$) isomorphous series.

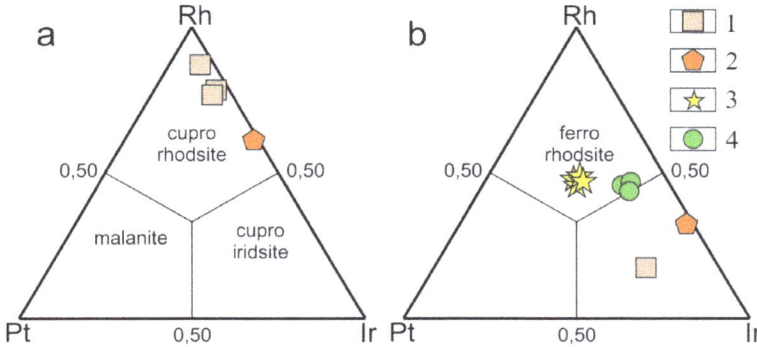

Figure 14. The composition (at. %) of Cu bearing (**a**) and Fe bearing (**b**) Pt-Ir-Rh thiospinels in isoferroplatinum from chromite-platinum ore zones of the Svetloborsky massif (1), the Vershinniy eluvial placer (2), the Travyanistyi proximal placer (3), and the Glubokinskoe distal placer (4).

In addition to cuprorhodsite, cuproiridsite (CuIr$_2$S$_4$) and ferrorhodsite (FeRh$_2$S$_4$) were found once only. They form complex intergrowths with various sulfides (mainly kashinite), located near the boundaries of Pt-Fe alloy grains (Figure 13a).

3.2.5. Other Platinum Group Minerals

One elongated grain of modified Ir (presumably iridium oxide, according to the measured oxygen content; Figure 15a) was found in isoferroplatinum from the Travyanistyi ravine, and a small aggregate consisting of Pt-Fe alloys and an unnamed Rh-Fe-Ni sulfide phase (Figure 15b; Table 6, No. 14) located at the isoferroplatinum grain boundary, also from the Travyanistyi ravine.

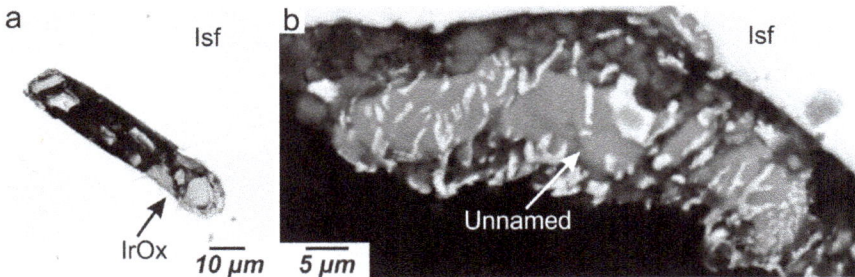

Figure 15. (**a**) the iridium oxide (IrOx) and (**b**) unnamed phase from the Travyanistyi proximal placer. Isf—isoferroplatinum. BSE images (SEM).

3.3. Mineral Assemblages—A Comparison

The present study established that the PGM assemblages of all studied placers are characterized by the absolute predominance of isoferroplatinum grains, with a small amount of ferroan platinum and minerals of the tetraferroplatinum group. For chromite-platinum mineralization, the similar characteristics of the assemblage composition are noted, along with slightly more common minerals of the tetraferroplatinum group (see Figure 6), whereas for the dunite-type mineralization, the pseudomorphs of tetraferroplatinum and tulameenite are noted in about half of the studied grains.

Similar features are observed when analyzing the nature of the impurity element distribution in isoferroplatinum (Table 7). According to the calculated average content of minor elements, the differences in isoferroplatinum contents, as a whole, do not go beyond the limits of calculation errors, and the compositions of isoferroplatinum from the placers and lode chromite-platinum ore zones are geochemically similar.

Table 7. Isoferroplatinum compositions from chromite-platinum ore zones of the Svetloborsky massif (1), the Vershinniy eluvial placer (2), the Travyanistyi proximal placer (3) and the Glubokinskoe distal placer (4).

No.	Fe	Cu	Ru	Rh	Pd	Ir	Pt	PGE
1	(19.6–27.8)	(0.4–2.6)	(0.0–0.3)	(0.0–2.2)	(0.2–1.6)	(0.0–12.5)	(55.2–75.4)	(70.3–78.7)
	24.29	1.34	0.05	0.71	0.51	1.19	72.27	74.28
2	(22.6–28.8)	(0.0–1.9)	(0.0–2.1)	(0.0–3.3)	(0.0–1.5)	(0.0–1.4)	(65.3–77.4)	(71.1–77.4)
	21.90	1.59	0.06	0.72	0.73	2.19	72.37	76.38
3	(21.9–27.4)	(0.6–3.2)	(0.0–1.0)	(0.0–1.9)	(0.0–1.5)	(0.0–6.7)	(64.9–74.0)	(70.6–75.2)
	23.39	1.82	0.23	1.06	0.57	0.97	71.59	74.52
4	(21.9–27.4)	(0.6–3.2)	(0.0–1.0)	(0.0–1.9)	(0.0–1.5)	(0.0–6.7)	(64.9–74.0)	(70.6–75.2)
	23.85	1.78	0.6	0.73	0.63	1.78	70.94	74.18

Note: The data are given in (minimum–maximum) and average. The number of analyses: No. 1—64; No. 2—56; No. 3—62; and No. 4—127.

The coincidence of the isoferroplatinum compositions from the proximal and distal placers and lode chromite-platinum mineralization is demonstrated in Figure 16, where the concentrations of impurity elements from the placers associated with other Ural-Alaskan type massif of the same region–Veresovoborsky–are given for greater clarity. In the Figure 16, the intermediate position of the isoferroplatinum from the Svetloborsky massif is noted regarding Ir, Pd, and Rh concentrations, whereas the isoferroplatinum of the Veresoborsky massif by Pd-Rh specifics.

Os-Ir-Ru alloy inclusions are found in all types of lode mineralization. They are characterized by the same wide compositional variation as the Os-Ir-Ru minerals from the placers. It should be noted that Ir nuggets weighing up to 12 g are found only in chromite-platinum ore zones, whereas osmium plates are more characteristic of dunite-type mineralization, and iridium is found in a subordinate amount [9].

The comparison of the peculiarities of the remaining PGM in this case is inexpedient due to their relative rarity in the studied objects, although the bowieite was established of dunite-type mineralization [27], whereas it was not found in the placers. The absence of a wide variety of rarer minerals in the studied assemblages emphasizes once again the difference of the PGM from the placers and dunite-type mineralization, where the Rh analogue of tolovkite, hollingworthite, irarsite, sperrylite, hexaferrum, etc. [9] were established, whereas such mineral and species diversity is not typical for the platinum ore zones of the Svetloborsky massif [14].

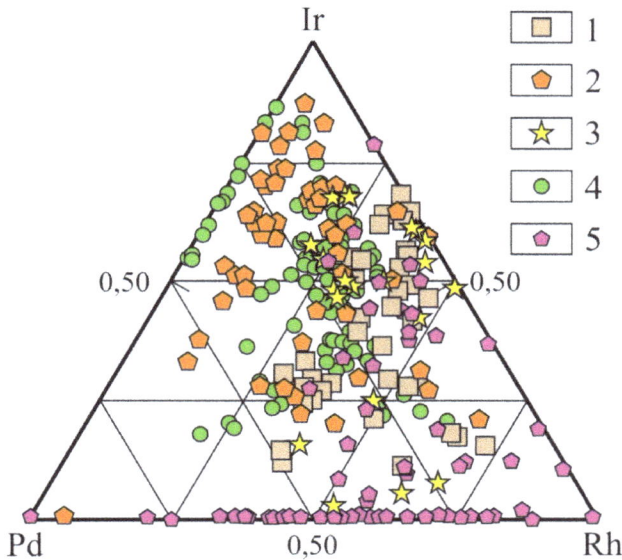

Figure 16. The concentrations of minor elements in isoferroplatinum from lode chromite-platinum mineralization (1), the Vershinniy eluvial placer (2), the Travyanistyi proximal placer (3), the Glubokinskoe distal placer (4), the lode and placer mineralization of the Veresovoborsky massif (5).

4. Discussion

The predominance of Pt-Fe alloys with inclusions of Os-Ir-Ru composition, as well as PGE sulfides (laurite-erlichmanite and kashinite-bowieite) in the studied objects is typical for the PGM assemblages from the placers and bedrocks of Ural-Alaskan type massifs in different regions of the world [18–24,28–35]. The presence of Pt-Ir-Rh thiospinels among the inclusions in Pt-Fe minerals from the placer assemblages is considered as a typomorphic feature of platinum placer objects associated with the Ural-Alaskan type massifs [36]. The placer assemblages studied, regardless of the distance from the lode source, completely correspond to the typical PGM assemblages of the lode sources, Ural-Alaskan type zonal massifs.

The analysis of the PGM assemblages of the proximal and distal placers allows the criteria for estimating the PGM transport distance to be singled out. The most important of these criteria is the nature of the morphology alteration of PGM individuals and aggregates. Pt-Fe alloys retain the primary morphological features characteristic of lode PGM in the placers with a small distance of clastic material transport. At a distance of more than 10 km, the degree of mechanical attrition of PGM becomes significant, and the morphological features characteristic of lode PGM assemblages are practically not preserved. Similar changes are established of gold and PGM grains from placers associated with different gold deposits of the world [37]. Another criterion for assessing the transport distance of PGM in placers is the degree of the rim integrity around isoferroplatinum, commonly composed of tetraferroplatinum group minerals. As a result of our research, it was found that with a long transport distance, these rims are completely destroyed due to mechanical wear during their transport to the placers, corroborating earlier studies of the placer systems associated with the Nizhnetagilsky massif [17].

A general analysis of the aggregate dimensions of the PGM made it possible to establish that in eluvial and distal placers with a short distance of material transport the average size of the PGM nuggets is larger than in distal placers located 10 km and more from the primary source. The nuggets of 0.5–1.5 mm in size prevailed in the Travyanistyi proximal placer comprising about 60% of the platinum in the entire volume of heavy concentrate. During early mining work nuggets were found weighing from the first grams to tens of grams [2]. In the Is river placer with transport distance of more than 10 km, the PGM grains with a size of 0.1–0.5 mm prevail. They comprise about 75% of the PGM in the entire volume of heavy concentrate. Throughout history of mining, nuggets were not found in these placers. A regular decrease in the dimensions of PGM grains with distance from the lode source was established for platinum placers of the Urals as a result of specialized investigations [16].

The comparative characteristics of the compositions of the Pt-Fe alloys and their inclusions from the placer assemblages and lode sources made it possible to establish their identity in many ways. Taking into account the presence of two types of mineralization—dunite and chromite-platinum—within the Svetloborsky massif, it is important to determine the contribution of each one to the platinum placer potential. The relationship of eluvial and proximal placers with chromite-platinum ore zones is out of question. This is confirmed by many facts, the main of which is the coincidence of the contours of the heavy concentrate anomaly (or the eluvial placer) and the designated zone of massive platinum-bearing chromitites, as well as the geomorphological features of the terrain. It is important to emphasize that the small size of platinum group minerals in the dunitic type of mineralization does not allow one to count on their significant contribution to the ore potential of the placer systems [38]. Proceeding from this, chromite-platinum mineralized zones appear to be the most likely lode source for the placers.

The volume of the PGM extracted from the Is river placers is more than 110 ton [39], which makes this placer unique and raises the question about the existence of other lode sources besides the Svetloborsky massif. A number of platinum-bearing rivers and ravines flows into the Is river, including the relatively large Prostakishenka and Pokap rivers. These rivers drain the dunites and the embedded chromite-platinum ore zones of the Veresovoborsky massif in the upper course. However, the contribution of the Veresovoborsky massif's bedrocks to the formation of the Isovsko-Turinskaya placer was not previously estimated. A different PGM assemblage is characteristic of the Veresovoborsky massif and the associated alluvial deposits as compared to the Svetloborsky massif. In the chromite-platinum ore zones of the Veresovoborsky massif, the ferroan platinum is largely observed, as well as the abundance of the tetraferroplatinum group minerals [14]. Along with this, the mineral inclusions in Pt-Fe alloys are not characteristic. The similar assemblages are also noted for the placers associated with the Veresovoborsky massif [40]. In addition, the difference in the trace element concentrations in the isoferroplatinum from the placers associated with the Svetloborsky and Veresovoborsky massifs is clearly demonstrated in Figure 16.

The Kachkanar gabbro-clinopyroxenite massif is considered as another lode source for the Isovskaya placer. The part of the ravines draining the clinopyroxenites of this massif flows into the Is river across the area from the Svetloborsky massif to the Glubokinskoe site. A number of works [26,41] provides the information that the clinopyroxenites of zonal Ural-Alaskan type massifs can serve as a source for the placer object formation. Similar placer objects, not related with the erosion of dunite bodies, were found within the Koryak Highlands [21,42]. However, the peculiarities of the chemical composition of the platinum group minerals from the placers associated with the clinopyroxenite fragments of zonal massifs or separate clinopyroxenite bodies differ in a number of features. The coincidence of most of the peculiarities of the PGM assemblages from the most remote part of the Isovskaya placer with the PGM assemblages from the chromite-platinum ore zones of the dunite core of the Svetloborsky massif indicates the insignificant influence of the bedrocks of the Kachkanar massif on the formation of the placer PGM assemblages.

Based on the coincidence of a number of the most significant features of the PGM mineralization of the Glubokinskoe site with the chromite-platinum ore zones of the Svetloborsky massif, combined with their significant difference from the PGM of the lode mineralization of the Veresovoborsky massif, it can be stated that the destruction of the chromite-platinum ore zones of the Svetloborsky clinopyroxenite-dunite massif provides the greatest contribution to the Isovsko-Turinskaya placer system.

The close chemical composition of Pt-Fe minerals from the placers located at different distances from the lode source can also be the evidence of the absence of pronounced zonality within the already destroyed part of the dunite core. However, for several PGM placer deposits, for example the deposits associated with the Galmoenan massif [21], a compositional variability of the Pt-Fe minerals is observed, which may be due to the existence of the vertical zonality of the lode platinum mineralization. The zonality of lode mineralization is also assumed proceeding from the existing models of Ural-Alaskan type massif formation and the experimental data [21,43–45].

5. Conclusions

The main transformation of PGM in placers is reduced to a regular change in the morphological features during their transport from a lode source to a placer. At a distance of more than 10 km, the degree of mechanical attrition of PGM becomes significant, and the morphological features characteristic of lode PGM are practically not preserved. In addition, in the placers with a significant transport distance, tetraferroplatinum group mineral rims on isoferroplatinum are completely destroyed. A general analysis of the aggregate dimensions made it possible to establish that in eluvial and alluvial placers with a short distance of material transport the average size of the platinum nuggets is larger than in distal placers located 10 km and more from the primary source. Composition of isoferroplatinum weakly change during of transport of detrital materials.

The results obtained indicate that the PGM bulk has entered the Isovsko-Turinskaya placer system as a result of the destruction of the chromite-platinum ore zones of the Svetloborsky massif. This proves that the chromite-platinum type of mineralization is the most important for the formation of the large placer systems associated with the Ural-Alaskan type massifs.

The coincidence of the compositional features of the PGM assemblages from the studied placers having different distances from the lode source may be the evidence of the absence of vertical zonality in the lode platinum mineralization within the destroyed part of the Svetloborsky massif.

Author Contributions: The authors (S.Y.S., R.S.P., A.V.K., D.A.K., D.A.V. and D.V.K.) discussed the obtained results and wrote the article together. Conceptualization, investigation and writing-original draft preparation, S.Y.S. and R.S.P.; writing-review and editing, supervision, A.V.K.; software, formal analysis, D.A.K. and D.A.V.; writing—review, editing and translation, D.V.K. All authors participated in the funding acquisition.

Funding: This research was funded by Russian Foundation for Basic Research, grant number 18-35-00151\18.

Acknowledgments: We are grateful to Anton Antonov (Saint-Petersburg) and Anton Kutyrev (Petropavlovsk-Kamchatsky) for analytical work.

Conflicts of Interest: The authors declare no conflict of interest. The funders had no role in the design of the study; in the collection, analyses, or interpretation of data; in the writing of the manuscript, or in the decision to publish the results.

References

1. Zaitsev, A.M. *Platinum Deposits in the Urals*; M.N. Kononov and I.F. Skulimovskiy: Tomsk, Russia, 1898; p. 74. (In Russian)
2. Vysotsky, N.K. Platinum deposits of Isovskoy and Nizhnetagilsky areas on the Urals. *Proc. Geol. Comm.* **1913**, *62*, 1–694. (In Russian)
3. Zavaritsky, A.N. *Lode Platinum Deposits on the Urals*; Publishing House of the Geological Committee: Saint Petersburg, Russia, 1928; p. 56. (In Russian)
4. Lazarenkov, V.G.; Malitch, K.N.; Sakhyanov, L.O. *Platinum Metal Mineralization of Zonal Ultrabasic and Comatiite Massifs*; Nedra: Leningrad, Russia, 1992; p. 217. (In Russian)
5. Ivanov, O.K. *Zoned Ultramafic Complexes of Ural (Mineralogy, Petrology, Genesis)*; Publishing House of Uralsky Mining University: Ekaterinburg, Russia, 1997; p. 488. (In Russian)
6. Garuti, G.; Pushkarev, E.; Zaccarini, F. Composition and paragenesis of Pt alloys from chromitites of the Uralian-Alaskan type Kytlym and Uktus complexes, Northern and Central Urals, Russia. *Can. Mineral.* **2002**, *40*, 357–376. [CrossRef]

7.   Auge, T.; Genna, A.; Legendre, C.; Ivanov, K.S.; Volchenko, Y.A. Primary platinum mineralization in the Nizhny Tagil and Kachkanar ultramafic complexes, Urals, Russia: A genetic model for PGE concentration in chromite-rich zones. *Econ. Geol.* **2005**, *100*, 707–732. [CrossRef]

8.   Pushkarev, E.V.; Anikina, E.V.; Garuti, G.; Zaccarini, F. Chrome-platinum mineralization of the Nizhny Tagil type in the Urals: Structural and material characteristics and the problem of genesis. *Lithosphere* **2007**, *3*, 28–65. (In Russian)

9.   Tolstykh, N.D.; Telegin, Y.M.; Kozlov, A.P. Platinum mineralization of the Svetloborsky and Kamenushinsky massifs (Urals Platinum Belt). *Russ. Geol. Geophys.* **2011**, *52*, 603–619. [CrossRef]

10. Zaccarini, F.; Garuti, G.; Pushkarev, E.V. Unusually PGE-rich chromitite in the Butyrin vein of the kytlym Uralian-Alaskan complex, Northern Urals, Russia. *Can. Mineral.* **2011**, *49*, 1413–1431. [CrossRef]

11. Tolstykh, N.D.; Kozlov, A.P.; Telegin, Y.M. Platinum mineralization of the Svetly Bor and Nizhny Tagil intrusions, Ural Platinum Belt. *Ore Geol. Rev.* **2015**, *67*, 234–243. [CrossRef]

12. Malitch, K.N.; Stepanov, S.Y.; Badanina, I.Y.; Khiller, V.V. Bedrock platinum-group elements mineralization of zonal clinopyroxenite-dunite massifs of the Middle Urals. *Dokl. Earth Sci.* **2017**, *476*, 1147–1151. [CrossRef]

13. Palamarchuk, R.S.; Stepanov, S.Y.; Khanin, D.A.; Antonov, A.V. PGE Mineralization of Massive chromitites of the Iov dunite body (Northern Urals). *Mosc. Univ. Geol. Bull.* **2017**, *72*, 68–76. [CrossRef]

14. Stepanov, S.Y.; Kozlov, A.V.; Malitch, K.N.; Badanina, I.Y.; Antonov, A.V. Platinum group element mineralization of the Svetly bor and Veresovy bor clinopyroxenite-dunite massifs, Middle Urals, Russia. *Geol. Ore Depos.* **2017**, *59*, 244–255. [CrossRef]

15. Malitch, K.N.; Thalhammer, O.A.R. Pt-Fe nuggets derived from clinopyroxenite-dunite massifs, Russia: A structural, compositional and osmium-isotope study. *Can. Mineral.* **2002**, *40*, 395–418. [CrossRef]

16. Barannikov, A.G.; Osovetsky, B.M. Platinum and platinum containing placers of the Urals, the criteria and signs of their spatial connection with primary sources. *News Ural State Min. Univ.* **2014**, *35*, 12–29. (In Russian)

17. Stepanov, S.Y.; Pilyugin, A.G.; Zolotarev, A.A., Jr. Comparison of the compositions of the platinum group minerals of the chromitites and placers of the Nizhny Tagil massif, Middle Urals. *J. Min. Inst.* **2015**, *211*, 22–28. (In Russian)

18. Tolstykh, N.D.; Krivenko, A.P.; Sidorov, E.G.; Laajoki, K.V.O.; Podlipskiy, M. Ore mineralogy of PGM placers in Siberia and the Russian Far East. *Ore Geol. Rev.* **2002**, *20*, 1–25. [CrossRef]

19. Tolstykh, N.D.; Sidorov, E.G.; Laajoki, K.V.O.; Krivenko, A.P.; Podlipskiy, M. The association of platinum-group minerals in placers of the Pustaya river, Kamchatka, Russia. *Can. Mineral.* **2000**, *38*, 1251–1264. [CrossRef]

20. Vildanova, E.Y.; Zaitsev, V.P.; Kravchenko, L.I.; Landa, E.A.; Litvinenko, A.F.; Markovskij, B.A.; Melkomukov, V.N.; Mochalov, A.G.; Nazimova, Y.V.; Popruzhenko, S.V.; et al. *Koryak-Kamchatka Region is A New Platinum-Bearing Province of Russia*; Publishing House of VSEGEI: St. Petersburg, Russia, 2002; p. 383. (In Russian)

21. Tolstykh, N.D.; Sidorov, E.G.; Kozlov, A.P. Platinum-group minerals in lode and placer deposits associated with the Ural-Alaskan-type Gal'moenan complex, Koryak-Kamchatka platinum belt, Russia. *Can. Mineral.* **2004**, *42*, 619–630. [CrossRef]

22. Sidorov, E.G.; Tolstykh, N.D.; Podlipsky, M.Y.; Pakhomov, I.O. Placer PGE minerals from the Filippa clinopyroxenite-dunite massif (Kamchatka). *Russ. Geol. Geophys.* **2004**, *45*, 1128–1144.

23. Oberthur, T.; Weiser, T.W.; Melcher, F. Alluvial and eluvial platinum-group minerals from the Bushveld complex. *S. Afr. J. Geol.* **2014**, *117*, 255–274. [CrossRef]

24. Oberthur, T. The fate of Platinum-Group Minerals in the Exogenic Environment–From Sulfide Ores via Oxidized Ores into Placers: Case Studies Bushveld Complex, South Africa, and Great Dyke, Zimbabwe. *Minerals* **2018**, *8*, 581. [CrossRef]

25. Palamarchuk, R.S.; Stepanov, S.Y.; Khanin, D.A.; Antonov, A.V.; Zolotarev, A.A., Jr. Comparative characteristics of platinum group minerals from the eluvial placer and chromitites of the Svetloborsky clinopyroxenite-dunite massif (Middle Urals). *Mineralogy* **2017**, *4*, 37–50. (In Russian)

26. Cabri, L.J.; Feather, C.E. Platinum-iron alloys: Nomenclature based on a study of natural and synthetic alloys. *Can. Mineral.* **1975**, *13*, 117–126.

27. Zaccarini, F.; Bindi, L.; Pushkarev, E.; Garuti, G.; Bakker, R.J. Multi-analytical characterization of minerals of the bowieite-kashinite series from the Svetly Bor complex, Urals, Russia, and comparison with worldwide occurences. *Can. Mineral.* **2016**, *54*, 461–473. [CrossRef]

28. Mertie, J.B. *Economic Geology of the Platinum Metals*; United States Government Printing Office: Washington, DC, USA, 1969.

29. Cabri, L.J.; Harris, D.C.; Weiser, T.W. The mineralogy and distribution of Platinum Group Mineral (PGM) placer deposits of the world. *Explor. Min. Geol.* **1996**, *5*, 73–167.

30. Weiser, T.W. Platinum-group minerals (PGM) in placer deposits. In *The Geology, Geochemistry, Mineralogy and Mineral Beneficiation of Platinum-Group Elements*; Cabri, L.J., Ed.; Canadian Institute of Mining, Metallurgy and Petroleum (CIM): Westmount, QC, Canada, 2002; CIM Special Volume 54, pp. 721–756.

31. Slansky, E.; Johan, Z.; Ohnenstetter, M.; Barron, L.M.; Suppel, D. Platinum mineralization in the Alaskan-type intrusive complexes near Fifield, N.S.W., Australia, Part 2. Platinum-group minerals in placer deposits at Fifield. *Mineral. Petrol.* **1991**, *43*, 161–180. [CrossRef]

32. Nixon, G.; Cabri, L.J.; Laflamme, J.H.G. Platinum-group-element mineralization in lode and placer deposits associated with the Tulameen Alaskan-type complex, British Columbia. *Can. Mineral.* **1990**, *28*, 503–535.

33. Legendre, O.; Auge, T. Alluvial platinum-group minerals from the Manampotsy area, East Madagascar. *Aust. J. Earth Sci.* **1992**, *39*, 389–404. [CrossRef]

34. Johan, Z.; Slansky, E.; Kelly, D.A. Platinum nuggets from the Kompiam area, Enga Province, Papua New Guinea: Evidence for an Alaskan-type complex. *Mineral. Petrol.* **2000**, *68*, 159–176. [CrossRef]

35. Johan, Z.; Ohnenstetter, M.; Fisher, W.; Amosse, J. Platinum-group minerals from the Durance River Alluvium, France. *Mineral. Petrol.* **1990**, *42*, 287–306. [CrossRef]

36. Podlipskii, M.Y.; Sidorov, E.G.; Tolstykh, N.D.; Krivenko, A.P. Cobalt-bearing malanite and other Pt-thiospinels from the Maior river placers (Kamchatka). *Russ. Geol. Geophys.* **1999**, *40*, 645–648.

37. McClenaghan, M.B.; Cabri, L.J. Review of gold and platinum group element (PGE) indicator minerals methods for surficial sediment sampling. *Geochemistry* **2011**, *11*, 251–264. [CrossRef]

38. Volchenko, Y.A.; Ivanov, K.S.; Koroteev, V.A.; Auge, T. Structural and materal evolution of the complexes of Ural Platinum-supporting Belts while forming the chromite-platinum deposits of the Ural type, Part I. *Lithosphere* **2007**, *3*, 3–27.

39. Mosin, K.I. *The History of Platinum Mining in the Urals*; Nizhne Turinskaya Printing House: Nizhnyaya Tura, Russia, 2002; p. 246. (In Russian)

40. Palamarchuk, R.; Stepanov, S.; Kozlov, A. The characteristics of placer platinum mineralization associated with Urakian-Alaskan massifs, Middle Urals, Russia. In Proceedings of the 13th International Platinum Symposium, Polokwane, South Africa, 30 June–6 July 2018; pp. 152–153.

41. Duparc, L.; Tikhonowitch, M. *Le Platine et Les Gites Platiniferes de l'Oural et du Monde*; Quarto: Geneve, Switzerland, 1920; p. 542.

42. Kutyrev, A.V.; Sidorov, E.G.; Antonov, A.V.; Chubarov, V.M. Platinum-group mineral assemblage of the Prizhimny Creek (Koryak Highland). *Russ. Geol. Geophys.* **2018**, *59*, 935–944. [CrossRef]

43. Amosse, J.; Dable, P.; Allibert, M. Thermochemical behavior of Pt, Ir, Rh, and Ru vs fO_2 and fS_2 in a basaltic melt. Implications for the differentiation and precipitation of these elements. *Mineral. Petrol.* **2000**, *68*, 29–62.

44. Nekrasov, I.Y.; Lennikov, A.M.; Oktyabrsky, R.A.; Zalishchak, B.L.; Sapin, V.I. *Petrology and Platinum-Bearingness of Ring Alkaline-Ultrabasic Complexes*; Laverov, N.P., Ed.; Nauka: Moscow, Russia, 1994; p. 381. (In Russian)

45. Kessel, R.; Beckett, J.R.; Stolper, E.M. Thermodynamic properties of the Pt-Fe system. *Am. Mineral.* **2001**, *86*, 1003–1014. [CrossRef]

Article

Mineralogy of Platinum-Group Elements and Gold in the Ophiolite-Related Placer of the River Bolshoy Khailyk, Western Sayans, Russia

Andrei Y. Barkov [1,*], Gennadiy I. Shvedov [2], Sergey A. Silyanov [2] and Robert F. Martin [3]

[1] Research Laboratory of Industrial and Ore Mineralogy, Cherepovets State University, 5 Lunacharsky Avenue, 162600 Cherepovets, Russia

[2] Institute of Mining, Geology and Geotechnology, Siberian Federal University, 95 Avenue Prospekt im. gazety "Krasnoyarskiy Rabochiy", 660025 Krasnoyarsk, Russia; g.shvedov@mail.ru (G.I.S.); silyanov-s@mail.ru (S.A.S.)

[3] Department of Earth and Planetary Sciences, McGill University, 3450 University Street, Montreal, QC H3A 0E8, Canada; robert.martin@mcgill.ca

* Correspondence: ore-minerals@mail.ru; Tel.: +7-8202-51-78-27

Received: 21 May 2018; Accepted: 5 June 2018; Published: 12 June 2018

check for updates

Abstract: We describe assemblages of platinum-group minerals (PGM) and associated PGE–Au phases found in alluvium along the River Bolshoy Khailyk, in the western Sayans, Russia. The river drains the Aktovrakskiy ophiolitic complex, part of the Kurtushibinskiy belt, as does the Zolotaya River ~15 km away, the site of other placer deposits. Three groups of alloy minerals are described: (1) Os–Ir–Ru compositions, which predominate, (2) Pt–Fe compositions of a Pt_3Fe stoichiometry, and (3) Pt–Au–Cu alloys, which likely crystallized in the sequence from Au–(Cu)-bearing platinum, Pt(Au,Cu), Pt(Cu,Au), and $PtAuCu_2$, to $PtAu_4Cu_5$. The general trends of crystallization of PGM appear to be: [Os–Ir–Ru alloys] → Pt_3Fe-type alloy (with inclusions of Ru-dominant alloy formed by exsolution or via replacement of the host Pt–Fe phase) → Pt–Au–Cu alloys. We infer that Rh and Co mutually substitute for Fe, not Ni, and are incorporated into the pentlandite structure via a coupled mechanism of substitution: $[Rh^{3+} + Co^{3+} + \square \rightarrow 3Fe^{2+}]$. Many of the Os–Ir–Ru and Pt–Fe grains have porous, fractured or altered rims that contain secondary PGE sulfide, arsenide, sulfarsenide, sulfoantimonide, gold, Pt–Ir–Ni-rich alloys, and rarer phases like Cu-rich bowieite and a Se-rich sulfarsenide of Pt. The accompanying pyroxene, chromian spinel and serpentine are highly magnesian, consistent with a primitive ultramafic source-rock. Whereas the alloy phases indicate a highly reducing environment, late assemblages indicate an oxygenated local environment leading to Fe-bearing Ru–Os oxide (zoned) and seleniferous accessory phases.

Keywords: platinum-group elements; gold; platinum-group minerals; placer deposits; ophiolite complexes; western Sayans; Russia

1. Introduction

It was L.A. Yachevskiy who first found grains of platinum-group minerals (PGM) identified as "osmian iridium" and "platinum" in a placer in 1910 in the Usinskiy area of the western Sayans [1]. These grains, enriched in platinum-group elements (PGE), were recognized in a heavy-mineral concentrate provided by Chirkov, a gold miner. Later, B.M. Porvatov, M.K. Korovin, and N.K. Vysotskiy [1–3] described the geology and occurrences of Au–PGM-bearing placers in the Usinskiy area and other regions of the western Sayans. Krivenko et al. [4] conducted large-scale investigations of associations of PGM in Au–PGE-bearing zones of placers of the Altai-Sayan folded region. Tolstykh et al. [5] reported the occurrence of a highly unusual association of PGM in a placer deposit of the

Zolotaya River ("Golden River") watershed, located ca. 15 km from the Bolshoy Khailyk placer. There, arsenotellurides and telluroarsenides of Ir–Os–Ru and an arsenoselenosulfide of Ir occur with a Pt–Au–Cu alloy phase in association with grains of Os–Ir–Ru and Pt–Fe alloys. These placer occurrences were attributed to lode serpentinite rocks of the Kurtushibinskiy ophiolite belt [5]. In addition, G.I. Shvedov and V.V. Nekos [6] found grains of placer PGM at the River Bolshoy Khailyk.

In the Aktovrakskiy complex [7], which forms part of the Kurtushibinskiy belt in Krasnoyarskiy kray (Figure 1), serpentitite bodies are fairly abundant at Bolshoy Khailyk, the site of our investigation. The area largely consists of volcanogenic and sedimentary rocks of the Chinginskaya suite of Lower Cambrian age, discordantly overlain by Ordovician "terrigenous" rocks. Dikes and intrusive bodies of Devonian rocks (granite, diorite, and metagabbroic rocks) are considered responsible for placer gold in the area.

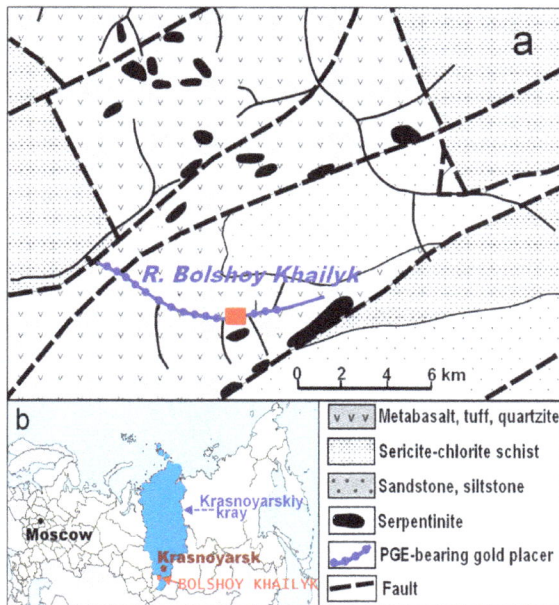

Figure 1. Regional geology of the Bolshoy Khailyk area (**a**) based on [6], with minor modifications, and a map (**b**) showing the location of the placer area in the Russian Federation. The red square (**a**) shows the sampling area (this study).

Our aims in the present article are to document the occurrences and mineralogical characteristics of assemblages of PGM and PGE–Au-rich phases in the alluvial placer associated with the River Bolshoy Khailyk (Figure 1), a tributary of the Urbun (Urgun) River, which flows into the River Yenisei. Our results and observations provide genetic implications and insights into the mineral-forming environments and trends of crystallization of the ophiolite-related PGE–Au mineralization at Bolshoy Khailyk.

2. Materials and Methods

We have examined numerous PGM grains and micrometric inclusions found in heavy-mineral concentrates collected at thirteen prospecting dug pits at a smooth bend along the riverbed of Bolshoy Khailyk, within the mining concession outlined (Figure 1). The concentrates are invariably enriched in grains of chromian spinel (~15–75 vol %), magnetite (up to ~30%), clinopyroxene and amphiboles (up to ~20% each), with subordinate amounts (<5%) of epidote, zoisite, and grossular-rich garnet.

Among the detrital grains of PGM, the Os–Ir–Ru alloy minerals strongly predominate (~80 vol %). Grains of these alloys vary in size from <0.5 to ~2 mm across. Commonly, they display a well-preserved hexagonal outline (Figure 2) and are composed of intergrowths of crystals of varying compositions (e.g., Figure 3b,f). Variously rounded grains also are present, however (Figure 3e).

Figure 2. Photo of a concentrate of platinum-group minerals collected in the Bolshoy Khailyk placer deposit. The brownish film present on the surface of some grains represents iron hydroxides.

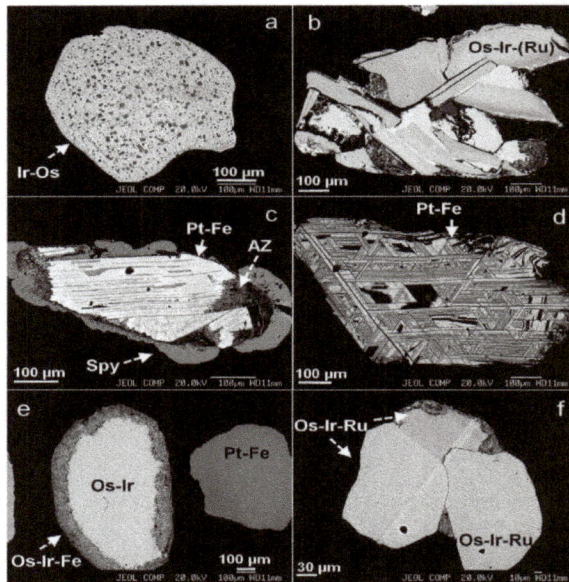

Figure 3. Back-scattered electron (BSE) images showing characteristic textures of placer grains of platinum-group minerals from the Bolshoy Khailyk placer. (**a**) A partly rounded grain of Ir–Os alloy contains inclusions of Pt–Fe alloy. (**b,f**) Intergrowths of subhedral crystals of Os–Ir–Ru alloys. (**c**) A placer grain of Pt–Fe alloy hosting abundant lamellae of Ru-rich alloy (gray) is surrounded by a rim of sperrylite, Spy. Note the existence of a zone of metasomatic alteration (labeled AZ). (**d**) A subhedral crystal of Pt–Fe alloy displays a perfectly developed rhombohedral cleavage, accompanied by voids. (**e**) A rounded grain of Os–Ir alloy consists of a rim of Fe-enriched alloy phase of Os–Ir–Fe; note that this rim is very porous and fractured.

Grains of Pt–Fe alloy with a Pt_3Fe-type stoichiometry (i.e., isoferroplatinum or ferroan platinum) account for up to 20 vol % in the samples examined; the prefix "ferroan" is used here to conform to the historically used name, without implication as to the valence of iron. The size of the Pt–Fe alloy grains, 2.5–3 mm across, generally exceeds that of the associated Os–Ir–Ru alloys. Commonly, these grains are anhedral. Nevertheless, subhedral to euhedral grains of Pt–Fe alloy are also encountered occasionally. The grain shown in Figure 3d has a perfectly developed rhombohedral cleavage, with rhombic patterns or voids. Lamellae of Ru-dominant alloy are hosted by the Pt–Fe alloy matrix (Figure 4b); this alloy phase of Ru can also display a rhombic shape, which conforms to the observed cleavage (Figure 5a), and is likely pseudomorphically formed as a result of late replacement of the Pt_3Fe host.

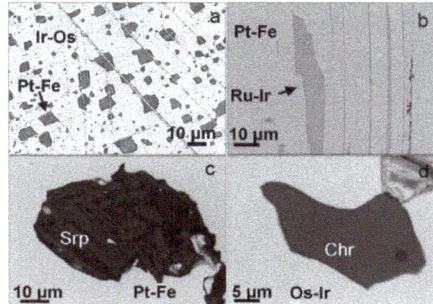

Figure 4. (**a**) BSE image showing exsolution phases of Pt–Fe alloy hosted by a grain of Ir–Os alloy. Note the presence of an Or–Ir alloy of late generation, cutting the exsolution lamellae of Pt–Fe alloy; this late alloy filled the cleavage space at a subsolidus stage. (**b**) BSE image of lamellar inclusions of a Ru-dominant alloy, labeled Ru–Ir, which has the composition $[Ru_{65.6}Ir_{12.5}Os_{9.6}Rh_{6.7}Pt_{5.7}]$, and is hosted by an isoferroplatinum-type alloy of Pt–Fe: $[Pt_{63.4}Fe_{24.8}Rh_{6.0}Cu_{3.6}Pd_{2.2}]$. (**c**) BSE image of inclusion of serpentine (Srp) hosted by a Pt–Fe alloy, $[Pt_{68.6}Fe_{23.9}Cu_{3.7}Pd_{2.8}Rh_{1.0}]$. (**d**) Inclusion of a chromian spinel, labeled Chr, which corresponds to magnesiochromite, and is hosted by a placer grain of Os–Ir alloy $[Os_{45.0}Ir_{39.7}Ru_{15.3}]$; BSE image.

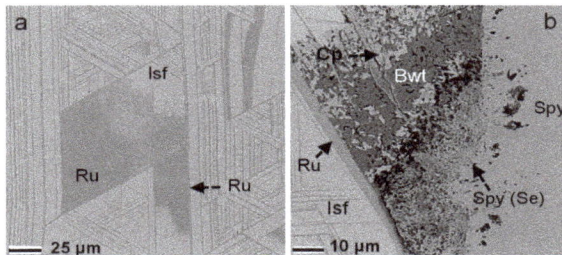

Figure 5. (**a**) BSE image showing exsolution lamellae of Ru-dominant alloy, labeled Ru $[Ru_{70.2-72.4}Rh_{7.1-7.7}Os_{7.5-7.8}Pt_{6.1-6.9}Ir_{5.9-6.1}Fe_{0.7-1.6}]$, hosted by a phase of Pt–Fe alloy related to isoferroplatinum (Isf): $Pt_{64.3}Fe_{24.3}Rh_{7.6}Cu_{2.3}Ni_{1.4}$. (**b**) BSE image displaying the texture and mineral assemblage developed in a zone of metasomatic alteration (labeled AZ in Figure 3c), which consists of secondary phases: bowieite rich in Cu, Bwt $[(Rh_{0.70}Pt_{0.65}Cu_{0.61}Ru_{0.05})_{\Sigma2.01}(S_{2.83}As_{0.16})_{\Sigma2.99}]$, cooperite, Cp $[(Pt_{0.88}Rh_{0.04})_{\Sigma0.92}S_{1.08}]$ and a seleniferous variety of sperrylite, Spy-(Se) $[(Pt_{0.54}Rh_{0.45})_{\Sigma0.99}(As_{1.29}Se_{0.40}S_{0.31})_{\Sigma2.00}]$. This alteration occurs locally, close to the margin of the placer grain of isoferroplatinum, Isf $[(Pt_{2.47}Rh_{0.25}Pd_{0.09})_{\Sigma2.8}(Fe_{0.99}Cu_{0.15}Ni_{0.05})_{\Sigma1.2}]$, and in contact with the outer rim of Se-free sperrylite, Spy $(Pt_{0.98}Rh_{0.01})_{\Sigma0.99}(As_{1.87}S_{0.12}Te_{0.02})_{\Sigma2.01}$. Note the presence of lamellae composed of a Ru-dominant alloy, labeled Ru $[Ru_{68.3}Ir_{12.0}Os_{7.2}Rh_{6.9}Pt_{5.7}]$.

Many of placer grains of Os–Ir–Ru and Pt–Fe alloys have a porous (Figure 3e), abundantly fractured or altered rim (Figure 6b,c), associated with the development of secondary phases of PGE sulfide, arsenide, sulfarsenide, native gold and a Pt–Ir–Ni-rich alloy. Such zones of metasomatic alteration (labeled AZ: Figures 3c and 5b) are developed close to grain margins; they are isolated and surrounded by an outer rim of sperrylite. The rare compounds formed in these zones include Cu-rich bowieite and a Se-rich sulfarsenide of Pt (Figure 5b).

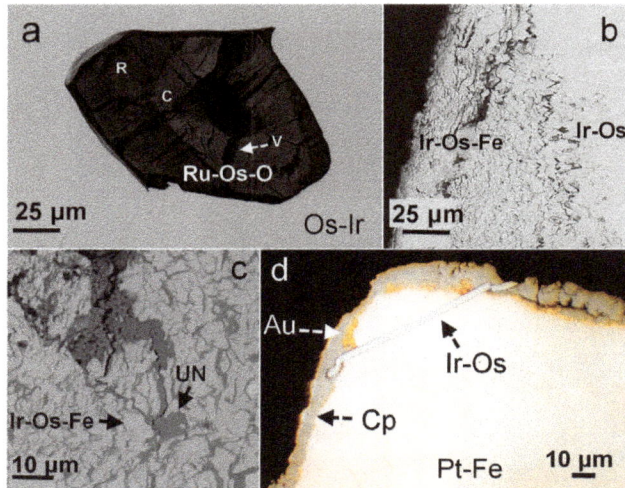

Figure 6. (**a**) BSE image of a zoned grain of Fe-bearing Ru–Os oxide, labeled Ru–Os–O, which is hosted by Os–Ir alloy [$Os_{45.9}Ir_{44.2}Ru_{9.9}$]. The letter C refers to the point analysis of the core zone, the letter R to that of the rim, and the letter V to the veinlet. (**b**) BSE image showing a porous and fractured rim (labeled Ir–Os–Fe) relatively enriched in Fe and Ni, that is developed around a placer grain of Ir–Os alloy poor in Fe (0.5 wt % Fe), with the following composition: [$Ir_{60.3}Os_{32.4}Ru_{3.8}Rh_{1.7}Fe_{1.7}Pt_{0.1}$]. The rim contains 5.3 wt % Fe, 0.88 wt % Ni, and corresponds to [$Ir_{48.6}Os_{29.8}Fe_{15.6}Ru_{3.4}Ni_{2.5}$]. (**c**) BSE image of veinlets composed of unnamed alloys, labeled UN [$(Ir,Pt)(Ni,Fe,Cu)_3–(Pt,Ir)(Ni,Fe,Cu)_3$], which are hosted by an Ir–Os–Fe alloy rich in Fe (8.0 wt %) and Ni (1.8 wt %) [$Ir_{48.4}Os_{24.7}Fe_{22.3}Ni_{4.7}$]; the latter alloy phase forms the rim (**b**). (**d**) Optical photograph showing the development of a composite rim, composed of cooperite (Cp) [$(Pt_{0.93}Pd_{0.02})S_{1.05}$] and Au–Ag alloy, labeled Au [$Au_{75.3}Ag_{22.5}Cu_{2.2}$]. Note the presence of a staple-like phase of Ir–Os alloy [$Ir_{50.7}Os_{26.3}Ru_{8.5}Fe_{6.5}Pt_{3.6}Ni_{2.9}Rh_{1.4}$]. These phases occur in the narrow rim around a grain of Pt–Fe alloy [$Pt_{65.4}Fe_{25.5}Rh_{3.2}Ir_{3.1}Pd_{2.5}Os_{0.2}Ni_{0.1}$].

In addition, our materials contain micrometric inclusions of clinopyroxene, chromian spinel, amphiboles, serpentine and base-metal sulfides. We also examined compositional variations in one unusual grain ~2 mm across, composed of the phase $PtAu_4Cu_5$ with inclusions of other intermetallic compounds of the system Pt–Cu–Au, of an unnamed Pt–Cu stannide and a Co–Rh-rich pentlandite (Figure 7a,b).

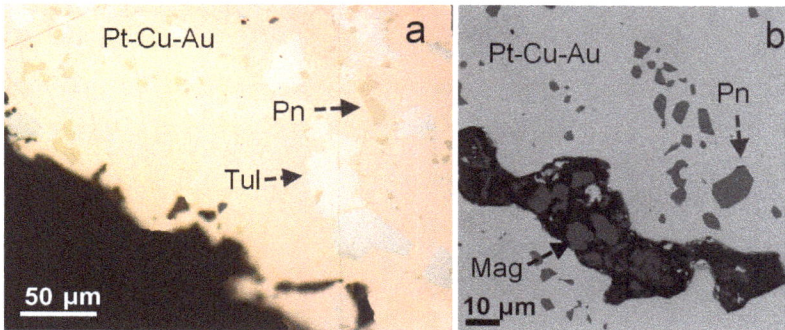

Figure 7. (**a**) Optical photograph showing inclusions of the tulameenite–ferronickelplatinum series (Tul) and of pentlandite (Pn), which are hosted by a placer grain of Pt–Au–Cu alloy. (**b**) BSE image of subhedral inclusions of Co–Rh-rich pentlandite (Pn) enclosed within the Pt–Au–Cu alloy. Mag is magnetite.

In this study, we used essentially the same approach and analytical facilities as in the complementary project, devoted also to associations of PGM of the Sayan region [8]. Compositions of various PGE alloys, PGM, PGE- and Au-rich phases, silicate minerals and hydrous silicates were investigated with wavelength-dispersive analysis (WDS) using a Camebax-micro electron microprobe (CAMECA SAS, Gennevilliers Cedex, France) at the Sobolev Institute of Geology and Mineralogy, Russian Academy of Sciences, Novosibirsk, Russia. The analytical conditions used for PGE-rich minerals were the following: 20 kV and 60 nA; the $L\alpha$ line was used for Ir, Rh, Ru, Pt, Pd, and As; the $M\alpha$ line was used for Os and Au, and the $K\alpha$ line was used for S, Fe, Ni, Cu, and Co. We used as standards pure metals (for the PGE and Au), $CuFeS_2$ (for Fe, Cu, and S), synthetic FeNiCo (for Ni and Co), and arsenopyrite (for As). The minimum detection limit is $\leq$0.1 wt % for results of the WDS analyses. The WDS analyses of chromite, clinopyroxene, and amphiboles were acquired at 20 kV and 40 nA, using $K\alpha$ lines, and the following standards: chromite (for Fe, Mg, Al, and Cr), ilmenite (Ti), manganiferous garnet (for Mn), and synthetic V_2O_5 (for V). The amphibole analyses were done using diopside (for Ca), albite (for Na), orthoclase (for K), pyrope (Mg, Fe, Al, and Si), a glass of diopside composition doped with 2 wt % TiO_2 for Ti, and manganiferous and chromiferous garnets (for Mn and Cr) as standards. For serpentine, we used an olivine standard (Mg, Si, Fe, and Ni), as well as diopside (Ca), and, as noted, garnets for Mn and Cr.

We employed scanning-electron microscopy (SEM) and energy-dispersive analysis (EDS) for the analysis of phases whose grain size is on the order of $\leq$2–5 μm, or larger. These phases were analyzed at 20 kV and 1.2 nA using a Tescan Vega 3 SBH facility combined with an Oxford X-Act spectrometer (Oxford Instruments, Abingdon, UK) at the Siberian Federal University, Krasnoyarsk, Russia. Pure elements (for the PGE, Fe, Cu), as well as FeS_2 (for S), InAs (for As), were used as standards. The $L\alpha$ line was used for As and the PGE, except for Pt and Au ($M\alpha$ line); the $K\alpha$ line was used for Fe, Cu, Ni, Co, and S.

3. Results

3.1. Grains of Os–Ir–Ru, Pt–Fe Alloys and Exsolution-Induced Phases

As noted, the Os–Ir–Ru and Pt–Fe alloy minerals account for ~80% and ~20 vol %, respectively, of the detrital PGM. In contrast to the Os–Ir–Ru alloys at Sisim, eastern Sayans [8], we detect no zonation in these grains of alloy at Bolshoy Khailyk. However, there are significant grain-to-grain variations (Table 1), reflected in differences in contrast in back-scattered electron (BSE) images of mutually intergrown grains (Figure 3f). Inclusions of Pt–Fe alloy are present in an Os–Ir–Ru alloy matrix, and a Ru–Ir–Os alloy occurs as lamellae hosted in a Pt–Fe alloy (Figure 4a,b and Figure 5a).

Table 1. Compositions of grains of Ru–Os–Ir alloy minerals from the Bolshoy Khailyk placer deposit.

#			Ru	Os	Ir	Rh	Pt	Pd	Fe	Ni	Cu	Total
1	Ru-dominant	Matrix	34.92	58.67	6.48	bdl	bdl	bdl	bdl	bdl	bdl	100.1
2			54.8	35.36	7.04	bdl	bdl	bdl	1.77	1.07	bdl	100
3			24.28	41.74	29.68	1.83	3.02	bdl	0.15	bdl	bdl	100.7
4			33.18	45.99	15.93	0.53	4.07	bdl	0.08	bdl	bdl	99.8
5			28.47	34.94	33.18	1.8	1.24	bdl	0.18	0.04	bdl	99.9
6		Inclusion	33.21	26.78	32.78	1.89	5.5	bdl	0.2	bdl	0.07	100.4
7			35.48	26.02	32.39	1.96	4.88	bdl	0.21	bdl	bdl	100.9
8			35.87	24.9	32.57	2.13	5.14	bdl	0.21	bdl	bdl	100.8
9			50.46	14.38	19.36	4.53	10.64	bdl	0.14	0.05	bdl	99.6
10	Os-dominant	Matrix	10.76	68.49	17.79	0.76	2.66	bdl	0.06	bdl	bdl	100.5
11			0.75	90.45	8.56	bdl	bdl	bdl	bdl	bdl	bdl	99.8
12			12.59	45.96	39.68	1.24	0.55	bdl	0.26	bdl	bdl	100.3
13			4.22	53.29	40.9	0.38	0.6	bdl	0.14	0.06	0.13	99.7
14			0.98	79.1	18.72	0.25	0.69	bdl	0.06	bdl	bdl	99.8
15		Inclusion	1.73	64.17	33.67	0.23	0.34	bdl	bdl	bdl	bdl	100.1
16			0.31	58.82	41.06	0.1	0.26	bdl	0.61	bdl	bdl	101.2
17			0.42	57.51	42.19	0.13	bdl	bdl	0.41	0.06	bdl	100.7
18	Ir-dominant	Matrix	0.43	33.37	61.59	0.19	4.23	bdl	0.23	0.04	bdl	100.1
19			2.88	13.89	76.02	1.01	3.52	bdl	0.84	0.12	bdl	98.3
20			0.87	6.06	81.71	0.45	8.91	bdl	1.77	0.18	bdl	100
21			0.27	27.92	68.5	bdl	2.08	0.05	0.28	0.04	0.04	99.2
22			1.74	36.73	58.39	0.59	1.91	bdl	0.28	0.06	bdl	99.7
23		Inclusion	1.26	38.36	54.7	0.55	4.41	bdl	0.33	bdl	bdl	99.6
24			5	29.02	56.48	0.82	4.02	bdl	2.11	1.02	bdl	98.5
25			5.82	30.05	51.04	0.64	6	bdl	2.83	2.03	bdl	98.4

		Atomic proportions (per a total of 100 at %)								
#			Ru	Os	Ir	Rh	Pt	Pd	Fe	Ni	Cu
---	---	---	---	---	---	---	---	---	---	---	---
1	Ru-dominant	Matrix	50.2	44.9	4.9	0	0	0	0	0	0
2			66.6	22.8	4.5	0.00	0	0	3.9	2.2	0
3			37.0	33.8	23.8	2.74	2.4	0	0.4	0	0
4			48.3	35.5	12.2	0.76	3.1	0	0.2	0	0
5			42.3	27.6	25.9	2.63	1.0	0	0.5	0.1	0
6		Inclusion	47.5	20.4	24.7	2.66	4.1	0	0.5	0	0.2
7			49.9	19.4	23.9	2.70	3.6	0	0.5	0	0
8			50.3	18.5	24.0	2.93	3.7	0	0.5	0	0
9			64.2	9.7	13.0	5.66	7.0	0	0.3	0.1	0
10	Os-dominant	Matrix	18.3	62.0	15.9	1.3	2.3	0	0.2	0	0
11			1.4	90.1	8.4	0.0	0.0	0	0.0	0	0
12			21.0	40.8	34.9	2.0	0.5	0	0.8	0	0
13			7.6	51.2	38.9	0.7	0.6	0	0.5	0.2	0.4
14			1.8	78.5	18.4	0.5	0.7	0	0.2	0	0
15		Inclusion	3.2	63.2	32.8	0.4	0.3	0	0.0	0	0
16			0.6	57.4	39.6	0.2	0.2	0	2.0	0	0
17			0.8	56.4	41.0	0.2	0.0	0	1.4	0.2	0
18	Ir-dominant	Matrix	0.8	33.2	60.6	0.3	4.1	0	0.8	0.1	0
19			5.3	13.5	73.0	1.8	3.3	0	2.8	0.4	0
20			1.6	5.8	77.2	0.8	8.3	0	5.8	0.6	0
21			0.5	28.0	68.1	0.0	2.0	0.1	1.0	0.1	0.1
22			3.2	36.0	56.7	1.1	1.8	0	0.9	0.2	0
23		Inclusion	2.3	37.9	53.4	1.0	4.2	0	1.1	0	0
24			8.5	26.3	50.7	1.4	3.6	0	6.5	3.0	0
25			9.5	26.2	44.0	1.0	5.1	0	8.4	5.7	<0.1

Note. Results of WDS analyses are listed in weight%; "bdl" indicates that amounts of elements are below detection limits.

Extremely narrow and linear "lamellae" of Os–Ir–Ru alloy cut the associated inclusions of Pt–Fe alloy and occupy the cleavage plane of some of the Ir–Os–Ru placer grains, thus implying subsolvus conditions of their deposition. Some Ru-dominant phases exhibit a rhombohedral shape (Figure 5a);

these are likely pseudomorphic and reproduce the characteristic forms of the rhombohedral cleavage in the matrix of Pt_3Fe-type phase. A wire-like inclusion of Ir–Os alloy is noteworthy (Figure 6d).

A strong enrichment in Ru is detected in lamellar phases, e.g., those in Figure 4b: $Ru_{65.6}Ir_{12.5}Os_{9.6}Rh_{6.7}Pt_{5.7}$, hosted by the isoferroplatinum-type alloy $[Pt_{63.4}Fe_{24.8}Rh_{6.0}Cu_{3.6}Pd_{2.2}]$, or in the lamellae shown in Figure 5a $[Ru_{70.2-72.4}Rh_{7.1-7.7}Os_{7.5-7.8}Pt_{6.1-6.9}Ir_{5.9-6.1}Fe_{0.7-1.6}]$, enclosed by a Pt–Fe alloy related to isoferroplatinum $[Pt_{64.3}Fe_{24.3}Rh_{7.6}Cu_{2.3}Ni_{1.4}]$. The observed Ru-enrichment is a reflection of relative abundance of Ru in the ophiolite environment, and points to the accumulation of levels of Ru during crystallization. It appears that Ru behaved somewhat incompatibly during the crystallization of the Pt–Fe alloy. In addition, a relative Rh-enrichment is common in compositions of Pt–Fe alloy at Bolshoy Khailyk (Table 2).

Table 2. Compositions of placer grains of isoferroplatinum and inclusions of minerals of the tulameenite-ferronickelplatinum series hosted by a Pt–Cu–Au alloy grain from Bolshoy Khailyk.

#		Pt	Ir	Rh	Pd	Ru	Os	Fe	Ni	Cu	Sb	Sn	Total
1	Isf	86.32	bdl	0.35	2.05	bdl	bdl	9.45	0.15	0.7	bdl	bdl	99.0
2		86.98	1.88	1.88	bdl	0.23	bdl	9.05	0.05	0.17	bdl	bdl	100.2
3		86.22	0.14	0.96	1.22	0.08	bdl	9.3	0.21	0.94	bdl	bdl	99.1
4		85.11	bdl	1.44	1.52	0.31	bdl	9.03	0.33	0.84	bdl	bdl	98.6
5		86.38	0.14	0.72	1.12	0.04	bdl	9.49	0.34	0.65	bdl	bdl	98.9
6		87.47	bdl	0.12	1.58	0.04	0.05	9.19	0.08	0.57	bdl	bdl	99.1
7		88.79	bdl	0.29	0.46	0.05	0.07	8.77	0.13	0.33	bdl	bdl	98.9
8		85.89	bdl	1.05	1.53	0.14	bdl	8.7	0.22	1.66	bdl	bdl	99.2
9		85.81	bdl	1.53	2.49	bdl	bdl	9.33	0.22	0.84	bdl	bdl	100.2
10		91.29	bdl	0.13	0.27	bdl	bdl	8.62	0.03	0.23	bdl	bdl	100.6
11	Tul-Fnp	74.81	bdl	bdl	bdl	bdl	bdl	11.9	5.6	7.03	bdl	bdl	99.3
12		74.28	bdl	bdl	bdl	bdl	bdl	11.38	4.21	8.62	bdl	bdl	98.5
13		76.03	bdl	bdl	bdl	bdl	bdl	9.11	2.38	12.18	bdl	0.9	100.6
14		77.31	bdl	bdl	bdl	bdl	bdl	11.85	4.67	8.01	bdl	bdl	101.8
15		77.25	bdl	bdl	bdl	bdl	bdl	11.62	5.12	7.41	bdl	bdl	101.4
16		75.17	bdl	bdl	bdl	bdl	bdl	9.85	2.66	9.62	0.77	0.95	99.0
17		76.63	bdl	bdl	bdl	bdl	bdl	11.86	3.77	7.97	bdl	bdl	100.2
18		75.7	bdl	bdl	bdl	bdl	bdl	11.56	4.48	7.22	bdl	bdl	99.0
19		76.3	bdl	bdl	bdl	bdl	bdl	10.52	1.91	9.15	1.44	0.7	100.0
20		76.89	bdl	bdl	bdl	bdl	bdl	11.1	2.74	9.51	0.32	0.47	101.0

Atomic proportions (per a total of 4 a.p.f.u.)

#	Pt	Ir	Rh	Pd	Ru	Os	Fe	Ni	Cu	Sb	Sn	Cu+Ni
1	2.73	0	0.02	0.12	0	0	1.04	0.02	0.07	0	0	–
2	2.78	0.06	0.11	0	0.01	0	1.01	0.01	0.02	0	0	–
3	2.72	0	0.06	0.07	0	0	1.03	0.02	0.09	0	0	–
4	2.69	0	0.09	0.09	0.02	0	1.00	0.03	0.08	0	0	–
5	2.74	0	0.04	0.07	0	0	1.05	0.04	0.06	0	0	–
6	2.80	0	0.01	0.09	0	0	1.03	0.01	0.06	0	0	–
7	2.90	0	0.02	0.03	0	0	1.00	0.01	0.03	0	0	–
8	2.70	0	0.06	0.09	0.01	0	0.96	0.02	0.16	0	0	–
9	2.66	0	0.09	0.14	0	0	1.01	0.02	0.08	0	0	–
10	2.97	0	0.01	0.02	0	0	0.98	0	0.02	0	0	–
11	1.91	0	0	0	0	0	1.06	0.48	0.55	0	0	1.03
12	1.92	0	0	0	0	0	1.03	0.36	0.69	0	0	1.05
13	1.97	0	0	0	0	0	0.82	0.20	0.97	0	0.04	1.17
14	1.95	0	0	0	0	0	1.04	0.39	0.62	0	0	1.01
15	1.96	0	0	0	0	0	1.03	0.43	0.58	0	0	1.01
16	1.99	0	0	0	0	0	0.91	0.23	0.78	0.03	0.04	1.02
17	1.98	0	0	0	0	0	1.07	0.32	0.63	0	0	0.95
18	1.98	0	0	0	0	0	1.05	0.39	0.58	0	0	0.97
19	2.02	0	0	0	0	0	0.97	0.17	0.74	0.06	0.03	0.91
20	1.98	0	0	0	0	0	1.00	0.23	0.75	0.01	0.02	0.99

Note. Numbers 1–10 pertain to results of WDS analyses, #11–20 to EDS analyses. The label Isf means isoferroplatinum-type alloy; Tul-Fnp are members of the tulameenite-ferronickelplatinum series; "bdl" indicates that amounts of elements are below detection limits.

Alloy phases of the tulameenite–ferronickelplatinum series (≤0.1 mm across) occur as inclusions in the matrix of a ternary Pt–Au–Cu phase (Figure 7a), and show strong variations of composition owing to Ni-for-Cu substitution (Table 2). An intermediate member contains ~0.5 Ni atoms per formula unit (hereafter a.p.f.u.) in this series. In addition, partly subhedral inclusions (≤20 μm in size) of a ferronickelplatinum-type alloy (#1–3, Table 3) were observed as inclusions in grains of Os–Ir–Ru alloy.

Table 3. Compositions of Ni-enriched alloys of Pt–Ir–Ni–Fe from the Bolshoy Khailyk deposit.

	Pt	Ir	Rh	Ru	Os	Fe	Ni	Cu	Total
1	76.79	bdl	0.46	bdl	bdl	8.37	12.65	2.02	100.29
2	73.89	bdl	0.36	bdl	bdl	10.08	13.54	2.28	100.15
3	78.08	bdl	0.29	bdl	bdl	8.63	12.61	2.01	101.62
4	51.64	5.44	bdl	bdl	bdl	12.89	27.61	2.55	100.13
5	51.64	5.4	bdl	bdl	bdl	13.3	28.29	1.62	100.25
6	49.55	6.1	bdl	bdl	bdl	13.48	27.79	1.56	98.48
7	52.17	5.77	bdl	bdl	bdl	12.85	27.01	1.56	99.36
8	6.28	55.12	bdl	bdl	2.23	11.01	25.96	bdl	100.60
9	21.28	28.74	0.38	bdl	bdl	13.88	34.83	1.9	101.01
10	24.26	30.4	0.43	bdl	bdl	12.91	29.87	2.05	99.92
11	10.41	38.72	0.27	bdl	bdl	14.17	35.42	bdl	98.99
12	14.86	32.31	0.51	bdl	bdl	13.7	36.27	0.98	98.63
13	13.43	48.18	0.52	bdl	bdl	14.44	23.84	bdl	100.41
14	23.86	28.35	0.44	bdl	bdl	14.97	31.03	bdl	98.65
15	26.86	28.14	0.22	bdl	2.06	13.45	30.55	bdl	101.28
16	48.03	6.03	0.47	0.06	0.88	13.11	26.03	1.63	96.24
17	49.08	6.65	0.45	bdl	0.3	12.91	24.5	1.44	95.33
18	48.69	9.41	0.43	0.43	1.74	13.17	23.47	1.49	98.83
19	13.09	36.82	0.34	bdl	1.62	11.57	33.38	1.71	98.53
20	12.84	36.71	0.34	bdl	1.76	11.58	33.33	1.69	98.25

					Atomic proportions (per a total of 100 at %)					
	Pt	Ir	Rh	Ru	Os	Fe	Ni	Cu	Ni/Fe	(Ni + Fe + Cu)/ΣPGE
1	49.5	0	0.6	0	0	18.8	27.1	4.0	1.44	1.00
2	45.7	0	0.4	0	0	21.8	27.8	4.3	1.28	1.17
3	49.8	0	0.4	0	0	19.2	26.7	3.9	1.39	0.99
4	25.6	2.7	0	0	0	22.3	45.5	3.9	2.04	2.53
5	25.5	2.7	0	0	0	22.9	46.4	2.5	2.02	2.55
6	24.8	3.1	0	0	0	23.5	46.2	2.4	1.96	2.59
7	26.4	3.0	0	0	0	22.7	45.5	2.4	2.00	2.40
8	3.3	29.6	0	0	1.2	20.3	45.6	0	2.24	1.93
9	9.6	13.2	0.3	0	0	21.9	52.3	2.6	2.39	3.32
10	11.7	14.9	0.4	0	0	21.8	48.1	3.0	2.20	2.69
11	4.8	18.1	0.2	0	0	22.8	54.1	0	2.38	3.33
12	6.8	14.9	0.4	0	0	21.7	54.8	1.4	2.52	3.53
13	7.0	25.3	0.5	0	0	26.1	41.1	0	1.57	2.05
14	11.4	13.8	0.4	0	0	25.0	49.4	0	1.97	2.91
15	13.0	13.8	0.2	0	1.0	22.8	49.2	0	2.16	2.56
16	24.8	3.2	0.5	0.06	0.5	23.7	44.7	2.6	1.89	2.45
17	26.1	3.6	0.5	0	0.2	24.0	43.3	2.4	1.81	2.30
18	25.6	5.0	0.4	0.44	0.9	24.2	41.0	2.4	1.70	2.08
19	6.3	17.8	0.3	0	0.8	19.3	53.0	2.5	2.74	2.97
20	6.1	17.8	0.3	0	0.9	19.4	53.0	2.5	2.74	2.98

Note. Numbers 1–10 are results of SEM/EDS analyses, and #16–20 of WDS analyses; "bdl" indicates that amounts of elements are below detection limits. Compositions #1–3 are close to ferronickelplatinum, #4–20 are unnamed members of the series (Pt,Ir)(Ni, Fe, Cu)$_{3-x}$ and (Ir, Pt)(Ni, Fe, Cu)$_{3-x}$.

We document the existence at Bolshoy Khailyk of extensive fields of solid solution in the system Os–Ru–Ir (Figure 8). The general enrichment in Ru is recognized in the coexisting pairs of matrix and exsolution-induced inclusions. As shown for comparison in Figure 8, the Os–Ir–Ru alloy grains from the neighboring placer at the River Zolotaya [5] also display the characteristic pattern of enrichment in Ru.

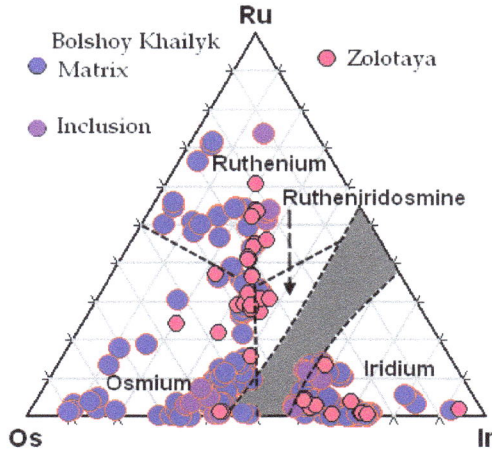

Figure 8. Compositional variations of grains of Os–Ir–Ru alloys from the Bolshoy Khailyk placer (this study), in comparison with alloys from the Zolotaya River deposit, western Sayans [5], in terms of the Os–Ir–Ru diagram (at %). The miscibility gap and nomenclature are on the basis of [9].

3.2. Fe–Ni-Enriched Rims and Unnamed Alloys of the Series $(Pt,Ir)(Ni,Fe,Cu)_{3-x}$–$(Ir,Pt)(Ni,Fe,Cu)_{3-x}$

The porous and fractured rim on grains of Ir–Os–Ru alloy (Figure 3e) is notably enriched in Fe and Ni relative to the adjacent phases analyzed beyond the rim. For example, the rim shown in Figure 6b has the typical composition: Fe 5.3, Ni 0.9, Ir 56.7, Os 34.4, and Ru 2.1, total 99.4 wt %, corresponding to $[Ir_{48.6}Os_{29.8}Fe_{15.6}Ru_{3.4}Ni_{2.5}]$ (in at %). This rim is developed around an alloy phase that contains: Fe 0.5, Ni <0.1, Ir 62.5, Os 33.2, Ru 2.1, Rh 0.9, Pt 0.1, total 99.3 wt %, or $[Ir_{60.3}Os_{32.4}Ru_{3.8}Rh_{1.7}Fe_{1.7}Pt_{0.1}]$. These variations indicate that Fe largely substitutes for Ir in the rim phase. The rim area shown in Figure 6c is even richer in Fe and Ni: Fe 8.0, Ni 1.8, Ir 60.1, Os 30.4, total 100.3 wt %, or $[Ir_{48.4}Os_{24.7}Fe_{22.3}Ni_{4.7}]$.

Interestingly, unnamed alloys of the series $(Pt,Ir)(Ni,Fe,Cu)_{3-x}$–$(Ir,Pt)(Ni,Fe,Cu)_{3-x}$ occur as veinlets of late phases associated with the Fe–Ni-enriched rim (Figure 6c). Their compositions reveal fairly strong variations in Pt–Ir and Ni–Fe–Cu contents (#4–20, Table 3). These alloy phases are Ni-dominant and contain essential Pt, which could be important to stabilize the structure. Indeed, the IrNi₃ phase has not been encountered in synthesis experiments, whereas the ordered phase PtNi₃ crystallizes at 580 °C in the system Pt–Ni [10].

3.3. Ni–Ir–Fe Alloy, Fe–Ni-rich Ir–Os Alloy, and Inclusions of Base-Metal Sulfides

Alloys of Ni–Ir–Fe and Fe–Ni–Ir–Os are associated with a local zone of alteration in a grain of Ir-dominant alloy, which contains Ir 72.2, Os 25.2, Fe 1.6, Rh 0.5, Ru 0.3, total 99.8 wt %, or $[Ir_{69.0}Os_{24.3}Fe_{5.4}Rh_{0.9}Ru_{0.5}]$. The Ni–(Fe)-rich alloy developed in this assemblage has the following composition: Ir 58.12, Os 9.44, Ni 18.99, Fe 11.47, total 98.0 wt %, or $[Ni_{36.7}Ir_{34.3}Fe_{23.3}Os_{5.6}]$. Its structure remains unknown; it is Ni-dominant and, thus, probably corresponds to garutiite (Ni,Fe,Ir), the hexagonal polymorph of native nickel cf. [11], or to its cubic polymorph, nickel. This phase is devoid of Pt, and differs from alloys of the series $(Pt,Ir)(Ni,Fe,Cu)_{3-x}$–$(Ir,Pt)(Ni,Fe,Cu)_{3-x}$ by a lower value of (Ni + Fe)/PGE (1.5). Nevertheless, we cannot exclude that it belongs to the latter solid solution.

One grain of Fe-rich alloy in this altered zone has an Ir-dominant composition: Ir 55.17, Os 30.27, Fe 8.17, Ni 4.46, Ru 0.77, total 98.8, or [$Ir_{42.5}Os_{23.5}Fe_{21.6}Ni_{11.2}Ru_{1.1}$]. Therefore, this phase is enriched strongly in the hexaferrum component, i.e., (Fe,PGE), cf. [12,13].

Micrometric and droplet-shaped inclusions of base-metal sulfides ≤10 μm in diameter occur close to margin in this grain of Ir-dominant alloy. Three types of phases are recognized on the basis of their compositions. The first is a monosulfide-type phase that contains: Ni 24.6, Fe 24.1, Cu 12.8, S 35.2, total 96.7 wt %, corresponding to the formula $(Fe_{0.40}Ni_{0.39}Cu_{0.19})_{\Sigma0.98}S_{1.02}$ calculated for a total of 2 a.p.f.u. A bornite-like phase is with the composition: Cu 52.1, Fe 16.5, S 28.8, total 97.4 wt %; its formula is $(Cu_{4.06}Fe_{1.47})_{\Sigma5.5}S_{4.5}$ (for 10 a.p.f.u.). And the third is represented by a godlevskite-like phase: Ni 70.2, S 30.2, total 100.4 wt %, or $Ni_{9.5}S_{7.5}$ (for 17 a.p.f.u.).

3.4. PGE Sulfides, Unnamed Ni[Ir(Co,Cu,Fe)]$_2$S$_4$, and Sperrylite

We analyzed several species of PGE-based sulfides found as inclusions (up to ~50 μm) hosted by grains of PGE alloys or as components of the rim or overgrowth on these grains. Results of analyses of ten grains (WDS) indicate that members of the laurite–erlichmanite series are relatively poor in Ir (up to ~10–15 wt %, Table 4). The exceptions are minute grains of Ir-rich laurite (≤5 μm across) deposited within a narrow rim that is developed around a grain of Ir–Os alloy; these are associated with members of the tolovkite–irarsite series and native gold [$Au_{74.1-91.1}Ag_{0.9-19.5}Cu_{6.4-8.1}$]. Note that the Au-rich alloy is enriched in Cu, and its composition ranges up to high-purity gold. The Ir-rich laurite is devoid of Os; it has the following composition: Ru 46.1, Ir 19.7, S 35.5, total 101.3 wt %, corresponding to $(Ru_{0.82}Ir_{0.18})_{\Sigma1.00}S_{1.99}$. The phase could possibly imply conditions of metastable crystallization at a low temperature as a representative of a late-stage assemblage at the rim. Compositions of five grains of cooperite are close to being stoichiometric (#3, 4, Table 4). A grain of bowieite in the zone of deuteric alteration (Figure 5b) is anomalously rich in Cu (#5, Table 4).

Table 4. Compositions of various PGE sulfides from the Bolshoy Khailyk deposit.

#		Ru	Os	Ir	Rh	Pt	Pd	Fe	Ni	Co	Cu	S	As	Total
1	Laurite	53.3	0.61	10.03	0.37	bdl	0.80	bdl	bdl	bdl	bdl	35.37	bdl	100.48
2		35.46	22.38	6.92	0.59	bdl	0.15	bdl	bdl	bdl	bdl	33.2	bdl	98.70
3	Cooperite	bdl	bdl	bdl	0.16	84.06	bdl	bdl	0.22	bdl	bdl	15.51	bdl	99.95
4		bdl	bdl	bdl	bdl	82.09	0.47	bdl	1.22	bdl	bdl	15.32	bdl	99.10
5	Bowieite (Cu-rich)	1.60	bdl	bdl	20.77	36.23	bdl	bdl	bdl	bdl	11.12	26.11	3.53	99.40
6	UN Tsp	bdl	bdl	42.33	bdl	bdl	bdl	4.13	12.17	6.57	4.31	26.55	bdl	100.11
7		bdl	bdl	41.93	bdl	bdl	bdl	4.15	13.93	6.38	5.5	27.62	bdl	99.51
8		bdl	bdl	43.59	bdl	bdl	bdl	3.98	12.25	6.09	5.87	27.37	bdl	99.15
9		bdl	bdl	43.84	bdl	bdl	bdl	4.52	12.86	6.70	5.41	27.81	bdl	101.14
						Atomic proportions								
#		Ru	Os	Ir	Rh	Pt	Pd	Fe	Ni	Co	Cu	S	As	ΣMe
1	Laurite	0.93	0.01	0.09	0.01		0.013	0	0	0	0	1.95	0	1.05
2		0.68	0.23	0.07	0.01		0.003	0	0	0	0	2.01	0	0.99
3	Cooperite	0	0	0	0.003	0.94	0	0	0.008	0	0	1.05	0	0.95
4		0	0	0	0	0.91	0.01	0	0.05	0	0	1.03	0	0.97
5	Bowieite	0.05	0	0	0.70	0.65	0	0	0	0	0.61	2.83	0.16	2.01
6	UN Tsp	0	0	1.00	0	0	0	0.34	0.94	0.51	0.31	3.76	0	3.09
7		0	0	0.96	0	0	0	0.33	1.05	0.48	0.38	3.80	0	3.20
8		0	0	1.02	0	0	0	0.32	0.94	0.46	0.42	3.84	0	3.16
9		0	0	1.00	0	0	0	0.36	0.96	0.50	0.37	3.81	0	3.19

Note. Numbers 1–4 pertain to results of WDS analyses, and #5–9 are SEM/EDS; "bdl" indicates that amounts of elements are below detection limits. The total of an.#6 includes 4.05 wt % Sb, which corresponds to 0.15 Sb a.p.f.u. The atomic proportions were calculated on the basis of a total of 3 a.p.f.u. (for laurite), 2 a.p.f.u. (cooperite), 5 a.p.f.u. (bowieite), and 7 a.p.f.u. for the unnamed phase related to thiospinels (UN Tsp).

We document the existence of an unknown sulfide that presumably is related to thiospinels, as it displays a Me$_3$S$_4$ stoichiometry, cf. [14], where Me represents the total content of metals (i.e., Ir, Co, Fe and Cu). This phase occurs in peripheral portions of irregular grains (≤50 μm), enclosed within a grain of Ir–Os alloy. Its composition corresponds to the formula $Ni[Ir(Co,Cu,Fe)]_2S_4$ (#6–9, Table 4). It seems likely that this phase is related to the synthetic thiospinel $NiIr_2S_4$ [15]; one of the two Ir atoms is presumably replaced by Co at Bolshoy Khailyk. The mixed-valence character of Ir is recognized in PGE thiospinels ([14] and references therein).

A total of 17 data-points (WDS) done on representative grains and inclusions of sperrylite yield normal levels of incorporation of Ir and Rh, up to ~3 wt % each, and up to 1.8 wt % S.

3.5. Zones of Deuteric Alteration

Zones of deuteric alteration (AZ: Figure 3c) are documented in a grain of placer Pt–Fe alloy of isoferroplatinum type $[(Pt_{2.47}Rh_{0.25}Pd_{0.09})_{\Sigma2.8}(Fe_{0.99}Cu_{0.15}Ni_{0.05})_{\Sigma1.2}]$. They occur near the grain margin, but are separated from it by the rim of S-bearing sperrylite $[(Pt_{0.98}Rh_{0.01})_{\Sigma0.99}(As_{1.87}S_{0.12}Te_{0.02})_{\Sigma2.01}]$. The matrix of the Pt–Fe alloy grain consists of abundant lamellae, oriented crystallographically, of Ru-dominant alloy $[Ru_{68.3}Ir_{12.0}Os_{7.2}Rh_{6.9}Pt_{5.7}]$.

In the altered zones, the late mineral assemblage is represented by bowieite rich in Cu (#5, Table 4), $[(Rh_{0.70}Pt_{0.65}Cu_{0.61}Ru_{0.05})_{\Sigma2.01}(S_{2.83}As_{0.16})_{\Sigma2.99}]$, cooperite $[(Pt_{0.88}Rh_{0.04})_{\Sigma0.92}S_{1.08}]$ and, interestingly, a seleniferous variety of sperrylite (Figure 5b) that contains Pt 36.99, Rh 16.07, As 33.79, Se 11.01, S 3.5, total 101.4 wt %. The stoichiometric formula, $(Pt_{0.54}Rh_{0.45})_{\Sigma0.99}(As_{1.29}Se_{0.40}S_{0.31})_{\Sigma2.00}$, points to an uncommon variety of seleniferous and rhodiferous sperrylite. Alternatively, though less likely, this phase may represent an unnamed species of arsenosulfoselenide: $(Pt, Rh)As_{1+x}(Se,S)_{1-x}$.

3.6. Tolovkite (IrSbS)—Irarsite (IrAsS)—Hollingworthite (RhAsS) Solid Solutions

Members of the tolovkite (IrSbS)—irarsite (IrAsS)—hollingworthite (RhAsS) solid solution form inclusions ($\leq$20 µm) enclosed within the Pt–Au–Cu alloy. The observed existence of Sb-for-As substitution is especially noteworthy (Figure 9, Table 5), as it was also documented in related PGM from the Svetlyi Bor complex (Svetloborskiy) of the Urals: $(Ir_{0.91}Rh_{0.08}Fe_{0.02})_{\Sigma1.01}(As_{0.59}Sb_{0.38})_{\Sigma0.97}S_{1.02}$ and $(Ir_{1.01}Rh_{0.03}Fe_{0.03})_{\Sigma1.07}(As_{0.73}Sb_{0.20})_{\Sigma0.93}S_{1.00}$ [16]. Tolovkite from its type locality, the alluvial placers of the Tolovka River zone associated with the Alpine-type Ust'-Belskiy complex, Magadanskaya oblast, Russia, is close to its end-member composition, IrSbS [17]. Another series of solid solution extends from IrSbS (tolovkite) toward its Rh counterpart, unnamed RhSbS [18]. Note that some compositions are Pt-rich (#16, Table 5), thus suggesting that the platarsite component (PtAsS [19]) also is involved.

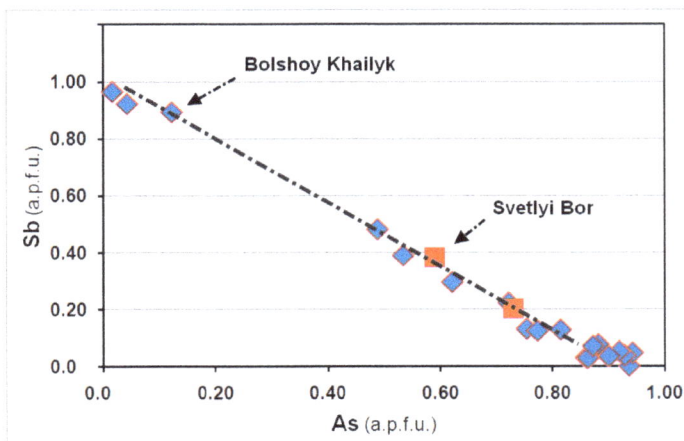

Figure 9. Compositional variations of phases of the tolovkite–irarsite solid solution from the Bolshoy Khailyk placer (this study) and the Svetlyi Bor (Svetloborskiy) complex, Urals, Russia [16].

Table 5. Compositions of the tolovkite–irarsite–hollingworthite solid solutions from the Bolshoy Khailyk deposit.

#	Ir	Rh	Ru	Os	Pt	Au	Fe	Ni	Co	Cu	Sb	As	S	Total
1	57.26	0.79	bdl	bdl	bdl	bdl	0.28	bdl	bdl	bdl	18.37	11.5	10.51	98.71
2	48.25	7.13	bdl	1.92	bdl	bdl	0.4	0.14	bdl	1.57	5.45	21.59	11.67	98.12
3	56.3	3.67	bdl	bdl	bdl	bdl	1.09	bdl	bdl	2.3	bdl	25.87	12.07	101.3
4	17.24	33.19	bdl	bdl	bdl	bdl	bdl	bdl	bdl	1.38	2.44	30.89	14.29	99.43
5	22.37	27.65	bdl	bdl	bdl	bdl	bdl	bdl	bdl	1.38	2.19	29.44	16.0	99.03
6	5.83	39.67	bdl	bdl	2.1	2.39	bdl	bdl	bdl	1.64	1.99	31.14	14.31	99.07
7	7.78	33.32	bdl	bdl	3.02	5.5	bdl	bdl	bdl	3.92	1.58	29.02	15.49	99.63
8	9.57	36.78	bdl	bdl	bdl	2.94	bdl	bdl	bdl	2.51	2.78	30.79	14.32	99.69
9	5.48	39.28	bdl	bdl	1.92	bdl	bdl	bdl	bdl	1.11	1.88	31.71	17.15	98.53
10	3.73	37.35	bdl	bdl	2.54	4.9	bdl	bdl	bdl	3.04	1.49	30.01	16.46	99.52
11	bdl	39.13	bdl	bdl	2.77	6.4	bdl	bdl	bdl	3.92	1.54	30.3	16.07	100.13
12	57.83	bdl	0.23	5.72	bdl	bdl	bdl	bdl	bdl	bdl	8.61	17.31	10.43	100.13
13	49.43	bdl	1.35	bdl	bdl	bdl	bdl	bdl	bdl	bdl	33.92	2.85	11.23	98.78
14	52.84	6.96	bdl	bdl	bdl	bdl	bdl	bdl	bdl	bdl	5.78	20.74	13.93	100.25
15	52.85	6.21	bdl	bdl	bdl	bdl	bdl	bdl	bdl	bdl	5.42	21.08	13.81	99.37
16	19.13	23.53	bdl	bdl	10.01	bdl	bdl	bdl	bdl	bdl	3.79	27.43	15.06	98.95
17	53.52	bdl	bdl	bdl	bdl	bdl	bdl	bdl	bdl	bdl	34.99	0.36	10.38	99.25
18	53.51	bdl	bdl	bdl	bdl	bdl	bdl	0.48	bdl	bdl	34.4	0.97	10.85	100.21
19	20.22	30.72	bdl	bdl	1.28	bdl	bdl	bdl	0.06	1.88	3.53	27.2	13.36	98.25
20	52.96	3.14	bdl	0.81	0.55	bdl	0.32	0.35	0.07	1.53	15.39	13.04	10.5	98.66
21	53.87	2.89	bdl	0.95	0.48	bdl	0.39	0.36	0.13	1.34	11.78	15.35	10.77	98.31

					Atomic proportions (per a total of 3 a.p.f.u.)										
	Ir	Rh	Ru	Os	Pt	Au	Fe	Ni	Co	Cu	Sb	As	S	ΣMe	As + Sb
1	0.95	0.02	0	0	0	0	0.02	0	0	0	0.48	0.49	1.04	0.99	0.97
2	0.71	0.20	0	0.03	0	0	0.02	0.01	0	0.07	0.13	0.81	1.03	1.03	0.94
3	0.79	0.10	0	0	0	0	0.05	0	0	0.10	0	0.94	1.02	1.04	0.94
4	0.21	0.74	0	0	0	0	0	0	0	0.05	0.05	0.94	1.02	0.99	0.99
5	0.27	0.61	0	0	0	0	0	0	0	0.05	0.04	0.90	1.14	0.93	0.94
6	0.07	0.86	0	0	0.02	0.03	0	0	0	0.06	0.04	0.93	1.00	1.04	0.97
7	0.09	0.72	0	0	0.03	0.06	0	0	0	0.14	0.03	0.86	1.07	1.04	0.89
8	0.11	0.80	0	0	0	0.03	0	0	0	0.09	0.05	0.92	1.00	1.03	0.97
9	0.06	0.81	0	0	0.02	0	0	0	0	0.04	0.03	0.90	1.14	0.93	0.93
10	0.04	0.78	0	0	0.03	0.05	0	0	0	0.10	0.03	0.86	1.10	1.01	0.89
11	0.00	0.81	0	0	0.03	0.07	0	0	0	0.13	0.03	0.86	1.07	1.04	0.89
12	0.94	0	0.01	0.09	0	0	0	0	0	0	0.22	0.72	1.02	1.04	0.94
13	0.82	0	0.04	0	0	0	0	0	0	0	0.89	0.12	1.12	0.87	1.01
14	0.75	0.18	0	0	0	0	0	0	0	0	0.13	0.75	1.18	0.93	0.88
15	0.76	0.17	0	0	0	0	0	0	0	0	0.12	0.77	1.18	0.92	0.90
16	0.24	0.55	0	0	0.12	0	0	0	0	0	0.07	0.88	1.13	0.91	0.96
17	0.93	0	0	0	0	0	0	0	0	0	0.96	0.02	1.09	0.93	0.98
18	0.91	0	0	0	0	0	0	0.03	0	0	0.92	0.04	1.10	0.93	0.96
19	0.25	0.72	0	0	0.02	0	0	0	0.002	0.07	0.07	0.87	1.00	1.06	0.94
20	0.85	0.09	0	0.01	0.01	0	0.02	0.02	0.004	0.07	0.39	0.53	1.00	1.07	0.92
21	0.85	0.09	0	0.02	0.01	0	0.02	0.02	0.007	0.06	0.29	0.62	1.02	1.07	0.91

Note. Number 1–18 are results of SEM/EDS analyses, and #19–21 are WDS analyses; "bdl" indicates that amounts of elements are below detection limits.

3.7. Zoned Oxide of PGE–Fe

Herein we describe an unusual grain of zoned oxide (or oxides), which consists of a core-like zone (marked with the letter C: Figure 6a), a rim-like zone (R) and a veinlet (V); these phases are rich in Ru, and, to a lesser extent, in Os and Ir, with substantial Fe and minor V (Table 6). The following observations are noteworthy: (1) The compositions are generally consistent with a simplified formula $Ru_6Fe^{3+}{}_2O_{15}$, in which the atomic Me:O ratio is close to 1:2; this grain could have arisen by desulfurization and oxidation of laurite (zoned), in which the Me:S proportion also is 1:2. If so, Fe and V were introduced externally via the oxidizing fluid. (2) Compositional variations (Table 6) indicate that levels of Os and Ir decreased with a relative increase in Fe and V from core to rim. (3) A further change is expressed inward from the rim, with a decrease in the total PGE with increase in contents of Fe and V; the low total suggests that the veinlet phase likely contains some water or a hydroxyl component. (4) The cavity observed in the center (Figure 6a) could represent the channelway of the oxidizing fluid.

Table 6. Compositions of PGE–Fe oxide phases in a zoned inclusion hosted by an Os–Ir–(Ru) alloy grain from the Bolshoy Khailyk placer deposit.

#		RuO_2	OsO_2	Ir_2O_3	Fe_2O_3	V_2O_5	Total
1	Core	64.89	9.97	6.45	18.35	0.97	100.63
2	Rim	65.95	4.61	4.42	19.82	1.27	96.07
3	Veinlet	50.35	bdl	1.91	24.78	1.95	78.99
Atomic proportions (per 15 oxygen atoms p.f.u.)							
#		Ru	Os	Ir	Fe^{3+}	V	ΣPGE
1		4.94	0.45	0.30	2.33	0.11	5.7
2		5.06	0.21	0.21	2.53	0.14	5.5
3		4.40	0	0.10	3.61	0.25	4.5

Note. Results of SEM/EDS analyses are quoted in weight %; "bdl" stands for not detected.

In various settings worldwide, grains of PGE-based oxides typically display similar features, e.g., [20–23]. These compounds remain poorly understood; in most occurrences, they could well represent cryptic mixtures rather than be single phases. Note that no bonding likely exists between platinum and oxygen in a Pt–Fe oxide grain studied by X-ray absorption spectroscopy [24]. The Pd-based oxide grains appear to be more diverse in their compositions than grains of Ru–Os–Ir-rich oxides. Indeed, the Pd-rich grains contain a variety of elements, e.g., Sb, Bi, Pb, and Tl [25–27], which could well reflect diverse precursor grains.

3.8. Pt–Au–Cu Alloys and Unnamed $(Pt,Pd)_3Cu_2Sn$

Four groups of clustered or individual data-points of Pt–Au–Cu alloys are recognized at Bolshoy Khailyk (#1 to 4 in Figure 10) (Table 7). Field #1 pertains to the phase $PtAu_4Cu_5$ [or Cu(Au,Pt)], the main portion of this grain (Figure 7a). The atomic proportions observed coincide with those of related intermetallic compounds reported from the Tulameen complex, Canada [28], an alluvial deposit at Sotajarvi, Finland [29], cf. [30], from the River Zolotaya [5], and from the Kondyor complex of Russia [31]. Note that the Pt–Au–Cu phases reported from Kondyor form a series of compositions toward CuAu, i.e., tetra-auricupride, first discovered in China [32]. Thus, the analyzed phase at Bolshoy Khailyk could represent a member in the solid-solution series of platiniferous tetra-auricupride, i.e., Cu(Au,Pt).

The compositional field #2 is clustered near the stoichiometry $PtAuCu_2$ (Figure 10). The composition #3 plots near PtCu, i.e., hongshiite [33]. Note that the auriferous hongshiite (Figure 10) is fairly close in composition to the phase reported from an alluvial occurrence at the River Durance, France [30]. The composition #4 plots close to PtAu, which is known from synthesis in nanometer-sized particles [34]. Field #5 likely reflects a subordinate extent of solid solution of Au in a Cu-bearing platinum alloy (Figure 10, Table 7). The likely trend of crystallization of the Pt–Au–Cu alloys is shown in Figure 10. In addition, inclusions of highly pure platinum (100.8 wt % Pt; EMP data) were encountered in this grain of the Pt–Au–Cu alloy.

An unnamed stannide of Pt and Cu with a subordinate content of Pd occurs as small inclusions (≤10 μm across) in the matrix of the Pt–Au–Cu alloy. Compositional data (#11–13, Table 7) indicate a (Pt + Pd + Cu):Sn proportion of 5:1, which points to an unnamed species of PGM. In the system $Pd_3Sn–Cu_3Sn$ [35], several ternary phases are known to form via solid-state transformations at low temperatures ≤550 °C, with a general increase in Cu with a drop in temperature in the system, e.g., Pd_5CuSn_2 [or $(Pd_{2.5}Cu_{0.5})_{\Sigma3.0}Sn$], $Pd_9Cu_3Sn_4$ [$Pd_{2.25}Cu_{0.75})_{\Sigma3.0}Sn$] and Pd_2CuSn, which are likely the synthetic equivalents of "stannopalladinite", taimyrite and cabriite [36], respectively. The Pt-for-Pd substitution is known in the taimyrite series, the Pt-dominant analogue of which is tatyanaite [37,38].

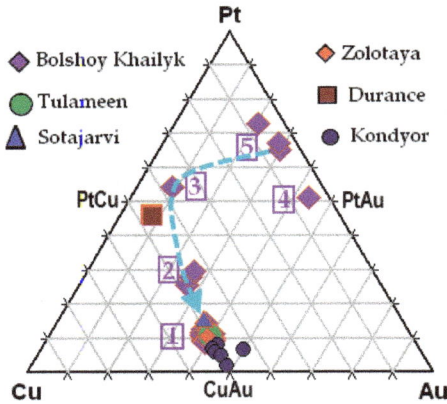

Figure 10. Compositions of alloys of Cu, Pt, and Au from the Bolshoy Khailyk placer (this study) on the diagram Cu–Pt–Au (at %). For comparison, compositions of Cu–Pt–Au phases are plotted from the Zolotaya River placer, western Sayans [5], the Tulameen complex, Canada [28], the Sotajarvi area, Finland [29], the Durance River area, France [30], and the Kondyor complex, Russian Far East [31]. The numbers 1 to 5, shown on this plot, pertain to compositional groups discussed in the text.

Table 7. Compositions of Pt–Au–Cu–(Sn) alloys from the Bolshoy Khailyk deposit.

#	Pt	Pd	Fe	Ni	Cu	Au	Sb	Sn	Total
1	14.66	bdl	bdl	bdl	25.23	59.63	bdl	bdl	99.52
2	16.03	bdl	bdl	bdl	24.1	60.5	bdl	bdl	100.63
3	36.07	bdl	bdl	bdl	22.17	38.14	2.55	0.77	99.7
4	38.41	bdl	1.74	bdl	20.39	36.34	0.91	3.13	100.92
5	39.07	bdl	1.95	bdl	19.08	35.2	1.18	3.82	100.3
6	65.36	bdl	1.05	bdl	14.7	10.85	6.45	1.26	99.67
7	53.17	bdl	bdl	bdl	1.65	46.15	bdl	bdl	100.97
8	75.55	bdl	0.45	bdl	2.26	21.61	bdl	bdl	99.87
9	67.09	bdl	bdl	bdl	1.74	31.02	bdl	bdl	99.85
10	68.51	bdl	bdl	bdl	1.49	29.22	0.46	bdl	99.68
11	58.82	7.69	1.01	0.37	14.64	1.37	bdl	15.23	99.13
12	59.63	7.37	0.99	0.36	14.78	1.39	bdl	15.77	100.29
13	59.66	7.96	1.32	0.43	14.71	0.55	bdl	15.3	99.93
Atomic proportions									
#	Pt	Pd	Fe	Ni	Cu	Au	Sb	Sn	
1	9.7	0	0	0	51.2	39.1	0	0	
2	10.7	0	0	0	49.3	40.0	0	0	
3	24.5	0	0	0	46.2	25.7	2.8	0.9	
4	25.7	0	4.1	0	41.8	24.0	1.0	3.4	
5	26.5	0	4.6	0	39.7	23.6	1.3	4.3	
6	47.6	0	2.7	0	32.9	7.8	7.5	1.5	
7	51.2	0	0	0	4.9	44.0	0	0	
8	71.6	0	1.5	0	6.6	20.3	0	0	
9	65.0	0	0	0	5.2	29.8	0	0	
10	66.7	0	0	0	4.5	28.2	0.7	0	
11	2.37	0.57	0.14	0.05	1.81	0.05	0	1.01	
12	2.38	0.54	0.14	0.05	1.81	0.05	0	1.03	
13	2.37	0.58	0.18	0.06	1.79	0.02	0	1.00	

Note. Analyses # 1–10 are compositions of Pt-Cu-Au alloy phases, which are listed in the order of increasing Pt content. Analyses #1 and 2 correspond to PtCu$_5$Au$_4$ [or Cu(Au,Pt)]. #11–13 pertain to unnamed (Pt,Pd)$_3$Cu$_2$Sn. An.#2 and #11–13 are results of WDS analyses; #1 and #3–10 are results of SEM/EDS; "bdl" indicates that amounts of elements are below detection limits. The atomic proportions are expressed per a total of 100 at %, except for #11–13 recalculated on the basis of a total of 6 a.p.f.u.

3.9. Rh–Co-Rich Pentlandite

It is known that pentlandite incorporates essential Rh or Co (e.g., [8,39–43]). In contrast to other occurrences, compositions of pentlandite at Bolshoy Khailyk display covariations of Co and Rh, both present in significant amounts in solid solution (Table 8). Thus, our results provide useful implications to establish a likely mechanism of the Rh incorporation. A total of 19 grains were analyzed, all of them subhedral to anhedral inclusions (≤20 μm across) hosted by the Pt–Au–Cu alloy (Figure 7a,b). The pair Co–Fe reveals a strongly negative correlation, whereas the pair Co–Ni is uncorrelated (Figure 11a,b). Interestingly, contents of Co display a significant and sympathetic correlation with Rh, a coefficient of correlation R of 0.8 (Figure 11c). The pair Fe–Rh is negatively correlated (R = −0.8). There are no definite correlations involving Ni with Rh or Fe (Figure 11e,f).

Table 8. Compositions of Co–Rh-bearing pentlandite from the Bolshoy Khailyk deposit.

	Fe	Ni	Co	Cu	Rh	Ir	Pt	S	Total
1	13.55	35.60	11.59	0.05	4.54	0.74	bdl	31.38	97.45
2	13.38	35.68	11.61	0.43	4.92	0.76	bdl	31.46	98.24
3	13.46	35.94	11.44	0.43	4.79	0.84	bdl	31.40	98.30
4	20.71	36.48	5.57	0.63	1.23	0.66	bdl	31.85	97.13
5	20.72	36.59	5.70	0.48	1.34	0.61	bdl	31.95	97.39
6	21.28	39.81	3.57	0.44	0.10	bdl	0.34	31.61	97.15
7	15.88	34.02	13.34	0.36	2.6	0.33	bdl	31.67	98.20
8	15.91	34.15	13.27	0.29	2.68	0.31	bdl	31.68	98.29
9	15.71	37.54	10.31	0.49	1.24	1.12	bdl	31.78	98.19
10	15.87	33.67	13.32	0.31	2.88	0.40	bdl	31.64	98.09
11	22.08	30.36	2.56	2.33	bdl	bdl	11.43	31.64	100.40
12	17.56	33.44	11.58	bdl	3.27	bdl	bdl	35.01	100.86
13	17.68	32.90	11.63	bdl	3.34	bdl	bdl	34.49	100.04
14	17.16	34.29	11.8	bdl	1.5	bdl	bdl	35.54	100.29
15	19.27	35.73	8.74	bdl	1.56	bdl	bdl	35.57	100.87
16	23.97	30.32	1.44	1.52	bdl	bdl	10.25	31.99	99.49
17	21.62	36.12	5.89	bdl	1.12	bdl	bdl	35.17	99.92
18	23.75	37.36	3.33	bdl	bdl	bdl	bdl	36.01	100.45
19	17.68	39.53	8.82	bdl	bdl	bdl	bdl	34.07	100.10

Atomic proportions (per a total of 17 a.p.f.u.)									
	Fe	Ni	Co	Cu	Rh	Ir	Pt	ΣMe	S
1	1.99	4.97	1.61	0.01	0.36	0.03	0	8.97	8.03
2	1.95	4.96	1.61	0.06	0.39	0.03	0	9.00	8.00
3	1.97	4.99	1.58	0.06	0.38	0.04	0	9.01	7.99
4	2.99	5.02	0.76	0.08	0.10	0.03	0	8.98	8.02
5	2.99	5.02	0.78	0.06	0.10	0.03	0	8.98	8.02
6	3.06	5.45	0.49	0.06	0.01	0	0.01	9.08	7.92
7	2.29	4.67	1.82	0.05	0.20	0.01	0	9.04	7.96
8	2.29	4.68	1.81	0.04	0.21	0.01	0	9.05	7.95
9	2.26	5.15	1.41	0.06	0.10	0.05	0	9.02	7.98
10	2.29	4.63	1.82	0.04	0.23	0.02	0	9.03	7.97
11	3.30	4.31	0.36	0.31	0	0	0.49	8.77	8.23
12	2.42	4.39	1.52	0	0.25	0	0	8.58	8.42
13	2.47	4.37	1.54	0	0.25	0	0	8.62	8.38
14	2.36	4.48	1.54	0	0.11	0	0	8.49	8.51
15	2.63	4.65	1.13	0	0.12	0	0	8.53	8.47
16	3.57	4.30	0.20	0.20	0	0	0.44	8.70	8.30
17	2.98	4.73	0.77	0	0.08	0	0	8.56	8.44
18	3.23	4.83	0.43	0	0	0	0	8.48	8.52
19	2.44	5.20	1.16	0	0	0	0	8.80	8.20

Note. Number 1–10 are results of WDS analyses. #11–19 are SEM/EDS analyses; "bdl" indicates that amounts of elements are below detection limits.

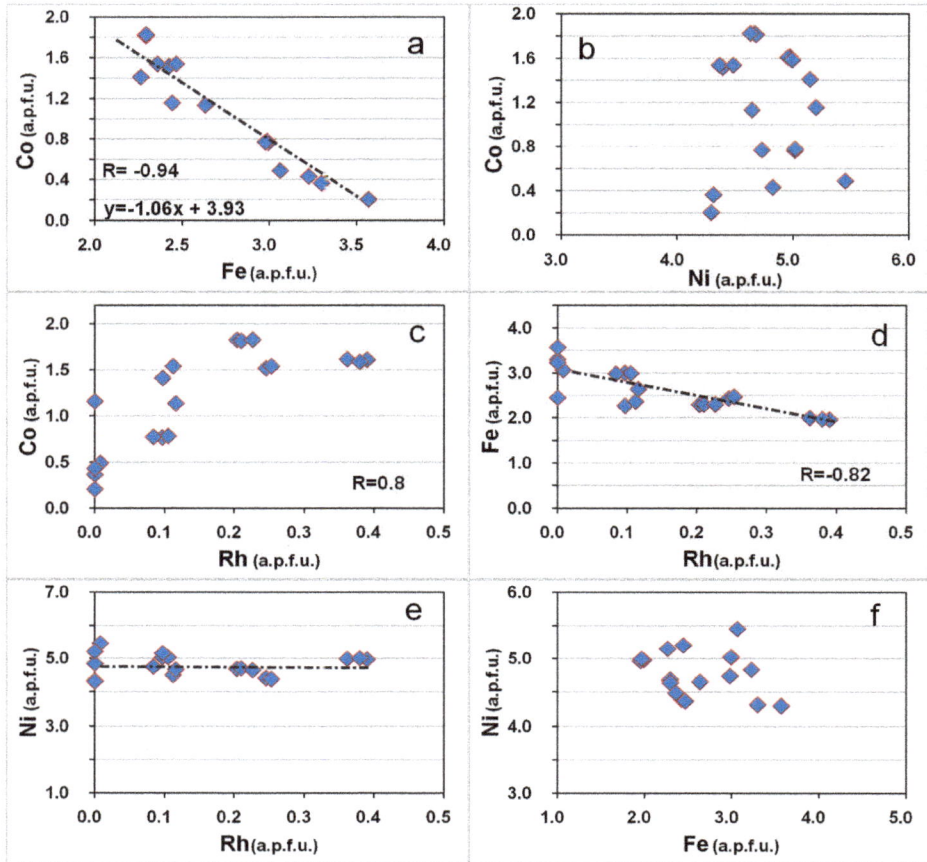

Figure 11. Variations and correlations observed in compositions of Rh–Co-rich pentlandite from the Bolshoy Khailyk placer in terms of binary plots Co vs. Fe (**a**), Co vs. Ni (**b**), Co vs. Rh (**c**), Fe vs. Rh (**d**), Ni vs. Rh (**e**), and Ni vs. Fe (**f**), with values expressed in values of atoms per formula unit, a.p.f.u., calculated for a total of 17 a.p.f.u. Values of the correlation coefficient R are shown, along with an equation of linear regression in (**a**).

3.10. Micrometer-Sized Inclusions in Grains of PGE Alloys

In general, placer grains of Os–Ir–Ru alloys contain inclusions more commonly than the Pt–Fe alloys. The analyzed inclusions of diopside (Table 9) are highly magnesian [$Wo_{48.3-48.6}En_{48.4-48.5}Fs_{2.6}Ae_{0.4-0.7}$; Mg# 96.9–97.9]. Inclusions of chromian spinel correspond to magnesiochromite; these are also highly magnesian, with values of Mg# up to 71. In addition, the chromian spinel displays unusually high values of Cr# (77–83.5: Table 9). Some grains (Figure 4d) possibly represent broken fragments of subhedral spinel from the lode rock.

Table 9. Compositions of inclusions of silicate minerals and chromian spinel hosted by placer grains of PGM from the Bolshoy Khailyk placer deposit.

#		SiO$_2$	TiO$_2$	Al$_2$O$_3$	Cr$_2$O$_3$	FeO (tot)	FeO (calc.)	Fe$_2$O$_3$ (calc.)	MnO	MgO	CaO	Na$_2$O	K$_2$O	Total
1	Cpx	53.3	0.02	1.82	0.84	1.63	0.57	1.18	0.02	17.5	24.3	0.18	bdl	99.61
2		53.4	0.02	1.74	0.81	1.62	0.07	1.72	0.05	17.7	24.7	0.12	bdl	100.16
3	Amp	53.9	0.03	3.62	0.12	9.05	bdl	bdl	0.18	16.8	11.5	0.46	0.05	95.71
4		49.5	0.05	8.43	0.52	9.67	bdl	bdl	0.09	15.1	10.8	0.54	0.29	94.99
5		49.2	0.09	8.85	0.36	8.43	8.17	0.30	0.12	16.6	11.0	0.55	0.03	95.23
6		49.7	0.08	9.85	0.09	8.83	bdl	bdl	0.13	16.5	10.2	0.61	0.04	96.03
7		51.1	0.04	7.67	0.11	8.92	bdl	bdl	0.05	17.9	9.3	0.82	0.18	96.09
8		50.4	0.02	8.4	0.14	7.09	bdl	bdl	0.01	17.9	11.1	0.93	0.22	96.21
9		51.7	bdl	8.5	0.55	9.83	bdl	bdl	bdl	15.6	12.3	bdl	0.31	98.79
10		51.3	bdl	8.2	0.58	10.7	bdl	bdl	bdl	16.4	9.7	bdl	0.28	97.16
11		51.8	bdl	11.5	0.20	5.44	bdl	bdl	bdl	18.3	11.8	0.9	bdl	99.94
12		49.0	bdl	12.2	bdl	4.0	2.34	1.84	bdl	17.3	13.2	bdl	bdl	95.7
13		49.4	bdl	10.9	0.41	4.32	3.00	1.46	bdl	17.8	12.4	bdl	bdl	95.23
14		54.3	bdl	6.63	bdl	8.46	7.57	0.99	bdl	18.5	12.2	bdl	bdl	100.09
15	Srp	45.3	bdl	bdl	bdl	4.79	bdl	bdl	bdl	38.1	bdl	bdl	bdl	88.19
16		46.2	bdl	bdl	bdl	5.03	bdl	bdl	0.32	39.0	bdl	bdl	bdl	90.55
17	Chr	bdl	0.11	7.88	57.6	14.3	10.28	4.47	0.28	13.7	bdl	bdl	bdl	94.32
18		bdl	0.09	7.67	57.9	14.4	10.05	4.83	0.27	13.9	bdl	bdl	bdl	94.71
19		bdl	0.08	7.64	57.5	19.4	17.91	1.66	1.27	8.0	bdl	bdl	bdl	94.06
20		bdl	0.07	7.74	57.8	19.3	18.08	1.36	1.16	8.0	bdl	bdl	bdl	94.21
21		bdl	0.10	7.61	57.4	19.4	17.20	2.44	0.52	9.0	bdl	bdl	bdl	94.27
22		bdl	0.09	8.2	56.6	18.4	15.18	3.58	0.40	10.5	bdl	bdl	bdl	94.55
23		bdl	bdl	10.5	52.1	22.9	20.22	2.98	0.49	7.1	bdl	bdl	bdl	93.39
24		bdl	bdl	8.3	61.4	17.9	15.67	2.48	bdl	11.4	bdl	bdl	bdl	99.25
25		bdl	bdl	8.5	61.3	17.7	15.78	2.13	bdl	11.3	bdl	bdl	bdl	99.01

							Atomic proportions										
	Si	Mg	Fe^{2+}	Ca	Mn	Na	K	Cr	Al	Fe^{3+}	Ti	Mg#	Cr#	Fe^{3+}#
1	1.94	0.95	0.02	0.95	<0.01	0.01	0	0.02	0.08	0.03	<0.01	96.9	–	–
2	1.93	0.95	<0.01	0.96	<0.01	0.01	0	0.02	0.07	0.05	<0.01	97.9	–	–
3	7.75	3.6	1.09	1.77	0.02	0.13	0.01	0.01	0.61	0	<0.01	76.4	–	–
4	7.23	3.29	1.18	1.69	0.01	0.15	0.05	0.06	1.45	0	<0.01	73.4	–	–
5	7.12	3.58	0.99	1.71	0.02	0.15	<0.01	0.04	1.51	0.03	0.01	78	–	–
6	7.11	3.52	1.06	1.56	0.02	0.17	0.01	0.01	1.66	0	0.01	76.5	–	–
7	7.3	3.81	1.07	1.42	0.01	0.23	0.03	0.01	1.29	0	<0.01	77.9	–	–
8	7.18	3.8	0.85	1.7	<0.01	0.26	0.04	0.02	1.41	0	<0.01	81.5	–	–
9	7.25	3.26	1.15	1.85	0	0	0.06	0.06	1.41	0	0	73.9	–	–
10	7.3	3.48	1.27	1.48	0	0	0.05	0.07	1.37	0	0	73.3	–	–
11	7.03	3.7	0.62	1.72	0	0.24	0	0.02	1.84	0	0	85.6	–	–
12	6.89	3.63	0.28	1.99	0	0	0	0	2.02	0.2	0	92.8	–	–
13	6.99	3.76	0.36	1.88	0	0	0	0.05	1.82	0.16	0	91.3	–	–
14	7.42	3.77	0.86	1.79	0	0	0	0	1.07	0.1	0	81.4	–	–
15	2.09	2.62	0.18	0	0.01	0	0	0	0	0	0	93	–	–
16	2.09	2.63	0.19	0	0.01	0	0	0	0	0	0	92.9	–	–
17	0	0.7	0.29	0	0.01	0	0	1.56	0.32	0.12	0.003	69.8	83.1	5.8
18	0	0.71	0.29	0	0.01	0	0	1.56	0.31	0.12	0.002	70.6	83.5	6.2
19	0	0.43	0.54	0	0.04	0	0	1.63	0.32	0.04	0.002	42.6	83.5	2.2
20	0	0.43	0.54	0	0.04	0	0	1.63	0.33	0.04	0.002	42.6	83.4	1.8
21	0	0.48	0.51	0	0.02	0	0	1.61	0.32	0.07	0.003	47.5	83.5	3.3
22	0	0.55	0.44	0	0.01	0	0	1.56	0.34	0.09	0.002	54.6	82.2	4.7
23	0	0.38	0.61	0	0.01	0	0	1.48	0.44	0.08	0	37.9	76.9	4
24	0	0.56	0.44	0	0	0	0	1.61	0.33	0.06	0	56.5	83.2	3.1
25	0	0.56	0.44	0	0	0	0	1.61	0.33	0.05	0	56.1	82.9	2.7

Note. These results of WDS analyses are expressed in weight%; "bdl" indicates that amounts of elements are below detection limits. The label Cpx is clinopyroxene (diopside), Amp is amphibole, Srp is serpentine, and Chr is chromian spinel (chromite-magnesiochromite series). An.#3–14 pertain to amphiboles. An.#3 corresponds to actinolite, #4–6, 8, 9, 11–14 to magnesio-hornblende, and #7, 10 are barroisite. FeO(tot) is all Fe as FeO. FeO(calc.) and Fe$_2$O$_3$ (calc.) are values calculated on the basis of stoichiometry and charge balance. The atomic proportions are based on O = 6 a.p.f.u. for Cpx, O = 23 for Amp, O = 7 for Srp, and O = 4 for chromian spinel. The index Mg# is 100Mg/(Mg + Fe^{2+} + Mn); Cr# is 100Cr/(Cr + Al); and Fe^{3+}# is 100Fe^{3+}/(Fe^{3+} + Cr + Al).

The amphibole inclusions are relatively abundant; they correspond to actinolite, magnesio-hornblende, and barroisite. Some of the inclusions of amphibole have Mg-rich compositions, with values of Mg# up to 93 (Table 9).

Some other inclusions crystallized from much more fractionated silicate melts. Among these are grains of nearly end-member albite (Ab$_{>90}$), hosted by a placer grain of Ru-dominant alloy

[$Ru_{56.1}Os_{39.8}Ir_{4.1}$], and inclusions of quartz (up to ~50 µm) hosted by a grain of a ternary alloy, $Ir_{35.5}Os_{32.6}Ru_{31.9}$. Euhedral grains of quartz were noted in inclusions hosted by Ir-dominant alloy from a placer deposit in British Columbia, Canada [44]. Plagioclase inclusions [$Ab_{58}An_{42}Or_{0.6}$] were found in the Pt–Au–Cu alloy grain at Bolshoy Khailyk. Serpentine (Table 9) and calcite are locally present in inclusions. A micrometric inclusion of grossular(?), with ~0.5% Cr_2O_3 and 3.7% MgO, displays a notable departure from normal stoichiometry, and likely is not single-phase. Compositions of five inclusions of magnetite (Figure 7b; WDS data) are relatively pure, with up to 0.52% MgO, 0.30% Cr_2O_3, and 0.31 wt % MnO.

4. Discussion

4.1. Provenance and Genetic Implications

We have pointed out the close association of the placer deposit with abundant bodies of ophiolitic serpentinite (Figure 1). In addition, we document a high extent of Ru-enrichment in the association of PGM at Bolshoy Khailyk. These characteristics clearly point to an ophiolitic source of the placer PGM, which is presumably related to the Aktovrakskiy complex, belonging to the Kurtushibinskiy belt of the western Sayans. The faceted morphology of the placer grains (Figure 3b,d,f) is consistent with a short distance of transport from the source rocks. The Aktovrakskiy complex was previously inferred to be the primary source for the associated PGM placers at the Zolotaya River, western Sayans [5]. We thus contend that this suite and related complexes of potentially mineralized ophiolites are the targeted PGM-producing source in other areas of the Altai-Sayan folded region.

We document the very high extents of Mg-enrichment in the inclusions of clinopyroxene (Mg# 96.9–97.9) and chromian spinel (Mg# 71) hosted by PGE alloys. In addition, serpentine is present, indicating the former presence of olivine in the system. Some of the amphibole inclusions also are highly magnesian (Mg# > 90), consistent with their crystallization in a primitive ultramafic rock. Therefore, the bulk of the investigated Os–Ir–Ru and Pt–Fe alloy minerals (with abundant exsolution of a Ru-dominant alloy) seem to have crystallized in a high-Mg chromitite or in a dunitic rock rich in magnesiochromite, with a subordinate role of pyroxenitic rocks inferred on the basis of the presence of quartz inclusions hosted by the PGE alloy.

We infer that subhedral grains of Os–Ir–Ru alloys crystallized first among the PGM and after the associated Mg-rich magnesiochromite and olivine, followed by subhedral grains of the isoferroplatinum-type alloy (Figure 3b,f); lamellar phases of Ru-dominant alloy (Figures 3b and 5a) formed largely by exsolution from the matrix phase upon cooling. They reflect a general enrichment in Ru existing in the lode ophiolite, and imply accumulation of levels of Ru during progressive crystallization of the Pt–Fe alloy. The presence of inclusions of amphiboles with lower values of Mg# (73–78; Table 9), and, especially, of Ab-rich plagioclase and quartz, imply that high degrees of fractionation were locally achieved in the ore-forming system.

The inferred incompatible behavior of minor constituents like S, As, Sb, Se, Sn, Au and base-metals (Cu, Ni, Co, Fe) was accompanied during the crystallization of Os–Ir–Ru–Pt alloys by a gradual buildup in the fugacity of S_2 and As_2. Consequently, a variety of sulfide, arsenide, antimonide, and intermetallic phases crystallized at this point, at a lower temperature, possibly at the expense of droplets of residual melt as micrometric inclusions, or as phases coexisting in altered zones near the rims. These implications are consistent with observations made at other placer deposits e.g., [8,18], and references therein. In addition, the porous texture of the Fe-enriched rims (Figure 3e) may reflect the abundance of volatile species; they are similar to rims developed in grains of Ir–Ru–Os alloys from a placer associated with the Trinity ophiolite complex in California [45,46].

The zones of metasomatic modification (Figures 3c and 5b) formed, presumably, as a consequence of (1) the expected buildup in levels of fS_2 and fAs_2, (2) an accumulation of levels of incompatible constituents, such as Cu and Se, in the isolated portions at the margins during the crystallization of the Pt–Fe alloy (the external rim of sperrylite is developed around the grain), and (3) the late deposition of the S–As–Se-bearing assemblage as a result of subsolidus reactions and deuteric alteration of the early formed Pt–Fe alloy phase.

4.2. PGE–Ni–Fe–Cu Alloy, PGE–Fe Oxide and Potential Mineral-Forming Reactions

Corresponding redox reactions likely proceeded at a low temperature and led to the deposition of the alloys of $(Pt,Ir)(Ni,Fe,Cu)_{3-x}$ type and PGE oxide (Figure 6a–c). These alloys occur as veinlets developed in porous and fractured zones at margins of placer grains. We suggest that they crystallized under reducing conditions at the expense of pre-existing pentlandite via a schematic reaction: $[(Ni,Fe)_9S_8 + 3Pt + 8H_2 \rightarrow 3Pt(Ni,Fe)_3 + 8H_2S]$. The Pt component was likely introduced and remobilized by a volatile phase from the associated PGE alloy. The other reactant, hydrogen, was likely a product of the serpentinization of olivine upon oxidation of Fe^{2+} in primary silicates to Fe^{3+} by a reaction involving H_2O ([47], and references therein).

The zoned grain of PGE–Fe oxide could form by combined oxidation and desulfurization of zoned grains of Ru–Os disulfide of the laurite-erlichmanite series. As noted, the observed zonal texture (Figure 6a) is likely a reflection of primary compositional zoning with respect to Ru–Os–(Ir), which is common in grains of laurite–erlichmanite worldwide. The likely schematic reaction at Bolshoy Khailyk is: $[6RuS_2 + Fe_2O_3 + 18O_2 \rightarrow Ru_6Fe_2O_{15} + 12SO_2]$, which involves a hematite component introduced by an oxidizing fluid phase.

4.3. Origin of Au–(Cu–Pt)-Rich Mineralization

The occurrence of Au–(Cu–Pt) alloys, i.e., $PtAu_4Cu_5$, $PtAuCu_2$, $Pt(Cu,Au)$, PtAu, and Au-bearing platinum alloy (Figure 10) appears to be related genetically with the Os–Ir–Ru–Pt mineralization at Bolshoy Khailyk. Thus, the inferred ophiolitic system was likely able to produce locally a high extent of Au-enrichment as a result of geochemical evolution and deposition from a fractionated Au–Cu-rich melt, which likely gained high levels of incompatible Cu + Au with subordinate Pt in a remaining volume during the bulk crystallization of Os–Ir–Ru alloy phases, after the preceding crystallization of chromian spinel (magnesiochromite) and olivine. The latter are presumably the main minerals in the inferred lode rocks (chromitite or magnesiochromite-rich dunite). As noted, PGE alloy grains hosting the quartz inclusions imply a pyroxenitic rock, which could be relevant to some other PGM grains containing inclusions of Ab-rich plagioclase and Fe-rich amphibole.

This suggestion is corroborated by the occurrence of an elevated Au content (up to 7.5 wt %) in the Fe-bearing Ir–Ru–Os alloy phases found associated with the Trinity ophiolite complex, California [46]. In addition, an Au–Ag alloy, ranging up to almost pure gold (Au_{99}), precipitated pseudomorphously by a subsolidus reaction between a fluid phase or a residual Au–Ag-rich melt and exsolution-induced inclusions of the Pt–Fe alloy phases [46].

Also, a residual melt enriched in Au likely appeared during the progressive crystallization of Pt–Fe alloy, leading to the formation of a rim of gold (Au_{80}) around placer grains of Pt–Fe alloy at Florence Creek, Yukon, Canada [48]. We presume that, at Bolshoy Khailyk, this mechanism was likely significant to produce the Au-rich phase precipitated as a narrow rim (Au_{75}) on the Pt–Fe alloy grain (Figure 6d). Note that the gold phase is deposited within the inner portion of the composite rim, being thus separated by the outer rim of cooperite (Figure 6d). Therefore, the gold alloy phase could not be formed by a surficial process in the placer environment.

In addition, the close association observed between Au and PGE is not unusual (e.g., in the Norilsk deposits, Russia; [37,38]; and references therein), because they are geochemically related and belong to the same group of highly siderophile elements, e.g., [49].

4.4. Mechanism of Incorporation of Rh and Co into the Structure of Pentlandite

Occurrences of Rh–Co-enriched grains of pentlandite hosted by the Pt–Au–Cu alloy (Figure 7b) have important crystallochemical implications. We infer, based on our results (Figure 11, Table 8), that Rh and Co display a strong sympathetic relationship. They mutually substitute for Fe, not Ni, and are incorporated into the pentlandite structure via a coupled mechanism of substitution: $[Rh^{3+} + Co^{3+} + \square \rightarrow 3Fe^{2+}]$.

4.5. Se-rich Compounds in Ophiolite-Related Associations, Western Sayans, and Their Implications

We document the occurrence of seleniferous and rhodiferous sperrylite, i.e., $(Pt_{0.54}Rh_{0.45})_{\Sigma0.99}(As_{1.29}Se_{0.40}S_{0.31})_{\Sigma2.00}$, which forms part of the deuterically deposited assemblage (Figure 5b). This phase contains 20 mol % of the sudovikovite component ($PtSe_2$); the latter species of PGM was also formed metasomatically in the Srednyaya Padma mine of the Velikaya Guba U-V deposit, Russian Karelia [50].

The occurrence of Se-rich sperrylite is highly unusual and even puzzling for the case of a highly primitive Os–Ir–Ru-dominant mineralization derived from an ophiolite. This is, in fact, the second occurrence of a PGE selenide in the area. The first report, from the Zolotaya River area [5], noted the presence of unnamed $Ir(As,Se,S)_2$ (~40 mol % $IrSe_2$) and Se-bearing sperrylite (2.35 wt % Se) as inclusions, hosted by an Ir-dominant alloy and located at the margin of a Pt_3Fe-type alloy grain, respectively.

Occurrences of compounds of Se and the iridium subgroup of the PGE (IPGE: Ir, Os, and Ru), or of phases of the PGE selenides in ophiolitic or other primitive ultramafic rocks, must be extremely rare. One such occurrence, i.e., members of the RuS_2 (laurite)—unnamed $RuSe_2$ series, was recently reported from chromitite of the Pados-Tundra ultramafic complex, Kola Peninsula, Russia [51]. Normally, laurite is depleted in Se [52]. Interestingly, all of the Se-rich compounds of PGE recognized presently in association with highly primitive rocks, display an AB_2-type stoichiometry: $Ir(As,Se,S)_2$, $Pt(As,Se)_2$, $(Pt,Rh)(As,Se,S)_2$, $Ru(S,Se)_2$, and $Ru(Se,S)_2$ ([5,51]; this study); thus, their structures are likely optimal to accommodate Se under the given conditions of crystallization.

The anomaly arises because unrealistically low values of S/Se are required to exist in the environment. The S/Se ratio of the mantle is 2850–4350, with a mean of ~3250 [53,54], and references therein. The chondritic value of this ratio is 2500 ± 270 [55]. Interestingly, the lowest values (190–700) were recorded in microglobules of sulfide in the Platinova Reef, Skaergaard layered complex [56]. Thus, a PGE sulfoselenide is unlikely to be stabilized instead of a pure sulfide as a primary magmatic phase in a highly primitive ultramafic rock. Besides, the PGE–Se-rich phases cannot be a consequence of a magmatic contamination, because crustal rocks have high values of S/Se: 3500 to 100,000 [54], and references therein.

At Pados-Tundra [51], the diselenide-disulfoselenide phases of Ru likely reflect a process of ultimate S-loss, causing a critical lowering of the S/Se value during a late-stage evolution of H_2O-bearing fluid involved in the formation of laurite–clinochlore intergrowths. An oxidizing character of this fluid was inferred cf. [57]. Sulfur is more mobile than Se in hydrothermal fluids, and is preferentially incorporated into aqueous fluids e.g., [58,59].

We believe that the occurrences of PGE selenides, derived from the ophiolitic rocks of western Sayans, result from the efficient fractional crystallization of Os–Ir–Ru–Pt–(Fe) alloys in a system that was initially reducing and closed. A buildup in levels of incompatible Se and S is inferred during the progressive crystallization, then followed by a locally effective removal of S, which is more mobile than Se in a fluid-saturated environment associated with the zones of metasomatic alteration, especially where the sulfur is oxidized to sulfate cf. [57].

Author Contributions: The authors (A.Y.B., R.F.M., G.I.S., and S.A.S.) discussed the obtained results and wrote the article together.

Acknowledgments: We are grateful to Liana Pospelova (Novosibirsk) for her expert assistance with the WDS EMP data. We thank two anonymous reviewers and the Editorial Board members, for their comments and improvements. A.Y.B. gratefully acknowledges a partial support of this investigation by the Russian Foundation for Basic Research (project # RFBR 16-05-00884).

Conflicts of Interest: The authors declare no conflicts of interest.

References

1. Vysotskiy, N.K. *Platinum and Areas of Its Mining*; Academy of Sciences of the USSR: Moscow, Russia, 1933; 240p. (In Russian)
2. Porvatov, B.M. The mountainous wealth of Uryankhay (Uriankhai). *Gold Platin.* **1914**, *3*, 64–67. (In Russian)
3. Korovin, M.K. *To the Question of the Existence of Deposits of Gold and Platinum in the Usinskiy Area*; Journal of the Society of Siberian Engineers: Tomsk, Russia, 1915; Volume 7. (In Russian)
4. Krivenko, A.P.; Tolstykh, N.D.; Nesterenko, G.V.; Lazareva, E.V. Types of mineral associations of platinum metals in auriferous placers of the Altai-Sayan folded region. *Russ. Geol. Geophys.* **1994**, *35*, 70–78. (In Russian)
5. Tolstykh, N.D.; Krivenko, A.P.; Pospelova, L.N. Unusual compounds of iridium, osmium and ruthenium with selenium, tellurium and arsenic from placers of the Zolotaya River (western Sayans). *Zap. Vsesoyuzn. Miner. Obshch.* **1997**, *126*, 23–34. (In Russian)
6. Shvedov, G.I.; Nekos, V.V. PGM of a placer at the R. Bolshoy Khailyk (western Sayans). *Geol. Resour. Krasnoyarskiy Kray* **2008**, *9*, 240–248. (In Russian)
7. Aleksandrov, G.P.; Gulyaev, Y.S. *The Geological Map of the USSR (Scale 1:200,000; M-46-III)*; The Western Sayan Series; NEDRA: Moscow, Russia, 1966.
8. Barkov, A.Y.; Shvedov, G.I.; Martin, R.F. PGE–(REE–Ti)-rich micrometer-sized inclusions, mineral associations, compositional variations, and a potential lode source of platinum-group minerals in the Sisim Placer Zone, Eastern Sayans, Russia. *Minerals* **2018**, *8*, 181. [CrossRef]
9. Harris, D.C.; Cabri, L.J. Nomenclature of platinum-group-element alloys: Review and revision. *Can. Mineral.* **1991**, *29*, 231–237.
10. Nash, P.; Singleton, M.F. The Ni-Pt (Nickel-Platinum) system. *Bull. Alloy Phase Diagr.* **1989**, *10*, 258–262. [CrossRef]
11. McDonald, A.M.; Proenza, J.A.; Zaccarini, F.; Rudashevsky, N.S.; Cabri, L.J.; Stanley, C.J.; Rudashevsky, V.N.; Melgarejo, J.C.; Lewis, J.F.; Longo, F.; et al. Garutiite, (Ni,Fe,Ir), a new hexagonal polymorph of native Ni from Loma Peguera, Dominican Republic. *Eur. J. Mineral.* **2010**, *22*, 293–304. [CrossRef]
12. Mochalov, A.G.; Dmitrenko, G.G.; Rudashevskii, N.S.; Zhernovskii, V.; Boldyreva, M.M. Hexaferrum (Fe,Ru), (Fe,Os), (Fe,Ir)—A new mineral. *Zap. Vseross. Miner. Obshch.* **1998**, *127*, 41–51. (In Russian)
13. Cabri, L.J.; Aiglsperger, T. A review of hexaferrum based on new mineralogical data. *Mineral. Mag.* **2018**, 1–16. [CrossRef]
14. Barkov, A.Y.; Martin, R.F.; Halkoaho, T.A.A.; Poirier, G. The mechanism of charge compensation in Cu-Fe-PGE thiospinels from the Penikat layered intrusion, Finland. *Am. Mineral.* **2000**, *85*, 694–697. [CrossRef]
15. Berlincourt, L.E.; Hummel, H.H.; Skinner, B.J. Phases and phase relations of the platinum-group elements. In *Platinum-group elements: Mineralogy, Geology, Recovery*; Cabri, L.J., Ed.; Canadian Institute of Mining, Metallurgy and Petroleum: Westmount, QC, Canada, 1981; Special Volume 23, pp. 21–43.
16. Tolstykh, N.D.; Telegin, Yu.M.; Kozlov, A.P. Platinum mineralization of the Svetloborsky and Kamenushinsky massifs (Urals Platinum Belt). *Russ. Geol. Geophys.* **2011**, *52*, 603–619. [CrossRef]
17. Razin, L.V.; Rudashevskiy, N.S.; Sidorenko, G.A. Tolovkite, IrSbS, a new sulfoantimonide of iridium from the northeastern USSR. *Int. Geol. Rev.* **1982**, *24*, 849–854. [CrossRef]
18. Barkov, A.Y.; Tolstykh, N.D.; Shvedov, G.I.; Martin, R.F. Ophiolite-related associations of platinum-group minerals at Rudnaya, western Sayans, and Miass, southern Urals, Russia. *Mineral. Mag.* **2018**. [CrossRef]
19. Cabri, L.J.; Laflamme, J.G.; Stewart, J.M. Platinum-group minerals from Onverwacht; II, Platarsite, a new sulfarsenide of platinum. *Can. Mineral.* **1977**, *15*, 385–388.
20. Augé, T.; Legendre, O. Platinum-group element oxides from the Pirogues ophiolitic mineralization, New Caledonia; origin and significance. *Econ. Geol.* **1994**, *89*, 1454–1468. [CrossRef]
21. Melcher, F.; Oberthür, T.; Lodziak, J. Modification of detrital platinum-group minerals from the eastern Bushveld complex, South Africa. *Can. Mineral.* **2005**, *43*, 1711–1734. [CrossRef]
22. Oberthür, T.; Melcher, F.; Gast, L.; Wöhrl, C.; Lodziak, J. Detrital platinum-group minerals in rivers draining the eastern Bushveld complex, South Africa. *Can. Mineral.* **2004**, *42*, 563–582. [CrossRef]
23. Oberthür, T.; Weiser, W.; Melcher, F. Alluvial and eluvial platinum-group minerals from the Bushveld complex, South Africa. *S. Afr. J. Geol.* **2014**, *117*, 255–274. [CrossRef]

24. Hattori, K.H.; Takahashi, Y.; Augé, T. Mineralogy and origin of oxygen-bearing platinum-iron grains based on an X-ray absorption spectroscopy study. *Am. Mineral.* **2010**, *95*, 622–630. [CrossRef]

25. Barkov, A.Y.; Shvedov, G.I.; Polonyankin, A.A.; Martin, R.F. New and unusual Pd-Tl-bearing mineralization in the Anomal'nyi deposit, Kondyor concentrically zoned complex, northern Khabarovskiy kray, Russia. *Mineral. Mag.* **2017**, *81*, 679–688. [CrossRef]

26. Barkov, A.Y.; Fleet, M.E.; Martin, R.F.; Halkoaho, T.A.A. New data on "bonanza"-type PGE mineralization in the Kirakkajuppura PGE deposit, Penikat layered complex, Finland. *Can. Mineral.* **2005**, *43*, 1663–1686. [CrossRef]

27. Tolstykh, N.D.; Krivenko, A.P.; Lavrent'ev, Y.G.; Tolstykh, O.N.; Korolyuk, V.N. Oxides of the Pd-Sb-Bi system from the Chiney massif (Aldan Shield, Russia). *Eur. J. Miner.* **2000**, *12*, 431–440. [CrossRef]

28. Cabri, L.J.; Laflamme, J.H.G. Analyses of minerals containing platinum-group elements. In *Platinum-group elements: Mineralogy, Geology, Recovery*; Cabri, L.J., Ed.; Canadian Institute of Mining, Metallurgy and Petroleum: Westmount, QC, Canada, 1981; Special Volume 23, pp. 151–173.

29. Törnroos, R.; Vuorelainen, Y. Platinum-group metals and their alloys in nuggets from alluvial deposits in Finnish Lapland. *Lithos* **1987**, *20*, 491–500. [CrossRef]

30. Johan, Z.; Ohnenstetter, M.; Fischer, M.; Amosse, J. Platinum-group minerals from the Durance River alluvium, France. *Mineral. Petrol.* **1990**, *42*, 287–306. [CrossRef]

31. Nekrasov, I.Y.; Lennikov, A.M.; Zalishchak, B.L.; Oktyabrsky, R.A.; Ivanov, V.V.; Sapin, V.I.; Taskaev, V.I. Compositional variations in platinum-group minerals and gold, Konder alkaline-ultrabasic massif, Aldan Shield, Russia. *Can. Mineral.* **2005**, *43*, 637–654. [CrossRef]

32. Chen, K.; Yu, T.; Zhang, Y.; Peng, Z. Tetra-auricupride, CuAu, discovered in China. *Sci. Geol. Sin.* **1982**. *17*, 111–116. (In Chinese, English Abstract)

33. Yu, Z. New data for daomanite and hongshiite. *Acta Geol. Sin.* **2001**, *75*, 458–466.

34. Shah, A.; Latif-Ur-Rahman, L.-U.; Qureshi, R.; Zia-Ur-Rehman, Z.-U. Synthesis, characterization and applications of bimetallic (Au-Ag, Au-Pt, Au-Ru) alloy nanoparticles. *Rev. Adv. Mater. Sci.* **2012**, *30*, 133–149.

35. Evstigneeva, T.L.; Nekrasov, I.Y. Conditions of the formation of tin-bearing platinum-group minerals in the system Pd-Cu-Sn and its partial cross sections. In *Tin in Magmatic and Postmagmatic Processes*; Nekrasov, I.Y., Ed.; Nauka Press: Moscow, Russia, 1984; pp. 143–169. (In Russian)

36. Evstigneeva, T.L.; Genkin, A.D. Cabriite Pd$_2$SnCu, a new species in mineral group of palladium, tin and copper compounds. *Can. Mineral.* **1983**, *21*, 481–487.

37. Barkov, A.Y.; Martin, R.F.; Poirier, G.; Tarkian, M.; Pakhomovskii, Y.A.; Men'shikov, Y.P. Tatyanaite, a new platinum-group mineral, the Pt analogue of taimyrite, from the Noril'sk complex (northern Siberia, Russia). *Eur. J. Mineral.* **2000**, *12*, 391–396. [CrossRef]

38. Barkov, A.Y.; Martin, R.F. Compositional variations in natural intermetallic compounds of Pd, Pt, Cu, and Sn: New data and implications. *Can. Mineral.* **2016**, *54*, 453–460. [CrossRef]

39. Makovicky, E.; Makovicky, M.; Rose-Hansen, J. The system Fe-Rh-S at 900° and 500 °C. *Can. Mineral.* **2002**, *40*, 519–526. [CrossRef]

40. Makovicky, M.; Makovicky, E.; Rose-Hansen, J. Experimental studies on the solubility and distribution of platinum group elements in base metal sulfides in platinum deposits. In *Metallogeny of Basic and Ultrabasic Rocks*; Gallagher, M.J., Ixer, R.A., Neary, C.R., Prichard, H.M., Eds.; Institution of Mining and Metallurgy: London, UK, 1986; pp. 415–425.

41. Genkin, A.D.; Laputina, I.P.; Muravitskaya, G.N. Ruthenium- and rhodium-containing pentlandite—An indicator of hydrothermal mobilization of platinum metals. *Int. Geol. Rev.* **1976**, *18*, 723–728. [CrossRef]

42. Merkle, R.K.W.; von Gruenewaldt, G. Compositional variation of Co-rich pentlandite: Relation to the evolution of the upper zone of the western Bushveld complex, South Africa. *Can. Mineral.* **1986**, *24*, 529–546.

43. Junge, M.; Wirth, R.; Oberthür, T.; Melcher, F.; Schreiber, A. Mineralogical siting of platinum-group elements in pentlandite from the Bushveld Complex, South Africa. *Miner. Depos.* **2015**, *50*, 41–54. [CrossRef]

44. Barkov, A.Y.; Fleet, M.E.; Nixon, G.T.; Levson, V.M. Platinum-group minerals from five placer deposits in British Columbia, Canada. *Can. Mineral.* **2005**, *43*, 1687–1710. [CrossRef]

45. Barkov, A.Y.; Fleet, M.E.; Martin, R.F.; Feinglos, M.N.; Cannon, B. Unique W-rich alloy of Os and Ir and associated Fe-rich alloy of Os, Ru, and Ir from California. *Am. Mineral.* **2006**, *91*, 191–195. [CrossRef]

46. Barkov, A.Y.; Martin, R.F.; Shi, L.; Feinglos, M.N. New data on PGE alloy minerals from a very old collection (probably 1890s), California. *Am. Mineral.* **2008**, *93*, 1574–1580. [CrossRef]
47. Klein, F.; Bach, W. Fe-Ni-Co-O-S phase relations in peridotite-seawater interactions. *J. Petrol.* **2009**, *50*, 37–59. [CrossRef]
48. Barkov, A.Y.; Martin, R.F.; LeBarge, W.; Fedortchouk, Y. Grains of Pt-Fe alloy and inclusions in a Pt-Fe alloy from Florence creek, Yukon, Canada: Evidence for mobility of Os in a Na-H_2O-Cl-rich fluid. *Can. Mineral.* **2008**, *46*, 343–360. [CrossRef]
49. O'Driscoll, B.; González-Jiménez, J.M. Petrogenesis of the platinum-group minerals. *Rev. Mineral. Geochem.* **2015**, *81*, 489–578. [CrossRef]
50. Polekhovskiy, Y.S.; Tarasova, I.P.; Nesterov, A.P.; Pakhomovskiy, Y.A.; Bakhchisaraitsev, A.Y. Sudovikovite $PtSe_2$–a new platinum selenide from a metasomic rock of southern Karelia. *Dokl. Akad. Nauk* **1997**, *354*, 82–85. (In Russian)
51. Barkov, A.Y.; Nikiforov, A.A.; Tolstykh, N.D.; Shvedov, G.I.; Korolyuk, V.N. Compounds of Ru–Se–S, alloys of Os–Ir, framboidal Ru nanophases and laurite–clinochlore intergrowths in the Pados-Tundra complex, Kola Peninsula, Russia. *Eur. J. Mineral.* **2017**, *29*, 613–622. [CrossRef]
52. Hattori, K.H.; Cabri, L.J.; Johanson, B.; Zientek, M.L. Origin of placer laurite from Borneo: Se and As contents, and S isotopic compositions. *Mineral. Mag.* **2004**, *68*, 353–368. [CrossRef]
53. Lorand, J.-P.; Alard, O.; Luguet, A.; Keays, R.R. Sulfur and selenium systematics of the subcontinental lithospheric mantle: Inferences from the Massif Central xenolith suite (France). *Geochim. Cosmochim. Acta* **2003**, *67*, 4137–4151. [CrossRef]
54. Smith, J.W.; Holwell, D.A.; McDonald, I.; Boyce, A.J. The application of S isotopes and S/Se ratios in determining ore-forming processes of magmatic Ni–Cu–PGE sulfide deposits: A cautionary case study from the northern Bushveld Complex. *Ore Geol. Rev.* **2016**, *73*, 148–174. [CrossRef]
55. Dreibus, G.; Palme, H.; Spettek, B.; Zipfel, J.; Wänke, H. Sulfur and selenium in chondritic meteorites. *Meteoritics* **1995**, *30*, 439–445. [CrossRef]
56. Holwell, D.A.; Keays, R.R.; McDonald, I.; Williams, M.R. Extreme enrichment of Se, Te, PGE and Au in Cu sulfide microdroplets: Evidence from LA-ICP-MS analysis of sulfides in the Skaergaard intrusion, east Greenland. *Contrib. Mineral. Petrol.* **2015**, *170*, 53. [CrossRef]
57. Simon, G.; Kesler, S.E.; Essene, E.J. Phase relations among selenides, tellurides, and oxides: II. Applications to selenide-bearing ore deposits. *Econ. Geol.* **1997**, *92*, 468–484. [CrossRef]
58. Ewers, G.R. Experimental hot water-rock interactions and their significance to natural hydrothermal systems in New Zealand. *Geochim. Cosmochim. Acta* **1977**, *41*, 143–150. [CrossRef]
59. Howard, J.H. Geochemistry of selenium: Formation of ferroselite and selenium behavior in the vicinity of oxidizing sulfide and uranium deposits. *Geochim. Cosmochim. Acta* **1977**, *41*, 1665–1678. [CrossRef]

Article

Micrometric Inclusions in Platinum-Group Minerals from Gornaya Shoria, Southern Siberia, Russia: Problems and Genetic Significance

Gleb V. Nesterenko [1], Sergey M. Zhmodik [1,2,*], Dmitriy K. Belyanin [1,2], Evgeniya V. Airiyants [1] and Nikolay S. Karmanov [1]

[1] Sobolev Institute of Geology and Mineralogy, Siberian Branch of the Russian Academy of Sciences, pr. Akademika Koptyuga 3, 630090 Novosibirsk, Russia; nesterenko@igm.nsc.ru (G.V.N.); bel@igm.nsc.ru (D.K.B.); jenny@igm.nsc.ru (E.V.A.); krm@igm.nsc.ru (N.S.K.)

[2] Faculty of Geology and Geophysics, Novosibirsk State University, 630090 Novosibirsk, Russia

* Correspondence: zhmodik@igm.nsc.ru; Tel.: +7-913-891-22-57

Received: 11 March 2019; Accepted: 22 May 2019; Published: 27 May 2019

Abstract: Micrometric inclusions in platinum-group minerals (PGMs) from alluvial placers carry considerable information about types of primary rocks and ores, as well as conditions of their formation and alteration. In the present contribution, we attempt to show, with concrete examples, the significance of the data on the composition and morphology of micrometric inclusions to genetic interpretations. The PGM grains from alluvial placers of the Gornaya Shoria region (Siberia, Russia) were used as the subject of our investigation. In order to determine the chemical composition of such ultrafine inclusions, high-resolution analytical methods are needed. We compare the results acquired by wavelength-dispersive spectrometry (WDS; electron microprobe) and energy-dispersive spectrometry (EDS) and scanning electron microscopy (SEM) methods. The results obtained have good convergence. The EDS method is multi-elemental and more effective for mineral diagnostics in comparison with WDS, which is its certain advantage. The possible conditions for the formation of inclusions and layers of gold, sulfoarsenides and arsenides in Pt_3Fe grains, which have an original sub-graphic and layered texture pattern, are discussed. They are the result of solid solution and eutectic decompositions and are associated with the magmatic stages of grain transformation, including the result of the interaction of Pt_3Fe with a sulfide melt enriched with Te and As.

Keywords: platinum-group elements; gold; platinum-group minerals; placer deposits; micrometric inclusions; Gornaya Shoria; Siberia; Russia

1. Introduction

The majority of investigations of the mineralogical and geochemical peculiarities of the platinum-group elements (PGEs) are faced with the problem of determining the composition of micrometric inclusions [1–4]. In the present contribution, we attempt to show, with concrete examples, the significance of the data on the chemical composition and morphology of micrometric and close to nanometric inclusions to elucidate conditions of the formation of the platinum-group minerals (PGMs) and mineral associations related to them. Determination of the chemical composition of such ultrafine inclusions requires application of high-resolution analytical methods. Scanning electron microscopy with energy-dispersive spectrometry (SEM-EDS) satisfies these requirements fully. When determining the composition of micro-inclusions whose size is smaller than the area of X-ray radiation generation, the composition of the surrounding mineral ("host mineral") exerts an influence. The impact of this effect may vary significantly. The accounting, appraisal, and removal of this impact constitute separate problems in extracting a "useful" signal against a background of an "interfering" one, or, in other words, ascertainment of the most realistic ("separated") composition of the mineral inclusions. In some cases, this problem is simple and can be solved with the help of modern tools. In other cases, it is difficult due to the superposition of analytical signals from elements of included and including mineral phases. We demonstrate present-day possibilities for analysis of micro-inclusions in PGM grains, and interpret the results of such analysis. Grains composed of platinum-group minerals were used as the subject of our investigation. The PGM grains originate in alluvial gold placers of Gornaya Shoria (Western Siberia) (Figure 1a–c). Previously, these objects were not investigated using high-spot resolution of the SEM-EDS analysis.

With the development of local methods of analysis (electron-microprobe (EMP), SEM, Secondary-Ion Mass Spectrometry (SIMS)), the number of papers in which great attention is paid to the study of micro- and nano-inclusions in platinum-group minerals has increased significantly. In many cases, data on the composition and morphology of inclusions and on metasomatic changes allow us to draw conclusions about the sources of PGEs and the processes of their transformation [2,5–11].

2. Geological Setting

The study area, in the Kuznetsk Alatau Ridge, Gornaya Shoria, and Salair Ridge, represent the western part of the Altai-Sayan folded area (ASFA) (Figure 1) formed during the Caledonian and Hercynian orogenies. Gornaya Shoria is represent a mountain region where the ranges of the North-Eastern Altai, Kuznetsk Alatau Ridges and Salair mountain range converge into a complex knot. Variations in the structure and lithology were controlled by geodynamic processes that led to the formation of oceanic and island-arc complexes, subsequent collision during accretion of the Siberian craton, and protracted plume-related magmatism over the Neoproterozoic–Mesozoic interval [12–16].

Little published information on the primary PGE mineralization in the region is available, and its potential bedrock source is still unknown. The geology of the region is dominated by complexes considered to have potential for PGE mineralization (e.g., Neoproterozoic–Lower Cambrian ophiolite complexes that have variably originated in mid-ocean ridge, back-arc, oceanic island, and island arc settings, or Lower-Middle Paleozoic bimodal volcanic complexes) [17–23]. The compositions of the placer PGM grains from this region were used to recognize a series of Uralian-Alaskan-type mafic–ultramafic complexes [24]. The Kaigadat massif, in the northwestern part of Kuznetsk Alatau (Figure 1b), was classified as a Uralian-Alaskan-type zoned mafic–ultramafic intrusion on the basis of its bulk composition and the widespread occurrence of Pt–Fe minerals as the dominant PGMs in the alluvium of nearby rivers and streams [25]. Although the ophiolitic nature of the Srednyaya Ters' massif (Figure 1b) has long been recognized [21,26,27], this has been disputed by some investigators, who argue that this massif represents a layered intrusion. The data of A.E. Izokh [28] show that dunites of the Srednyaya Ters' massif have high contents of Pd (up to 1 ppm) and Pt (up to 0.6 ppm) (atomic absorption spectrometry). The dunites are relatively enriched in disseminated sulfides and PGMs, represented by a wide variety of Pt and Pd compounds with Sb, As, and Te. Low-grade PGE mineralization (Ru-Ir-Os alloys) was found in serpentinites from the Seglebir massif of Gornaya Shoria [29] and rodingites from the Togul-Sungai massif of the Central Salair Ridge [30,31]. In addition, high Pt and Pd values were identified in early Cambrian chromite-rich ultramafic rocks, several layered peridotite–gabbro massifs, and carbonaceous schists of some Late Riphean, Late Vendian and Early Cambrian complexes of the Kuznetsk Alatau Ridge, Gornaya Shoria, and Salair Ridge [19]. Small mafic intrusions and dikes were also regarded by some investigators as the most probable source of the PGEs in placer deposits. Gold mineralization has long been considered to be genetically related to the dikes of the Middle-Upper Cambrian gabbro–diorite–diabase complex [32].

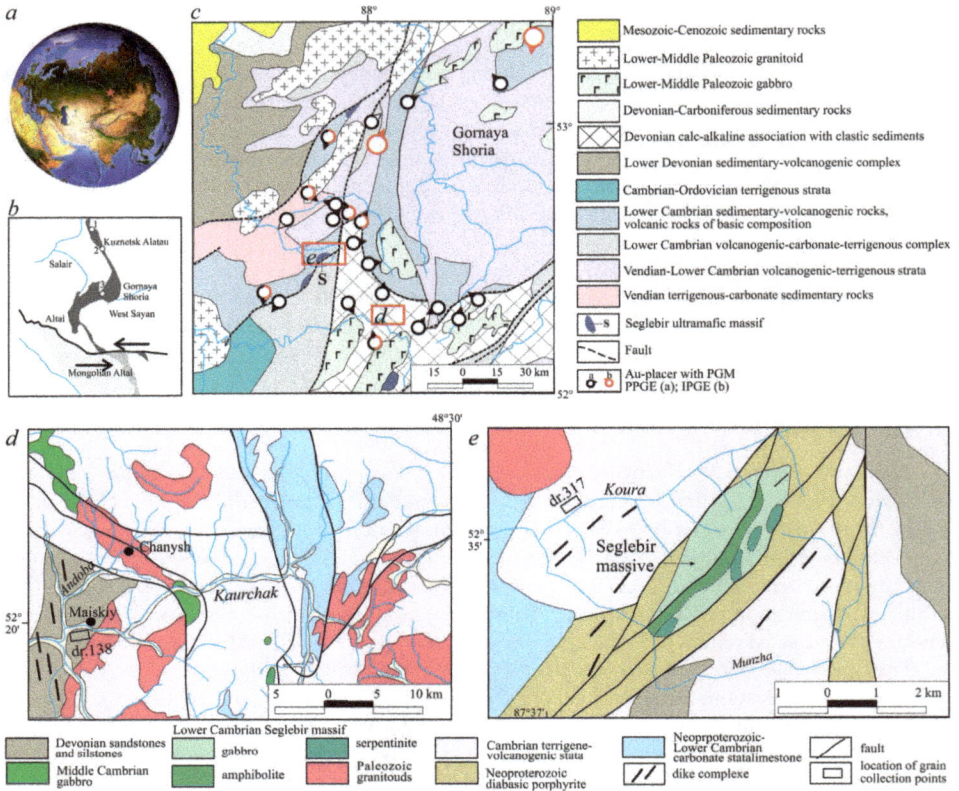

Figure 1. (a) Location of the study area; (b) the Kuznetsk Alatau–Gornaya Ghoria–Altai platinum-bearing belt is indicated by black color in the inset (modified from [20]): 1, Kaigadat massif; 2, Srednyaya Ters' massif; 3, Seglebir massif and placers of Gornaya Shoria; (c) geological map showing the location of PGM and gold placers in the western Altai-Sayan folded area (Gornaya Shoria); (d,e) geological maps of catchment areas r. Kaurchak (d) and r. Koura (e).

Most alluvial gold–PGM placer occurrences are related to Quaternary sediments [33,34]. Some of these occurrences were re-explored and revived for exploitation. The irregular distribution of placers within the study area is largely controlled by bedrock sources and geomorphology. Most placers are typically found at medium altitudes in river valleys, formed by erosional and depositional processes. A few placers occur at lower elevations, and they are virtually absent in high-altitude areas. No published information on the PGM content of most gold placer occurrences is available, but PGMs are generally present in low-grade placers, ranging from 0.03%–0.05% to a few percent of native gold (from 0.5–10.0 mg/m^3 to 500–800 mg/m^3 of rock). At some localities; however, the proportion of PGM makes up as much as 10–30 vol.% of the particles of gold [35–38].

One of the sources of the platinum-group minerals (PGMs) in the alluvial placers of Gornaya Shoria are the basic–ultrabasic massifs of the Seglebir complex, which belong to the Moskovkinsk group of lower Cambrian stratiform intrusions of peridotite–pyroxenite–gabbro [19]. The largest basic–ultrabasic Seglebir massif measures from 0.5 × 1.5 to 3 × 12 km, and extends in a northeasterly direction along the fault zone (Figure 1c).

The main source of PGM in the alluvial sediments of the River Koura and its tributaries, at the sampling site, should be considered rocks of the Seglebir massif (Figure 1e). The massif is composed of massive fine- to medium-grained gabbros, clinopyroxenites, diorites, and antigorite- and chrysotile-bearing serpentinites. Gabbro are characterized by average TiO_2 contents (0.75–1.07 wt.%). Dikes of the basic composition are common within the massif and in the enclosing strata. Nickel and copper mineralization are localized in serpentinized rocks of the Seglebir massif. Several models/interpretations are accepted in the literature about the formation of the Seglebir massif.

The source of the platinum-group minerals in the alluvial placers of the River Kaurchak and its tributaries can be considered rocks of the basic composition, which are attributed to the Middle Cambrian gabbro–diorite complex [19,39,40] (Figure 1d). The gabbro–diorite massifs are located upstream from the sampling site. The areas of intrusions do not exceed a few square kilometers. The arrays are composed of gabbros, olivine gabbros, gabbro-norites, hornblendites, and clinopyroxenites. Moderate to very high levels of TiO_2 and high P_2O_5 characterizes clinopyroxenites and gabbros. Portions of the complex are promising for ilmenite – Ti-magnetite mineralization.

3. Samples and Methods

All samples were taken using gold dredgers. The material was washed at the gravity contents plant using sluice boxes and wash pans to obtain concentrates. The PGMs were extracted from the concentrates after they were panned to recover gold grains. The final volume of the concentrate was 5–10 dm^3. The concentrate consisted of a heavy black sand. Substantial quantities of native gold and PGMs (the degree of enrichment ranging from 5 to 100,000 times) were present in the concentrate sample. Large-scale bulk sampling was employed to obtain an initial sand volume ranging from hundreds to a few thousand cubic meters. The sampling of the heavy-mineral concentrate was conducted at several placer deposits. The initial sand volume was 50–400 dm^3. The final treatment of all samples comprised hand-panning using a stepwise procedure [41] to minimize loss of precious metals. The PGM grains were hand-picked from the final concentrates under a binocular microscope and then examined for grain size, morphology, and surface texture. Selected PGM grains were mounted in epoxy blocks and polished with a diamond paste for further analysis. Microtextural observations of PGM were performed by means of reflected-light microscopy with a Zeiss AxioScope A1 microscope (Carl Zeiss MicroImaging GmbH, Germany, www.zeiss.de).

The composition and morphology of the PGM grains were investigated using a MIRA 3 LMU (Tescan Orsay Holding, Brno, Czech Republic) scanning electron microscope with an attached INCA Energy 450 XMax 80 (Oxford Instruments Nanoanalysis, Wycombe, UK) microanalysis energy-dispersive system at the X-ray Laboratory of the Institute of Geology and Mineralogy, Siberian Branch, Russian Academy of Sciences (analysts N.S. Karmanov, M.V. Khlestov). We employed an accelerating voltage of 20 kV, a beam current of 1600 pA, an energy resolution (MIRA) of 126–127 eV at the Mn $K\alpha$ line, and a region (3–5 μm), depending on the average atomic number of the sample and the wavelength of the analytical line.

The live time of spectrum acquisition was 30 s; in some cases, it reached 150 s. The standards used were FeS_2 (S), $PtAs_2$ (As), HgTe (Hg), PbTe (Pb and Te), and pure metals (Fe, Co, Ni, Cu, Ru, Rh, Pd, Ag, Sb, Os, Ir, Pt, and Au). For the analytical signal of S, Fe, Co, Ni, and Cu, the K-family of radiation was used, and for the remaining elements, the L family. The use of the L family for Os, Ir, Pt, Au, Hg, and Bi avoids the mutual overlaps of M families of these elements. Minimum detection limits (3σ criterion) of the elements (wt.%) were found to be 0.1–0.2 for S, Fe, Co, Ni, Cu; 0.2–0.4 for As, Ru, Rh, Pd, Ag, Sb, Te; 0.4–0.7 for Os, Ir, Pt, Au, Hg, Bi. The analytical error for the main components did not exceed 1–2 relative % and satisfied the requirements for quantitative analysis. Energy-dispersion spectrometry at the elementary conditions of the analysis is a quantitative method, as shown by Newbury et al. [42,43].

Lavrent´ev and co-workers [44] showed, on the same suite of rocks as in the present work, good reproducibility between EDS and wavelength-dispersive spectrometry (WDS) analyses of a large number of garnet, pyroxene, olivine, spinel, and ilmenite grains. We performed an additional test by analyzing some PGM grains on a Camebax Micro microanalyzer using WDS. The analytical conditions were accelerating voltage 20 kV, 20–30 nA beam current, beam size <2 μm, and 10 s counting time. The following X-ray lines and standards were used: PtLα, IrLα, OsMα, PdLα, RhLα, RuLα, AgLα, AuLα (pure metals), AsLα (synthetic InAs), SbLα (synthetic CuSbS$_2$), SKα, FeKα, CuKα (synthetic CuFeS$_2$), NiKα, CoKα (synthetic FeNiCo), BiMα (synthetic Bi$_2$Se$_3$). Element interference was corrected using experimentally measured coefficients [45]. Detection limits of the elements (wt.%) were 0.17 for Pt, 0.15 for Ir; 0.04 for Os, 0.04 for Pd, 0.04 for Rh, 0.04 for Ru, 0.03 for Fe, 0.06 for Cu, 0.07 for Ni, 0.05 for Co, 0.02 for S, 0.05 for As, and 0.06 for Sb.

Analysis by WDS (EMP) and EDS (SEM) methods was performed on grains from the same sites (Table 1). The coincidence of the results obtained is quite satisfactory (Table 1), especially if we take into account the possible heterogeneity of the grains under study, as well as the incomplete coincidence of the position of the analyzed points. Our findings suggest that the SEM-EDS method provides quantitative data in the study of the composition of the PGMs. It should be noted that identification of sample heterogeneity and high-spot resolution of the EDS analysis make it preferable over WDS for analysis of assemblages and aggregates (e.g., PGM exsolutions) in the nanometer-size range. In these cases, the microprobe data can be used to characterize bulk compositions of these nanoscale polymetallic aggregates.

Table 1. Comparison of sample results of WDS (microprobe) and EDS (SEM) of PGM analyzes of rivers Koura and Kaurchak (Gornaya Shoria, southern Siberia, Russia) (wt.%).

Placer	Grain	Analys	Pt	Ir	Os	Pd	Rh	Ru	Cu	Ni	Fe	Total
Koura	82-44	WDS	11.48	16.61	45.09	0.0	0.75	25.42	0.13	0.0	0.48	99.97
Koura	82-44	EDS	10.3	19.2	43.8	0.0	1.0	25.6	0.0	0.0	0.5	100.4
Koura	82-44	EDS	9.1	18.2	45.2	0.0	1.3	26.2	0.0	0.0	0.3	100.3
Kaurchak	66-12	WDS	87.01	3.92	0.91	0.44	1.0	0.43	0.32	0.16	6.29	100.39
Kaurchak	66-12	EDS	85.8	3.8	1.2	0.0	0.8	0.0	0.3	0.0	6.3	98.2
Kaurchak	66-12	EDS	88.2	3.9	0.0	0.5	1.0	0.5	0.3	0.0	6.2	100.6
Koura	82-16	WDS	90.77	0.20	0.93		0.04	0.27	0.92	0.1	6.94	100.18
Koura	82-16	EDS	87.0		1.0				0.9		6.4	98.2
Koura	82-16	EDS	89.7		0.8				0.8		6.4	97.7
Koura	82-32	WDS	83.46	1.08	0.25	0.21	1.3	0.19	0.59	0.1	11.69	98.89
Koura	82-32	EDS	83.1	1.4			1.2				11.6	97.4
Koura	82-32	EDS	82.3	0.8			1.7				10.7	95.5
Koura	82-32	EDS	82.4	0.6			1.3				11.1	95.1
Kaurchak	66-4	WDS	89.73		0.68	0.37	0.07	0.25	0.16	0.1	9.05	100.48
Kaurchak	66-4	EDS	89.7		0.5						9.1	98.8
Kaurchak	66-4	EDS	91.6								8.6	100.3
Koura	82-23	WDS	88.13					0.74	1.08		7.83	97.83
Koura	82-23	EDS	88.6					0.5	1.2		7.2	97.5
Koura	82-23	EDS	90.5					0.3	1.0		7.4	99.2
Koura	82-23	EDS	89.6					0.6	1.1		7.1	98.4
Koura	82-35	WDS	87.65	1.39	0.93			0.19	3.74		5.45	99.35
Koura	82-35	EDS	87.9	1.9					3.7		5.3	98.8
Koura	82-35	WDS	74.95	2.1	2.7			0.62	17.95		0.27	98.64
Koura	82-35	EDS	76.9	1.7	1.3				20.0		0.2	100.1
Koura	82-86	WDS	71.30	14.48	3.74		0.15	0.36	0.49		6.9	97.44
Koura	82-86	EDS	71.1	19.1	3.4			0.5	0.5		6.2	100.8
Koura	82-86	EDS	72.9	14.9	2.5			0.6			7.1	98.0
Kaurchak	66-20	WDS	89.21	0.06	0.91	1.27	1.48	0.58	0.54		5.51	99.61
Kaurchak	66-20	WDS	88.63		1.01	1.13	2.00	0.52	0.3		5.91	99.54
Kaurchak	66-20	EDS	86.4		0.9	1.3	1.9	0.7	0.5		6.0	97.6
Kaurchak	66-20	EDS	85.9		1.2	1.3	1.9	0.6	0.4		5.5	96.7

Notes: Zero = under detection limit; blank = not analyzed.

4. Results and Discussion

Several hundred grains of platinum-group minerals derived from placers of the rivers Koura and Kourchak in the Gornaya Shoria territory (Figure 1) were examined. For the purpose of this investigation, we selected three grains for a detailed description.

4.1. Grain No. 1

4.1.1. Results

Multiphase grain No. 1, less than diameter of 1 mm (Figure 2a), was derived from the Koura River placer, dredge 317. The grain consisted of cuprous isoferroplatinum $Pt_3(Fe,Cu)$, copper-bearing platinum $Pt_{1.4}Cu_{0.6}$, hongshiite $Pt_{1.1}Cu_{0.9}$, and rhodarsenide $(RhPdPt)_2As$. The last two minerals had sizes close to the nanoscale. Their sizes were a few hundredths (from 0.03 to 0.07 μm), and at most copper platinum a few tenths of a micrometer (from 0.3 to 0.7 μm) (Figure 2c). Lamellae in cuprous isoferroplatinum and in copper-bearing platinum was abundant. They were determined in more than 50% of analyses (Table 2). Such Os-Ir-Ru lamellae were recognized in hongshiite much less commonly.

Table 2. The frequency of occurrence of PGMs containing Ru-Os-Ir lamellae in grain No. 1.

PGM	Number of Analyses	All Lamellae	Lamellae Thickness 0.1–0.2 μm *
Cuprous isoferroplatinum	20 (100%)	13 (65%)	9 (45%)
Cuprous platinum	15 (100%)	11 (73%)	9 (60%)
Hongshiite	42 (100%)	5 (12%)	1 (2.4%)

* Correspond to analyses in which the content of the sum (Ru + Os + Ir) exceeds 4 at.%.

The contents of Os, Ir, and Ru in many analyses of the Pt-Cu-Fe and Pt-Cu solid solutions were below the detection limit. In general, the contents of Ru, Os, and Ir is directly related to the thickness of lamellae. The proportion of Os/Ir/Ru remained more-or-less constant; in terms of a sum equal to 100% it was close to: $Os_{0.4}Ir_{0.4}Ru_{0.2}$ (Table 3). As a rule, slight deviations from the constant proportion of these elements were detected where the content of Ru, Os, and Ir is low (Table 3). Such discrepancy with respect to the average contents reflects the insufficiency of the method employed for analysis of contents close to the detection limit [45]. In grain No. 1, in addition to the lamellae of Os, Ir, and Ru, we discovered an isometric roundish nodule of osmium ($Os_{80}Ir_{15}Ru_5$), with a diameter of about 6 μm (see Figure 2a).

Table 3. Composition of copper platinum and copper isoferroplatinum of grain No. 1, "pure", with micro-inclusions of Ru-Os-Ir lamellae and "separated" composition of lamellae, according to SEM-EDS analysis, at.%.

	Fe	Cu	Ru	Os	Ir	Pt	Sum Ru + Os + Ir	Ru	Os	Ir
		Measured Composition						**"Separated" Lamella Composition**		
Copper Platinum	0.0 *	27.2	1.1	2.2	2.1	66.6	5.4	21	40	39
	0.0	25.2	3.0	7.1	6.8	57.8	17.0	18	42	40
	0.0	28.3	1.3	1.7	1.3	67.5	4.2	30	40	30
	0.0	23.9	4.6	13.3	9.2	49.0	27.1	17	49	34
	0.0	30.0	0.0	0.0	0.0	70.0	0.0	-	-	-
	0.0	30.1	0.0	0.0	0.0	69.9	0.0	-	-	-
	0.0	31.3	2.5	4.8	4.9	56.5	12.3	20	39	40
	0.0	28.3	0.5	1.6	1.7	68.0	3.8	-	-	-
	0.0	29.8	0.0	0.0	0.0	70.2	0.0	-	-	-
	0.0	22.1	4.8	11.5	10.6	51.1	26.9	18	43	39
	0.0	22.2	5.1	11.9	9.3	51.6	26.3	19	45	35
	0.0	30.0	0.0	0.0	0.0	71.1	0.0	-	-	-
	0.0	28.5	1.2	2.5	2.2	65.6	5.9	21	42	37
	0.0	28.4	0.8	1.5	2.1	67.2	4.4	18	34	48
Copper Isoferroplatinum	0.0	29.9	0.0	0.0	0.0	70.1	0.0	-	-	-
	16.2	9.1	0.0	0.0	0.0	74.7	0.0	-	-	-
	15.1	8.9	1.2	1.4	2.2	71.3	4.7	25	29	46
	15.5	10.9	0.0	0.0	0.0	73.6	0.0	-	-	-
	14.8	9.7	1.3	2.5	2.2	69.5	6.0	22	41	37
	13.5	9.5	2.5	6.8	5.9	61.8	15.2	16	45	39
	11.9	10.9	4.5	9.6	8.3	54.8	22.4	20	43	37
	15.1	11.4	0.0	0.0	0.0	73.5	0.0	-	-	-
	15.4	11.1	0.0	0.0	0.0	73.5	0.0	-	-	-
	12.7	9.0	4.2	8.0	8.6	57.6	20.7	20	38	41
	15.3	12.0	0.0	0.0	0.0	72.8	0.0	-	-	-
	15.0	9.9	1.0	1.3	0.0	72.8	2.3	-	-	-
	15.5	11.3	0.0	0.0	0.0	73.2	0.0	-	-	-
	16.1	9.8	0.0	0.0	0.0	73.3	0.0	-	-	-
	15.6	10.9	0.0	0.0	0.0	73.5	0.0	-	-	-
	14.6	9.2	0.9	1.6	2.2	71.5	4.7	19	35	46
	14.7	9.7	0.8	1.9	2.0	71.1	4.6	17	40	43
	14.1	10.3	0.9	1.3	1.3	72.2	3.5	-	-	-
	14.5	9.4	0.9	1.6	2.3	71.3	4.8	19	34	47
	13.4	11.7	0.6	1.7	1.5	71.2	3.8	-	-	-
	15.0	10.0	0.8	0.7	1.2	72.3	2.7	-	-	-

* bold highlights analyses with the sum of Ru + Os + Ir > 4.0%. Zero = not detected.

4.1.2. Discussion

In copper isoferroplatinum, in copper-containing platinum and hongshiite there are lamellae $(Os_{0.4}Ir_{0.4}Ru_{0.2})$, which have a constant composition $(Os_{0.4}Ir_{0.4}Ru_{0.2})$. It is an indication that all lamellae were formed as a result of decomposition of a single original solid solution. They could be close to Pt_3Fe or $Pt_3(Fe,Cu)$ (with minor elements Os, Ru, Ir, and others), which is more consistent with the data presented in Figure 2. It is completely replaced by minerals in the following succession: isoferroplatinum (?) —> cuprous isoferroplatinum —> copper-rich platinum —> hongshiite. The process of metasomatic replacement itself, judging from preservation of all three phases and the excess platinum in hongshiite $Pt_{1.1}Cu_{0.9}$, was relatively brief and incomplete.

The existence of Os, Ir, and Ru alloys in the form of small lamellae and relatively large isometric inclusions substantially different in composition indicates that they formed at different times under different conditions.

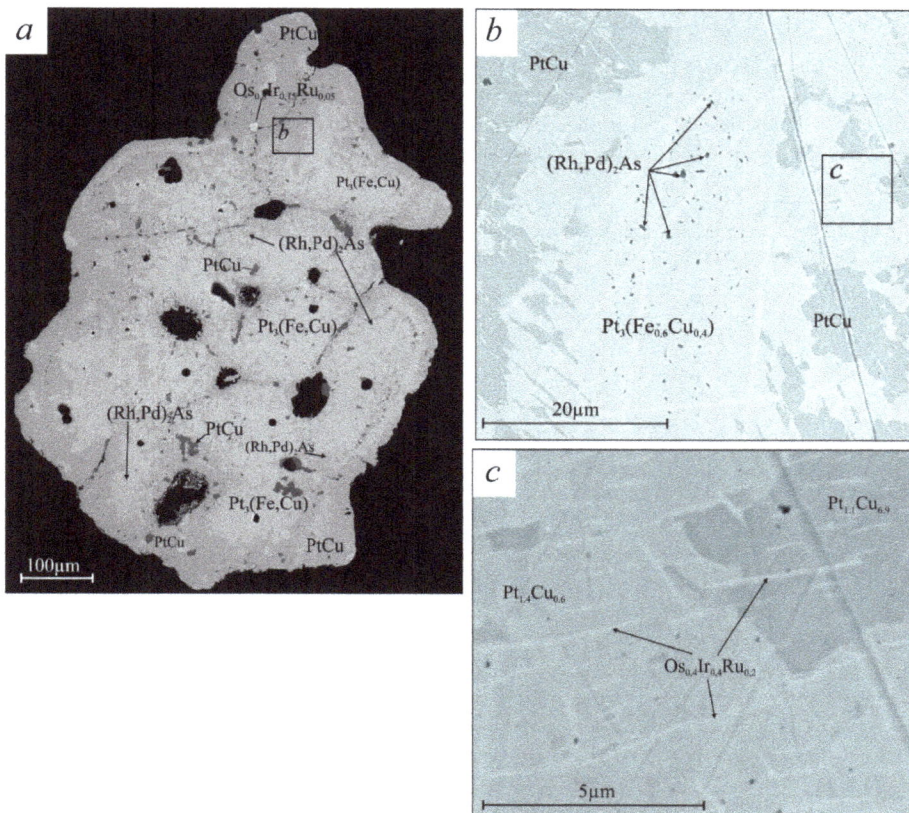

Figure 2. Back-scattered electron (BSE) image showing multiphase Grain No. 1 (**a**) A placer grain of $Pt_3(Fe,Cu)$ consists of a rim of Cu-enriched alloy phase (hongshiite PtCu) with inclusions of rhodarsenide $(Rh,Pd)_2As$. (**b,c**) The exsolution micro-lamellae of rutheniridosmine (the decomposition of a solid solution) in isoferroplatinum and platarsite (PtAsS). The composition of measurement points is listed in Table 4. White—isoferroplatinum (Pt,Fe) and osmium (Os); light gray and gray—platarsite (PtAsS), cooperite (PtS), and sperrylite (PtAs); rectangular contours—the position of the microsites "b" and "c". Explanation in the text.

The conditions were as follows: A multicomponent melt of platinum-group metals changed its composition as the high-temperature phases crystallized. The presence of nodules differing in composition ($Os_{0.8}Ir_{0.15}Ru_{0.05}$) and lamellae ($Os_{0.4}Ir_{0.4}Ru_{0.2}$) in the same PGM grain bears witness to a discontinuity of their formation in time and conditions and about the complex long history of its formation. This corresponds to the position that the Seglebir basic–ultrabasic massif served as the most probable source of PGMs for placers. Micro-emulsion rhodarsenide impregnation is present on micro-areas composed of cuprous isoferroplatinum (Figure 2b). The sizes of these micro-areas are mainly less than 0.5 μm, more rarely these increase to 1 μm. The micro-emulsion impregnations are randomly distributed (Figure 2b). The determination of the precise composition of distinct inclusions causes difficulties. The reason for this is their very small size and superposition of elements of the cuprous isoferroplatinum and rhodarsenide (Table 4).

The elements of cuprous isoferroplatinum (Fe,Cu,Pt) and Ir-Os-Ru alloys (Fe,Cu,Pt) affect the analyzes. The "separated" composition measured for the inclusions of ($Rh_{1.4}Pd_{0.3}Pt_{0.2}$)As and ($Rh_{1.1}Pd_{0.8}Pt_{0.1}$)As is close to the composition of rhodarsenide from the grain, which was represented by two varieties: ($Rh_{1.4}Pd_{0.3}Pt_{0.2})_{1.9}As_{1.1}$ and ($RhPd_{0.8}Pt_{0.2})_2As$ [46]. This similarity of compositions of micro-emulsion rhodarsenide impregnation and rhodarsenide from the grain confirms their genetic proximity. The emulsion impregnation has a metasomatic origin and it was formed contemporaneously with the main rhodarsenide inclusions. Thus, in this case, a relatively rare metasomatic structure is observed, which differs from the widely-developed emulsion structure of the solid solution decomposition [47].

Table 4. The result of determining the composition of emulsion inclusions of rhodarsenide in grain No. 1, according to SEM-EDS analysis.

No	Fe	Cu	As	Ru	Rh	Pd	Os	Ir	Pt	Total	%
1	2.8	2.0	11.0	0.5	22.0	5.3	1.0	1.3	53.8	99.7	wt.
2	3.1	2.3	9.1	0.7	14.2	10.8	0.0	1.6	57.0	98.7	wt.
1	6.3	4.0	18.7	0.7	27.2	6.3	0.7	0.9	35.2	100.0	at.
2	7.3	4.8	15.9	0.9	18.2	13.4	0.0	1.1	38.5	100.0	at.

4.2. Grain No. 2

4.2.1. Results

The diameter of the grain No. 2 is 0.3 mm. It also was derived from dredge 317 on the Koura River. The grain consisted of isoferroplatinum with the unclear zonation replaced by platarsite (PtAsS), cooperite (PtS), sperrylite (PtAs$_2$), and osmium laths ($Os_{0.7}Ir_{0.2}Ru_{0.1}$) (Figure 3). Micro-lamellae of Os-Ir-Ru are developed in isoferroplatinum and platarsite. Lamellae form a common lattice for both minerals (Figure 3c). The thickness of the lamellae of Os-Ir-Ru was about 0.1 μm (i.e., it was much less than the thickness of osmium laths (up to 2–6 μm) developed in the same grain). The lattice represents a structure decomposition of solid solution isoferroplatinum. Lamellae affected the composition of isoferroplatinum and platarsite enclosing them. In the latter, there was a high content of chemical elements forming (Os, Ir, Ru) lamellae (Table 5). The cooperite and sperrylite in the peripheral micro-zones contained no Ru, Os, and Ir, and Os-Ir-Ru lamellae were absent in these minerals (Figure 3b,c). The boundary of cooperite and sperrylite with platarsite is clear. Cooperite and sperrylite were formed at the later stage of replacement. The compositions of individual lamellae included into isoferroplatinum and platarsite, and separated by the technique described earlier, were quite close to each other. The scatter of either of the three elements (Os, Ir, Ru) contents did not exceed 20%, and more often 10%, and it varied within the range $Os_{0.7-0.5}Ir_{0.2-0.3}Ru_{0.1-0.2}$ (Table 6).

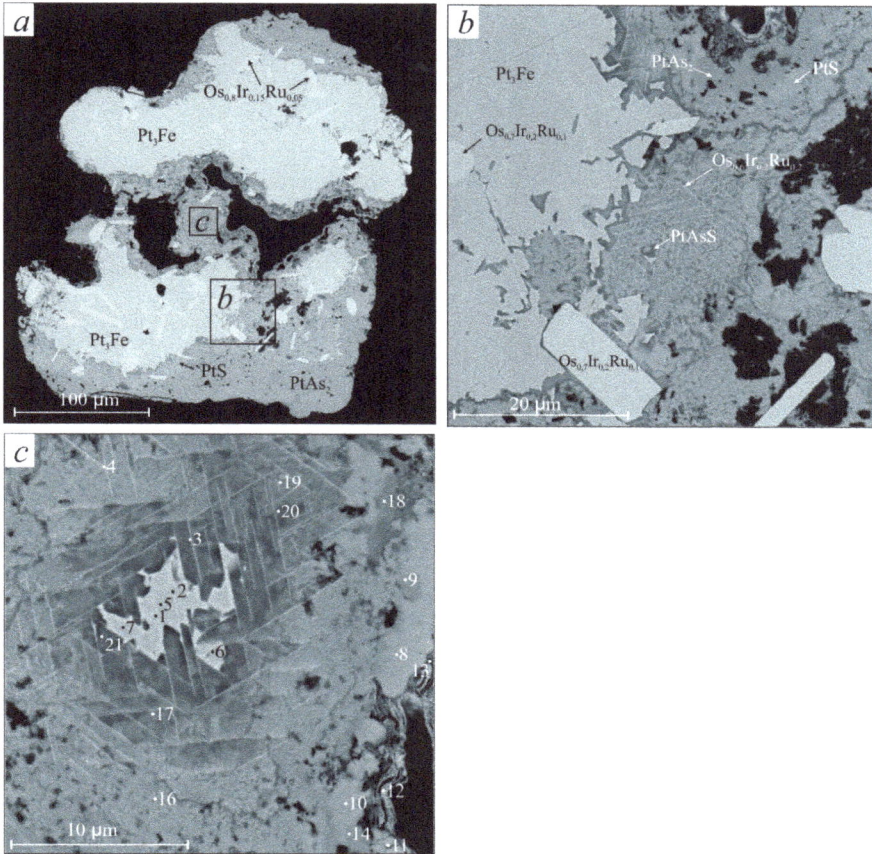

Figure 3. BSE image showing multiphase Grain No. 2. (**a**) A placer grain of Pt$_3$(Fe,Cu) consists of a rim of Cu-enriched alloy phase (hongshiite PtCu) with inclusions of rhodarsenide (Rh,Pd)$_2$As. (**b,c**) The exsolution micro-lamellae of rutheniridosmine (the decomposition of a solid solution) in isoferroplatinum and platarsite (PtAsS). For the composition of measurement points see Table 5. White—isoferroplatinum (Pt$_3$Fe) and osmium (Os); light gray and gray—platarsite (PtAsS), cooperite (PtS), and sperrylite (PtAs); rectangular contours—the position of the microsites "b" and "c".

Table 5. The composition of isoferroplatinum (1–10) and platarsite (11–22) of grain No. 2, "clean" and with inclusions of Ru-Os-Ir lamellae, according to SEM-EDS analysis, at.%.

No [1]	S	Fe	Cu	As	Ru	Rh	Sb	Os	Ir	Pt	Total [2], wt.%
1	0.0 [3]	19.8	3.3	0.0	1.6	1.8	0.0	3.6	2.2	67.7	97.2
2	0.0	18.2	0.0	0.0	3.1	0.0	1.9	12.2	5.7	58.9	95.7
3	0.0	13.4	0.0	0.0	7.8	1.1	0.0	28.7	10.1	38.9	93.6
4	0.0	22.6	0.0	0.0	0.0	0.0	1.8	0.0	0.0	75.6	96.9
5	0.0	20.4	0.0	0.0	1.5	0.0	2.5	0.0	2.8	72.8	99.6
6	0.0	21.6	0.0	0.0	1.2	0.0	2.1	1.2	1.6	72.4	96.3
7	0.0	20.9	0.0	1.1	0.0	1.1	1.6	1.9	2.5	70.9	98.4
8	0.0	21.8	1.0	0.0	1.1	0.0	1.6	0.0	2.2	72.4	97.6
9	0.0	19.5	0.0	0.0	2.3	0.0	1.5	10.5	2.5	63.8	96.4
10	0.0	19.5	0.0	0.0	2.5	0.0	0.0	9.8	2.4	65.9	94.3

Table 5. *Cont.*

No [1]	S	Fe	Cu	As	Ru	Rh	Sb	Os	Ir	Pt	Total [2], wt.%
11	37.7	0.0	0.0	11.5	1.9	4.0	0.0	10.4	3.7	30.8	96.3
12	33.2	0.0	0.0	17.0	2.5	5.7	0.0	10.5	5.2	26.0	95.3
13	40.6	0.0	0.0	16.5	1.4	5.5	0.0	0.0	2.6	33.5	100.2
14	39.6	0.0	0.0	10.2	1.0	2.9	0.0	5.1	3.0	38.1	96.6
15	28.7	2.9	0.0	29.1	1.4	18.5	0.0	0.7	2.7	16.1	101.5
16	33.0	0.0	0.0	26.6	1.8	11.1	0.0	2.3	5.4	19.8	98.4
17	20.4	0.5	0.0	37.5	1.4	13.1	0.0	9.6	3.7	13.8	102.8
18	17.5	0.0	0.0	46.8	1.1	8.3	0.5	2.0	2.5	21.4	97.0
19	26.6	0.0	0.0	34.9	2.6	5.3	0.0	1.8	7.2	21.6	100.5
20	12.3	0.0	0.0	50.6	0.8	4.4	0.7	2.2	2.9	26.3	98.4
21	24.4	0.4	0.0	39.5	1.0	15.6	0.0	1.9	3.4	13.8	100.1
22	23.6	0.6	0.0	39.7	1.2	15.3	0.0	2.0	2.9	14.7	97.4

[1] The binding of points is given in Figure 3c. [2] Total atom = 100%. [3] Analyses with the sum Ru + Os + Ir > 4.0% are highlighted in bold. Zero = not detected.

Table 6. A "separated" composition (at.%) of Ru-Os-Ir lamellae of grain No. 2.

No [1]	Ru	Os	Ir	Σ	PGM [2]
1	21	49	30	100	Isoferroplatinum
2	15	58	27	100	Isoferroplatinum
3	17	62	22	100	Isoferroplatinum
9	15	69	16	100	Isoferroplatinum
10	17	67	16	100	Isoferroplatinum
11	12	65	23	100	Platarsite
12	14	58	29	100	Platarsite
17	10	65	25	100	Platarsite

[1] The binding of points is given in Figure 3c. [2] PGM is a mineral containing Ru-Os-Ir lamellae.

4.2.2. Discussion

Weak variations in the composition of lamellae indicate that Os-Ir-Ru lamellae in platarsite are relict ones (as opposed to those in isoferroplatinum). This fact confirms the great stability of Os-Ir-Ru alloys in the process of alteration compared to isoferroplatinum. The composition of lamellae outlined above is quite close to the composition of "osmium" laths ($Os_{0.7}Ir_{0.2}Ru_{0.1}$) present in the grain. Probably it indicates an insignificant difference in time of formation of the osmium laths and the inclusions of isoferroplatinum and, as a consequence, the relatively rapid cooling of the ore-forming body and its small size. Such ore-forming body could be one of the numerous dikes of the Seglebir complex. Detection of grains of native isoferroplatinum in one of the samples of dyke gabbro on the adjacent territory [30] indirectly confirms this probability.

4.3. Grain No. 3

4.3.1. Results

Grain No. 3 was derived from the placer of Kaurchak River (dredge 138). The grain of PGM (200 × 500 μm) was of interest in view of detection of micro-inclusions of native gold in it, as well as sulfides, sulfoarsenides, and arsenides of platinum group elements. Three types of gold were distinguished (Au-I–Au-III). In the grain section (Figure 4), three zones were distinguished I — core, II — Rim-I, and III — Rim-II.

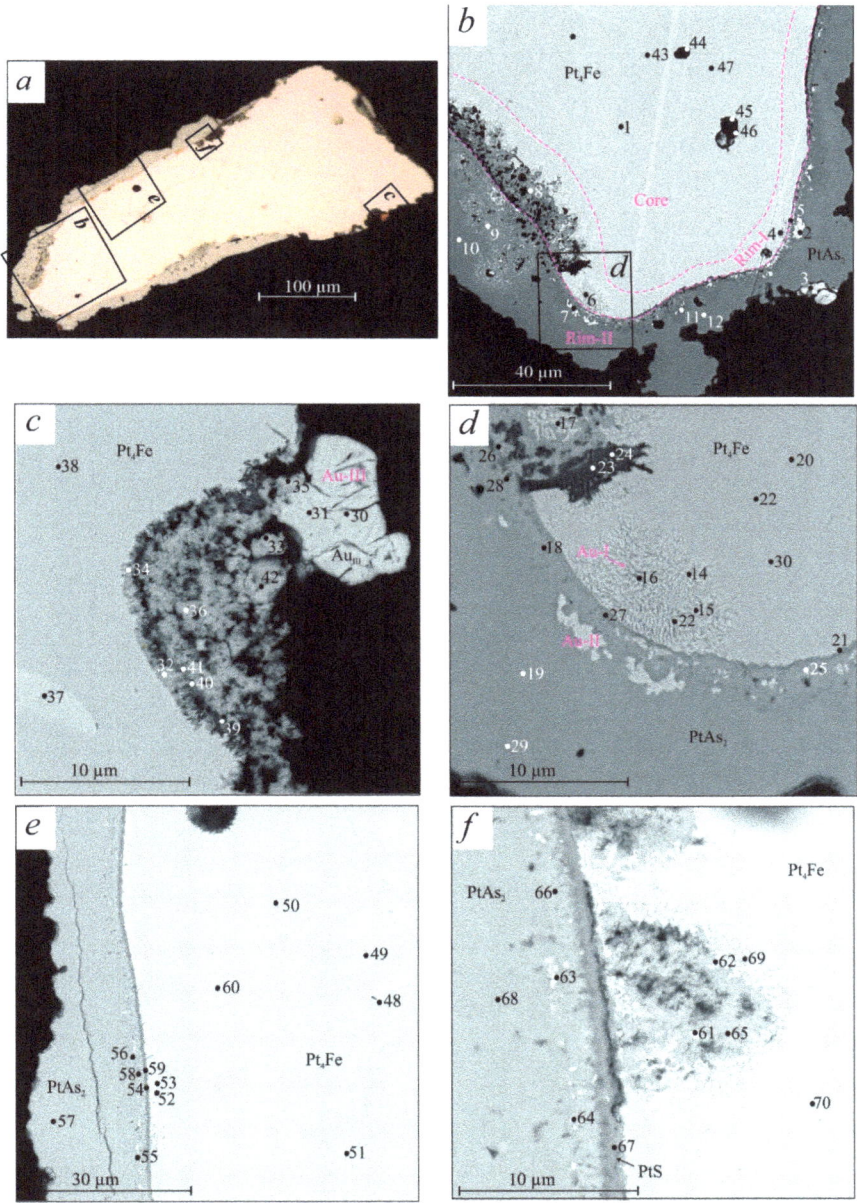

Figure 4. Grain No. 3 was derived from the placer of Kaurchak River (dredge 138). (**a**) Reflected light micrograph of grain No. 3 of the Pt-Fe alloy with alteration sperrylite composition rim (gray) and inclusions and overgrowths of native gold (reddish). (**b–f**) BSE images showing the structure and morphology of the local areas. Light gray background—matrix Pt_4Fe; gray and dark gray rim—sperrylite and platarsite; white inclusions and overgrowths—native gold; the compositions of the Pt-Fe alloy and inclusions at these points are given in Tables 7 and 8.

Minerals **2019**, 9, 327

Table 7. The composition of the core in grain No. 3 (Figure 4) and its inclusions and micro lamellae (at.%), determined by the SEM-EDS method.

Site [1]	No [2]	S	Fe	Co	Cu	Ru	Rh	Pd	Os	Ir	Pt	Formula	Mineral
4b	1	0.0	16.8	0.0	1.1	1.0	3.1	2.1	0.0	0.0	75.8	(Pt,Fe)	Pt-Fe alloy
4b	47	0.3	17.3	0.3	1.3	1.1	3.3	2.1	1.0	0.0	74.5	(Pt,Fe)	Pt-Fe alloy
4c	38	0.0	17.5	0.0	1.3	1.1	3.3	2	0.0	0.0	74.5	(Pt,Fe)	Pt-Fe alloy
4d	20	0.0	18.0	0.0	1.2	1.2	3.1	2	0.0	0.0	74.6	(Pt,Fe)	Pt-Fe alloy
4e	60	0.0	16.6	0.4	1.1	1.3	3.3	2.3	1.3	0.8	73	(Pt,Fe)	Pt-Fe alloy
4f	70	0.0	17.4	0.0	1.6	1.1	3.2	1.9	1.1	0.0	73.8	(Pt,Fe)	Pt-Fe alloy
4b [3]	43	37.0	2.0	2.0 [3]	4.4	0.6	21.8	24.3	0.0	0.0	7.7	$(Rh,Pd)_3S_2$	Untitled
4b	44	62.0	0.0	0.0	0.0	0.3	33.5	1.0	0.6	0.5	1.5	RhS_2	Untitled
4b	45	30.0	0.0	0.0	11.3	0.5	0.0	55.6	0.3	0.0	1.6	Pd_2S	Untitled
4b	46	57.8	0.0	0.0	0.0	3.6	36.1	0.0	0.4	0.3	1.2	Rh_2S_3	Bowieite
4c	37	0.0	0.0	0.0	0.0	1.3	3.5	0.0	87.5	3.4	4.3	$Os_{0.95}Ir_{0.04}Ru_{0.01}$	Osmium
4e	48	0.0	4.3	0.0	0.0	5.0	1.9	0.0	69.0	0.9	18.9	$Os_{0.92}Ru_{0.07}Ir_{0.01}$	Osmium
4e	49	0.0	6.1	0.0	0.0	6.6	2.6	0.9	57.4	0.0	26.4	$Os_{0.90}Ru_{0.10}$	Osmium
4e	50	0.0	4.9	0.0	0.0	4.2	2.4	0.5	65.3	1.3	21.4	$Os_{0.92}Ru_{0.06}Ir_{0.02}$	Osmium

[1] Site in Figure 4; [2] The binding of points is given in Figure 4; [3] At point 43 (Figure 4b). Zero = not detected.

The core occupies the large inner part of the grain. This zone has a homogeneous texture and it was comprised of a Pt-Fe solid solution. Its average composition, according to six analyses, was as follows (Table 7): 75 at.% Pt and 18 at.% Fe, which corresponds to platinum Pt_4Fe. Fe-Pt compounds with an iron content of 18 to 41 at.% correspond to isoferroplatinum, according to the Fe-Pt phase diagram [48,49]. The alloy in zone 1 (core) with 18 at.% Fe could be attributed to both isoferroplatinum and platinum. We used the term platinum. Minor elements (average grade) were as follows: 1.3 at.% Cu, 3.2 at.% Rh, 2.1 at.% Pd, and 1.3 at.% Ru. The scatter of values of each of them was insignificant (Table 7). The uniformity of the elements distribution in the core is shown in Figure 5. In the homogeneous mass of platinum, two rare micro-inclusions of two types observed, micro-lamellae of osmium and PGE-sulfide micro-nodules. The thickness of the micro-lamellae varies from 0.5 μm (data points No. 48–51 in Figure 4e) to 2 μm (data point No. 37 in Figure 4c). The composition of micro-lamellae was as follows: 92–95 at.% Os, 6–1 at.% Ru, 0–4 at.% Ir (Table 7, data points No. 37–50). The sizes of the micro-nodules were few μm. They were comprised of Rh and Pd sulfides: bowieite and unnamed varieties (data points No. 43–46 in Figure 4b, Figure 5, and in Table 7). PGE-sulfide micro-nodules were formed during liquation. Cavities in micronodules appeared as a result of the subsequent dissolution of sulfide minerals (Figure 4a,b,f).

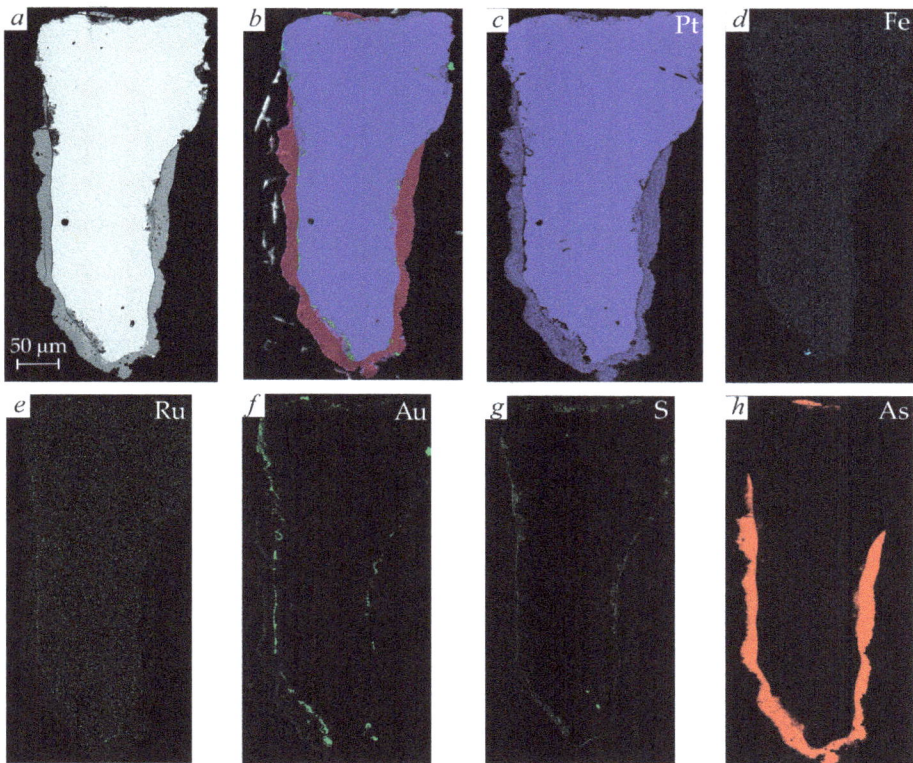

Figure 5. Backscattered-electron image (BSE) (**a**) and element distribution maps (**b**—polyelement; **c**—Pt; **d**—Fe; **e**—Ru; **f**—Au; **g**—S; **h**—As) in the grain No. 3 (see Figure 4) platinum nugget.

The Rim-I of grain No. 3 covered the edge part of the core composed of platinum (see Figure 4b). From the edge of the nucleus inward to 10 microns, micro-areas of unusual structure and composition were developed. Among them are two types of micro-areas. The first variety was the zone of alteration represented by platarsite-cooperite (data point No. 8 in Figure 4b). This has not been reviewed by us in detail. The second variety was investigated in detail (data point No. 6 in Figure 4b, data points No. 14–17 and 22 in Figure 4d, data points No. 61, 62, 65, 69 in Figure 4g).

These micro-areas had a peculiar dendrite-like microstructure, sub-graphic (eutectoid), represented by curved nanosized gold lamellae (from 100–150 nm up to 700–750 nm) (Figures 5 and 6). These nanoscale areas consisted of native gold (Au-I), rhodium, and platinum sulfoarsenides, with minor elements Os, Ru, and Ir (Table 8). The inclusions of native gold had the following composition: 83.8 at.% (84.7–82.6 at.%) Au; 15.9 at.% (15.3–17.4 at.%) Ag, and the average formula corresponded to Au_4Ag. The fineness of Au-I corresponded to 900–910 ‰. Gold in Rim-I also formed a micron layer, sharply bounded by Rim-II on the one hand and uneven on the other. Its power ranged from less than 1 micron to 3 microns.

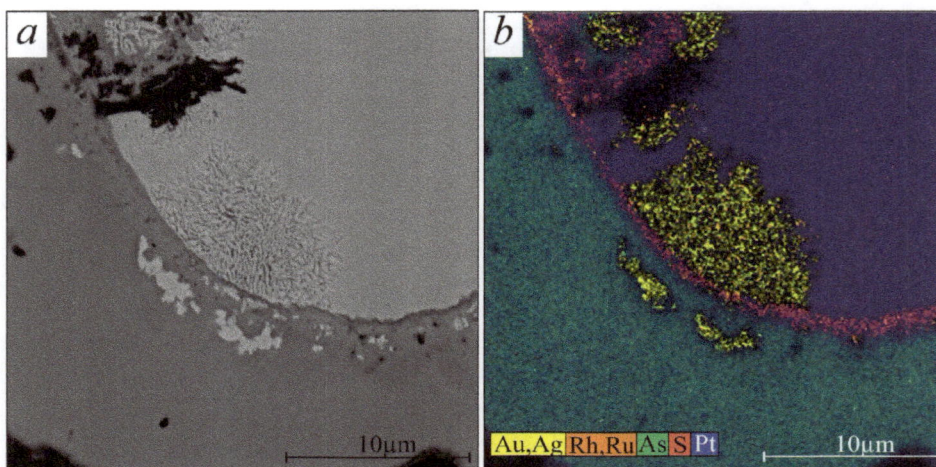

Figure 6. The areas of detail 4d of grain No. 3. Distribution patterns obtained in the mapping mode: Au, Ag, Rh, Ru, As, S, Pt.

Table 8. Full and "separated" chemical composition (at.%) (SEM-EDS method) Rim-I in areas with eutectic microstructure.

Site [1]	No [2]	S	Fe	As	Ru	Rh	Pd	Ag	Os	Ir	Pt	Au
				Full composition in areas with eutectic microstructure								
4b	6	0.0	0.0	14.6	0.0	6.2	0.0	11.3	1.6	0.0	5.2	61.1
4d	14	11.0	1.0	6.9	0.5	2.1	0.0	11.3	1.6	0.5	6.1	58.1
4d	15	15.6	0.4	12.5	1.0	4.9	0.0	8.8	1.4	0.7	5.5	48.6
4d	16	17.3	0.4	15.9	0.9	7.2	0.0	8.0	1.3	0.7	6.0	41.7
4d	17	16.5	1.7	9.6	0.5	3.1	0.0	9.7	1.4	0.6	6.3	49.9
4d	22	15.8	1.1	13.1	0.9	6.3	0.0	8.4	1.4	0.5	8.1	43.5
4f	69	15.2	0.6	12.4	0.9	4.9	0.0	9.7	1.4	0.7	8.1	46.1
		Separated composition of the sulfoarsenides in areas with eutectic microstructure										
4b	6	0.0	0.0	52.9	0.0	22.5	0.0		5.8	0.0		18.8
4d	14	37.0	3.4	23.2	1.7	7.1	0.0		5.4	1.7		20.5
4d	15	37.1	1.0	29.8	2.4	11.7	0.0		3.3	1.7		13.1
4d	16	34.8	0.8	32.0	1.8	14.5	0.0		2.6	1.4		12.1
4d	17	41.6	4.3	24.2	1.3	7.8	0.0		3.5	1.5		15.9
4d	22	33.5	2.3	27.8	1.9	13.3	0.0		3.0	1.1		17.2
4f	69	34.4	1.4	28.1	2.0	11.1	0.0		3.2	1.6		18.3
		Separated composition of inclusions of native gold (Au-I) in areas with eutectic microstructure										
4b	6							9.0				91.0
4d	14							10.0				90.0
4d	15							9.0				91.0
4d	16							9.0				91.0
4d	17							9.0				91.0
4d	22							10.0				90.0
4f	69							10.0				90.0

[1] Site in Figure 4; [2] The binding of points is given in Figure 4.

The Rim-II was the outer rim. The rim had an irregular thickness up to 20 μm and covered more than half of the grain contour. The Rim-II consisted of sulfide, sulfoarsenide, and arsenide layers (Figures 4 and 7). The sulfide layer Rim-II texture was homogeneous. The sulfoarsenide layer and the adjacent part of the arsenide layer had a spotty texture due to the inclusions of gold Au-II. The sulfide layer consisted of cooperite and sulfides Rh, Os, Ru, and Ir (Table 9). The sulfoarsenide layer represented by platarsite with minor elements of Rh, Fe, Os, Te, and small inclusions of gold (Au-II) (less than 1 micron). The composition of this gold Au-II determined by the relationship between Au and Ag was relatively consistent: 84.1 at.% (84.7–83.5 at.%) Au and 15.9 at.% (15.3–16.5 at.%) Ag. The gold fineness corresponded to 900–910 ‰. The arsenide layer was the most powerful. It consisted of sperrylite. On the border with the sulfoarsenide layer were larger (up to 5 microns) inclusions of gold (Au-II). Inclusions had an elongated shape along the border with the sulfoarsenide layer. The composition of gold was the same as the composition of gold in the sulfoarsenide layer.

Table 9. Full and "separated" chemical composition Rim-II.

	S	Fe	Co	Cu	As	Ru	Rh	Pd	Ag	Te	Os	Ir	Pt	Au	Hg
					Full composition of the sulfide layer										
Mean	40.6	2.2	0.0	0.2	13.5	3.3	7.6	0.1	1.1	0.3	4.8	1.8	19.6	5.1	0.0
Min	33.7	0.2	0.0	0.0	7.7	1.4	5.3	0.0	0.0	0.2	1.0	0.5	2.8	0.0	0.0
Max	46.8	4.5	0.0	0.4	18.2	6.8	11.4	0.3	3.4	0.4	9.2	4.2	31.0	17.0	0.0
					Separated composition of the sulfide layer										
Mean	43.2	2.3	0.0	0.2	14.4	3.5	8.1	0.1		0.3	5.1	1.9	20.9		
Min	33.7	0.2	0.0	0.0	7.7	1.4	5.3	0.0		0.2	1.0	0.5	9.2		
Max	43.5	4.5	0.0	0.4	18.2	5.8	10.2	0.3		0.3	9.2	3.2	31.0		
					Full composition of the sulfoarsenide layer										
Mean	15.7	0.2	0.3	0.0	50.3	0.0	3.7	0.1	0.0	0.5	0.1	0.0	28.4	0.5	0.0
Min	10.3	0.0	0.0	0.0	37.4	0.0	1.1	0.0	0.0	0.3	0.0	0.0	26.1	0.0	0.0
Max	26.4	0.9	1.7	0.0	56.8	0.0	6.00	0.6	0.0	0.7	0.7	0.0	31.4	3.3	0.0
					Separated composition of the sulfoarsenide layer										
Mean	15.8	0.2	0.1	0.0	50.9	0.0	3.8	0.0		0.5	0.1	0.0	28.7		
Min	10.6	0.0	0.0	0.0	37.5	0.0	1.1	0.0		0.3	0.0	0.0	26.2		
Max	26.7	1.0	0.1	0.0	55.9	0.0	6.0	0.2		0.7	0.5	0.0	31.5		
					Full composition of the arsenide layer										
Mean	7.3	0.7	1.4	0.1	57.5	0.1	0.5	0.9	0.3	1.0	0.0	0.0	29.1	1.0	0.0
Min	4.1	0.0	0.0	0.0	54.8	0.0	0.0	0.0	0.0	0.0	0.0	0.0	28.6	0.0	0.0
Max	11.8	1.4	1.9	0.6	58.7	0.5	2.5	1.7	1.4	1.6	0.0	0.0	30.2	6.1	0.0
					Separated composition of the arsenide layer										
Mean	7.4	1.4	1.4	0.1	58.3	0.1	0.5	0.9		1.0	0.0	0.0	29.5		
Min	4.1	0.6	0.0	0.0	54.8	0.0	0.0	0.0		0.0	0.0	0.0	28.5		
Max	11.8	2.2	1.9	0.6	61.9	0.5	2.5	1.7		1.7	0.0	0.0	30.9		
					Separated composition of inclusions of native gold (Au-II)										
Mean									9.9					90.3	0.0
Min									9.0					90.0	0.0
Max									10.0					91.0	0.0
					Separated composition of inclusions of native gold (Au-III)										
Mean									1.7					91.7	6.5
Min									1.0					87.0	4.0
Max									2.0					94.0	12.0

Inclusions of Au-III 5–10 μm in size were on the outer surface of grain No. 3. Forms of Au-III aggregates were as follows: Pads and smooth isometric microcrystals or their intergrowths (Figure 4b,c; data points No. 3, 30, and 31). The peculiarity of their compositions were the low Ag content: 3 at.%, 3 at.%, and 3 at.% Ag and the presence of Hg: 12at.%, 3.6 at.%, and 6 at.% (Table 10).

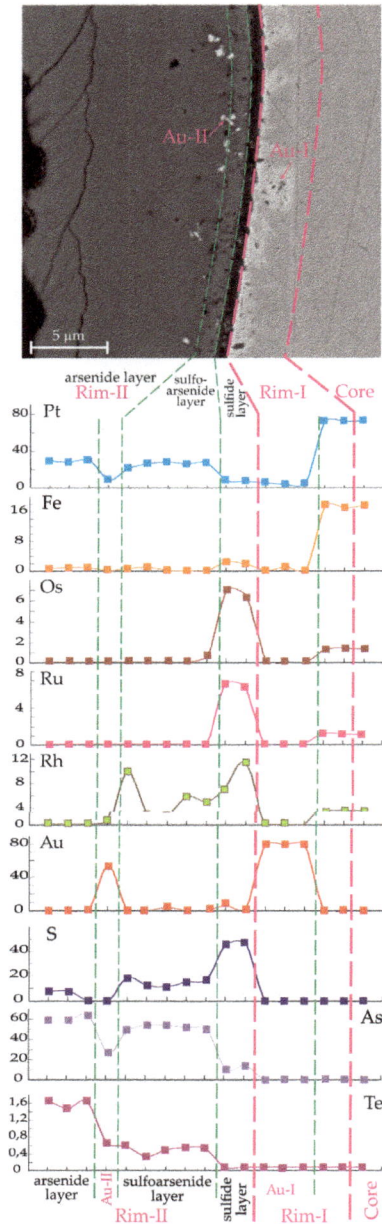

Figure 7. Backscattered-electron (BSE) image (top) and the graphs of changes of the main elements content (at.%) in the Core, Rim-I, Rim-II, with micro-layers in grain No. 3 (see also Figure 4e).

Table 10. "Separated" composition of inclusions of native gold (wt.%) in PGM grain No. 3 (Figure 4).

Site [1]	No [2]	Au	Ag	Hg	Formula	Au Generation
4b	4	90	10	0	$Au_{0.83}Ag_{0.17}$	Au-I
4b	6	91	9	0	$Au_{0.84}Ag_{0.16}$	Au-I
4d	14	90	10	0	$Au_{0.84}Ag_{0.14}$	Au-I
4d	15	91	9	0	$Au_{0.85}Ag_{0.15}$	Au-I
4d	16	91	9	0	$Au_{0.84}Ag_{0.16}$	Au-I
4d	17	91	9	0	$Au_{0.84}Ag_{0.16}$	Au-I
4d	22	90	10	0	$Au_{0.84}Ag_{0.16}$	Au-I
4e	52	90	10	0	$Au_{0.83}Ag_{0.17}$	Au-I
4e	53	90	10	0	$Au_{0.82}Ag_{0.18}$	Au-I
4f	61	91	9	0	$Au_{0.85}Ag_{0.15}$	Au-I
4f	62	90	10	0	$Au_{0.84}Ag_{0.16}$	Au-I
4f	69	90	10	0	$Au_{0.83}Ag_{0.17}$	Au-I
4b	2	90	10	0	$Au_{0.84}Ag_{0.16}$	Au-II
4b	5	90	10	0	$Au_{0.83}Ag_{0.17}$	Au-II
4b	7	90	10	0	$Au_{0.84}Ag_{0.16}$	Au-II
4b	12	91	9	0	$Au_{0.85}Ag_{0.15}$	Au-II
4e	54	90	10	0	$Au_{0.83}Ag_{0.17}$	Au-II
4f	63	91	9	0	$Au_{0.84}Ag_{0.6}$	Au-II
4b	3	87	1	12	$Au_{0.86}Ag_{0.03}Hg_{0.12}$	Au-III
4c	30	94	2	4	$Au_{0.93}Ag_{0.03}Hg_{0.04}$	Au-III
4c	31	92	2	6	$Au_{0.91}Ag_{0.03}Hg_{0.06}$	Au-III
4c	32	94	2	4	$Au_{0.92}Ag_{0.03}Hg_{0.04}$	Au-III

[1] Site in Figure 4, from 4b to 4f; [2] The binding of points is given in Figure 4. Zero = not detected.

4.3.2. Discussion

Grain No. 3 has a complex structure, which reflects the crystallization history of PGM and the distribution of PGE and other elements among the mineral phases. Grain No. 3 consists of Core, Rim-I, and Rim-II. The Rim-II consists of sulfide, sulfoarsenide and arsenide layers. The core has a composition of 75 at.% Pt and 18 at.% Fe. Minor elements (Rh, Pd, Cu) are distributed uniformly in the bulk of the core. This indicates their isomorphic occurrence. Alloys of this composition are rare for Western Siberia varieties [48], but not rare for the World [1]. According to the Rt-Fe state diagram, the temperature of formation of such an alloy is below 1110 °C [49,50]. Various crystallization mechanisms of the Pt-Fe alloy are described in the literature. One of them is associated with the crystallization of Pt-Fe alloys from MSS (monosulfide solid solution) [51]. Another mechanism of crystallization of Pt-Fe alloys is not associated with the decomposition of sulfide liquid. Crystallization of the Pt-Fe alloy occurred before immiscible sulfide formed [52,53]. The core contains inclusions of Rh and Pd sulfides in the central part. On the periphery of the core in rim-I, we observe areas with a complex dendrite structure with inclusions of Au-I and sulfoarsenides (Rh, Pt, ± Os, Ru, Ir). They correspond to the structure of the eutectic-decomposition multi-principle component alloys (MPCAs) [54]. The presence of sulfide globules and areas with a multicomponent eutectic melt indicates the possible crystallization of the Pt-Fe alloy from a multicomponent melt containing Pt-Fe-Pd-Rh-Au-Ag-As-S. The presence of sulfide globules and areas with a multicomponent eutectic melt indicates the possible crystallization of the Pt-Fe alloy at a temperature below 1100 °C from a multicomponent melt containing Pt-Fe-Pd-Rh-Au-Ag-As-S. The eutectic temperature of the multicomponent melt was below 1050 °C, which is indicated to us by the Au-Ag state diagram [55]. Another possible formation of Rim-I is the substitution mechanism of the core (Pt-Fe) by the Au-Ag-Rh-Ru-As-S melt. During the cooling of the system, the solid solution disintegrated with the formation of this sub-graphic structure.

We observed mineral zonality in Rim-II, which reflects a sequential change of crystallization conditions from residual melt (containing Pt, Rh, Pd, S, As, Te). The residual melt is separated from the MMS according to the model proposed by Dare et al. [51]. Initially, the melt had a high f_{S2} (fugacity of sulfur), which gradually decreased. The content of As and Te in the residual melt, on the contrary, increased. This is confirmed by the successive change in sulfide layers of sulfoarsenide and arsenide layers in the grain. The residual melt interacted with the platinum grain. The Pt-Fe alloy was replaced n the interaction zone with the enlargement of Au-I and the formation of Au-II. Formation of Au-III is associated with low-temperature processes of hydrothermal or supergene processes. The formation of Au-III occurred later and was associated with low-temperature hydrothermal processes or supergenic alteration.

5. Conclusions

The article shows, with specific examples, the importance of data on the composition of microscopic, close to nanoscale, mineral inclusions in order to ascertain conditions for the formation of associations of platinum-group minerals. In turn, the determination of the composition, close to the true one, of such inclusions and the host mineral, causes certain difficulties, even when using local methods of analysis (EMP, SEM). The reason is that the region of X-ray generation captures both the matrix and an inclusion. There is a mutual influence of the compositions (overlapping spectra). The complexity of the task is governed by a number of factors, and first of all by the ratio of elemental compositions of included and including minerals and the degree of overlap of spectral lines involved in the SEM-EDS analysis. The EDS method can be used on a par with WDS to study the chemical composition of minerals, including micro-inclusions in PGMs. This method is multi-elemental and more effective for mineral diagnostics in comparison with WDS, which is its certain advantage. With the EDS approach, it is possible to obtain quantitative data on the composition of micro-inclusions with a size of 3–5 μm. An EDS analysis of PGM grains ranging in size from less than 3 to 1 μm provides semiquantitative data. The EDS analysis of grains with a size of less than 1 μm gives, at best, a qualitative composition of PGMs or suggests the presence of PGM nanoparticles.

The main results of the study are as follows:

In the processes of metasomatic transformations of PGMs, the stability of the Os-Ir-Ru lamellae substantially exceeds that of isoferroplatinum, and the products of its alteration. This fact is confirmed by the identical composition of the lamellae included in cuprous isoferroplatinum, as well as in copper-bearing platinum and hongshiite in the first case (grain No. 1), and in isoferroplatinum and platarsite in the second case (grain No. 2).

Based on the significant difference in the Os-Ir-Ru compositions of the lamellae in isoferroplatinum and nodules, a conclusion is drawn about the great hiatus in time and the difference in conditions of formation of the indicated phases of grain No. 1. This corresponds to the position that the Seglebir basic–ultrabasic massif served as the most probable driving force.

On the contrary, it has been established that grain No. 2 solidified relatively swiftly. It was formed in one of the numerous small magmatic bodies of basic composition, which are widely distributed in the area. The role of such a body is very suitable for one of the numerous dikes of the Seglebir complex. Detection of grains of native isoferroplatinum in one of the samples of a dyke of gabbro on the adjacent territory [30] indirectly confirms this probability. The deposition of sperrylite and native gold from the basalt melt in layered basic–ultrabasic complexes are an established fact [56,57].

The source of grain No. 3 was intrusive massifs of the middle Cambrian gabbro. It is very difficult to restore the history of crystallization on a single grain. However, by the structural relationships of minerals, we can confidently speak of a complex sequence of the formation of the PGMs.

1. The formation of platinum with melt sulfide inclusions occurred from the primary monosulfide solid solution (MSS) [51,53] or sulfide liquids [52]. There was a liquation department of the Pt-Fe alloy with a small amount of Au, Ag, Rh, Ir, Os, S, and As. Os microcrystals crystallized first from the Pt-rich melt. As the system cooled down and platinum crystallized, a lower-temperature melt enriched in Au, Ag, S, As, and Rh (with minor elements of Ru, Os, Ir) accumulated and separated. The separated multicomponent melt during cooling formed a eutectic/eutectoid multi-principle component alloy.

2. The formation of layered Rim-II is associated with the interaction of platinum grains with the residual melt.

The formation of eutectoid structures and a border on the surface of the grains occurred in the magmatic stage. This is consistent with the findings of Badanina et al. [58] (according to the ^{187}Os/^{188}Os and ^{187}Re/^{188}Os systematics) about a single source of PGEs in the Pt-Fe alloy and arsenide rims, as well as experimental data [59]. Previously, the formation of arsenide, sulfoarsenide, and sulfide rims on Pt-Fe alloy grains was associated with post-magmatic hydrothermal processes ([24], etc.).

Variations in the content of basic chemical elements reflect the processes and conditions for the formation of PGMs. Sharp fluctuations in the local areas of the contents of elements that are not part of the mineral structure may indicate that nano-inclusions of a different composition fall into the region of X-ray generation. Such cases require detailed study using local methods of analysis with a resolution of several nanometers (HR-SEM, HR-TEM).

Author Contributions: Conceptualization, G.V.N. and S.M.Z.; methodology, N.S.K. and D.K.B.; software, D.K.B. and E.V.A.; validation, G.V.N., S.M.Z. and E.V.A.; formal analysis, E.V.A.; investigation, G.V.N. and S.M.Z.; resources, G.V.N. and N.S.K.; data curation, D.K.B. and E.V.A.; writing—original draft preparation, G.V.N. and S.M.Z.; writing—review and editing, G.V.N., S.M.Z. and E.V.A.; visualization, E.V.A. and N.S.K.; supervision, S.M.Z.; project administration, S.M.Z.; funding acquisition, S.M.Z.

Funding: This research was funded by Russian Fond Basic Research Grant No. 19-04-00464 and Ministry of Science and Higher Education of the Russian Federation (Project No. 0330-2016-0011).

Acknowledgments: The authors are grateful to the Guest Editors of the Special Issue and to the anonymous reviewer Canadian geologist for the benevolent attitude and significant improvement in the English language of this article. The authors express their sincere thanks to L.P. Boboshko for substantial assistance in processing the factual material and preparing the article for publication. The work was carried out at the Analytical Center for multi-elemental and isotope research SB RAS.

Conflicts of Interest: The authors declare no conflicts of interest.

References

1. Cabri, L.J.; Harris, D.C.; Weiser, T.W. Mineralogy and distribution of Platinum group mineral (PGM) in placer deposits. *Explor. Min. Geol.* **1996**, *5*, 73–167.

2. Barkov, A.Y.; Shvedov, G.I.; Silyanov, S.A.; Martin, R.F. Mineralogy of Platinum-Group Elements and Gold in the Ophiolite-Related Placer of the River Bolshoy Khailyk, Western Sayans, Russia. *Minerals* **2018**, *8*, 247. [CrossRef]

3. Oberthür, T.; Weiser, W.; Melcher, F. Alluvial and eluvial platinum-group minerals from the Bushveld complex, South Africa. *S. Afr. J. Geol.* **2014**, *117*, 255–274. [CrossRef]

4. Zaccarini, F.; Garuti, G.; Pushkarev, E.; Thalhammer, O. Origin of Platinum Group Minerals (PGM) Inclusions in Chromite Deposits of the Urals. *Minerals* **2018**, *8*, 379. [CrossRef]

5. Barkov, A.Y.; Shvedov, G.I.; Martin, R.F. PGE–(REE–Ti)-rich micrometer-sized inclusions, mineral associations, compositional variations, and a potential lode source of platinum-group minerals in the Sisim Placer Zone, Eastern Sayans, Russia. *Minerals* **2018**, *8*, 181. [CrossRef]

6. Melcher, F.; Oberthur, T.; Lodziak, J. Modification of detrital platinum-group minerals from the Eastern Bushveld complex, South Africa. *Can. Mineral.* **2005**, *43*, 1711–1734. [CrossRef]

7. McClenaghan, M.B.; Cabri, L.J. Review of gold and platinum group element (PGE) indicator minerals methods for surficial sediment sampling. *Geochem. Explor. Environ. Anal.* **2011**, *11*, 251–263. [CrossRef]

8. Oberthür, T.; Melcher, F.; Weiser, T. Detrital platinum-group minerals and gold in placers of south-eastern Samar Island, Philippines. *Can. Mineral.* **2017**, *54*, 45–62. [CrossRef]

9. Oberthür, T. The Fate of Platinum-Group Minerals in the Exogenic Environment—From Sulfide Ores via Oxidized Ores into Placers: Case Studies Bushveld Complex, South Africa, and Great Dyke, Zimbabwe. *Minerals* **2018**, *8*, 581. [CrossRef]

10. Garuti, G.; Fershtater, G.B.; Bea, F.; Montero, P.; Pushkarev, E.; Zaccarini, F. Platinum-group elements as petrological indicators in mafic-ultramafic complexes of the central and southern Urals. *Tectonophysics* **1997**, *276*, 181–194. [CrossRef]

11. González-Jiménez, J.M.; Proenza, J.A.; Martini, M.; Camprubí, A.; Griffin, W.L.; O'Reilly, S.Y.; Pearson, N.J. Deposits associated with ultramafic–mafic complexes in Mexico: The Loma Baya case. *Ore Geol. Rev.* **2017**, *81*, 1053–1065. [CrossRef]

12. Buslov, M.M.; Geng, H.; Travin, A.V.; Otgonbaatar, D.; Kulikova, A.V.; Ming, C.; Stijn, G.; Semakov, N.N.; Rubanova, E.S.; Abildaeva, M.A.; et al. Tectonics and geodynamics of Gorny Altai and adjacent structures of the Altai-Sayan folded area. *Russ. Geol. Geophys.* **2013**, *54*, 1250–1271. [CrossRef]

13. Dobretsov, N.L. Evolution of structures of the Urals, Kazakhstan, Tien Shan, and Altai-Sayan region within the Ural-Mongolian fold belt (Paleoasian Ocean). *Russ. Geol. Geophys.* **2003**, *44*, 5–27.

14. Dobretsov, N.L.; Buslov, M.M.; De Grave, J.; Sklyarov, E.V. Interplay of magmatism, sedimentation, and collision processes in the Siberian craton and the flanking orogens. *Russ. Geol. Geophys.* **2013**, *54*, 1135–1149. [CrossRef]

15. Kuzmin, M.I.; Yarmolyuk, V.V. Mantle plumes of Central Asia (Northeast Asia) and their role in forming endogenous deposits. *Russ. Geol. Geophys.* **2014**, *55*, 120–143. [CrossRef]

16. Rudnev, S.N.; Babin, G.A.; Kovach, V.P.; Kiseleva, V.Y.; Serov, P.A. The early stages of island-arc plagiogranitoid magmatism in Gornaya Shoriya and West Sayan. *Russ. Geol. Geophys.* **2013**, *54*, 20–33. [CrossRef]

17. Alabin, L.V.; Kalinin, Y.A. *Gold Metallogeny of Kuznetsk Alatau*; Izd. SO RAN: Novosibirsk, Russia, 1999. (In Russian)

18. Buslov, M.M. Tectonics and geodynamics of the Central Asian Foldbelt: The role of Late Paleozoic large-amplitude strike-slip faults. *Russ. Geol. Geophys.* **2011**, *52*, 52–71. [CrossRef]

19. Babin, G.A.; Gusev, N.I.; Yur'ev, A.A.; Uvarov, A.N.; Dubskii, V.S.; Chernykh, A.I.; Shchigrev, A.F.; Chusovitina, G.D.; Korableva, T.V.; Kosyakova, L.N.; et al. *Explanatory Note to the State Geological Map of the Russian Federation, Scale 1:1,000,000*, 3rd ed.; Sheet N-45, Novokuznetsk; Izd. Kartfabriki VSEGEI: St. Petersburg, Russia, 2007. (In Russian)

20. Izokh, A.E. Layered Mafic–Ultramafic Associations as Indicators of Geodynamic Settings (on the Example of the Central Asian foldbelt). Ph.D. Thesis, UIGGM SB RAS, Novosibirsk, Russian, 1999. (In Russian).

21. Kurenkov, S.A.; Didenko, A.N.; Simonov, V.A. *Geodynamics of Paleospreading*; GEOS: Moscow, Russia, 2002. (In Russian)

22. Pinus, G.V.; Kuznetsov, V.A.; Volokhov, I.M. *Hyperbasites of the Altai-Sayan Folded Area*; Izd. AN SSSR: Moscow, Russia, 1958. (In Russian)

23. Plotnikov, A.V.; Stupakov, S.I.; Babin, G.A.; Vladimirov, A.G.; Simonov, V.A. Age and geodynamic setting of the Kuznetsk Alatau ophiolites. *Dokl. Earth Sci.* **2000**, *372*, 608–612.

24. Tolstykh, N.D. Mineral Assemblages from Pt-Bearing Placers and Genetic Correlations with Their Bedrock Sources. Ph.D. Thesis, UIGGM SB RAS, Novosibirsk, Russia, 2004. (In Russian).

25. Podlipsky, M.Y.; Krivenko, A.P. New data on geological structure, lithology, and formational type of the Kaigadat massif as a primary source of Pt- and Fe-bearing PGM in placers. In *Topical Problems of Geology and Minerageny of Southern Siberia, 31 October–2 November 2001*; Elan'; Kemerovo District: Novosibirsk, Russian, 2001; pp. 126–132. (In Russian)

26. Gertner, I.F.; Krasnova, T.S. Geochemistry of ophiolitic rock paragenesis from Mts. Severnaya, Zelenaya, and Barkhatnaya (Kuznetsk Alatau). In *Petrology of Magmatic and Metamorphic Complexes*; TsNTI: Tomsk, Russian, 2000; Volume 4, pp. 35–41. (In Russian)

27. Krasnova, T.S.; Gertner, I.F. Ophiolite association of Mts. Severnaya–Zelenaya–Barkhatnaya (Kuznetsk Alatau). In *Petrology of Magmatic and Metamorphic Complexes*; TsNTI: Tomsk, Russian, 2000; pp. 28–34. (In Russian)

28. Polyakov, G.V.; Bognibov, V.I. *Platinum Potential of Ultramafic-Mafic Complexes of Southern Siberia*; Izd. SO RAN, NITS UIGGM SB RAS: Novosibirsk, Russia, 1995. (In Russian)

29. Gusev, A.I.; Grinev, R.O.; Chernyshev, A.I. Petrology and ore potential of the Seglebir ophiolite association (northeastern Gorny Altai and southern Gornaya Shoria). In *Petrology of Magmatic and Metamorphic Complexes*; TsNTI: Tomsk, Russian, 2004; Volume 4, pp. 130–133. (In Russian)

30. Agafonov, L.V.; Velinskii, V.V.; Loskutov, I.Y. Unusual Mineral Assemblage of the Noble Metals in the Dikes within Togul-Sungai Ultramafic Massif, Salair Ridge. *Dokl. Akad. Nauk* **1996**, *351*, 505–508. (In Russian)

31. Agafonov, L.V.; Borisenko, A.S.; Bedarev, N.V.; Loskutov, I.Y.; Akimtsev, V.A. PGE and other native element minerals in primary and placer deposits of Central Salair. In *Petrology of Magmatic and Metamorphic Complexes, 300th Anniversary of Mining and Geological Survey of Russia, Tomsk, 29–30 March 2000*; TsNTI: Tomsk, Russian, 2000; pp. 105–110. (In Russian)

32. Bulynnikov, A.Y. *Gold Ore Formations and Gold-Bearing Provinces of the Altai-Sayan Mountain System*; Tomsk. Gos. Univ.: Tomsk, Russia, 1948. (In Russian)

33. Butvilovskii, V.V.; Avakumov, A.E.; Gutak, O.Y. *The Gold Placer Potential of southern West Siberia. Overview of the History and Geology and Potential Assessment*; Kuzbass State Pedagogical Academy: Novokuznetsk, Russian, 2011. (In Russian)

34. Nesterenko, G.V. *Prediction of Gold Mineralization by Placers (on the Example of Southern Siberia)*; Nauka: Novosibirsk, Russia, 1991. (In Russian)

35. Kyuz, A.K. Platinum potential of Kuznetsk Alatau. *Sov. Gold Ind.* **1935**, *5*, 23–25. (In Russian)

36. Syrovatskii, V.V. *Perspective Lines of PGE Studies*; Zapsibgeologiya: Novokuznetsk, Russia, 1991. (In Russian)

37. Vysotskiy, N.K. *Platinum and Areas of Its Mining*; Academy of Sciences of the USSR: Moscow, Russia, 1933. (In Russian)

38. Zhmodik, G.V.; Nesterenko, E.V.; Airiyants, E.V.; Belyanin, D.K.; Kolpakov, V.V.; Podlipsky, M.Y.; Karmanov, N.S. Alluvial platinum-group minerals as indicators of primary PGE mineralization (placers of southern Siberia). *Russ. Geol. Geophys.* **2016**, *57*, 1437–1464. [CrossRef]

39. Studenikin, V.P.; Smirnova, A.I. *Geological Map of 1: 200 000 Scale, Gorno-Altai Series. Sheet N-45-XXXIV. Explanatory Note*; Gosgeoltehizdat: Moscow, Russia, 1963. (In Russian)

40. Fominsky, V.I. *Geological Map of 1: 200 000 Scale, Altai Series. Sheet N-45-XXXV. Explanatory Note*; Gosgeoltehizdat: Moscow, Russia, 1961; 95p. (In Russian)

41. Boitsov, V.E.; Surkov, A.V.; Akhapkin, A.A. Methodology for studying native gold from waste rock stockpiles. *Izv. Vuzov. Geol. Razved.* **2005**, *2*, 42–45. (In Russian)

42. Newbury, D.E.; Ritchie, N.W.M. Is Scanning Electron Microscopy/Energy Dispersive X-ray Spectrometry (SEM/EDS) Quantitative? *Scanning* **2013**, *35*, 141–168. [CrossRef] [PubMed]

43. Newbury, D.E.; Ritchie, N.W.M. Performing elemental microanalysis with high accuracy and high precision by scanning electron microscopy/silicon drift detector energy-dispersive X-ray spectrometry (SEM/SDD-EDS). *J. Mater. Sci.* **2015**, *50*, 492–518. [CrossRef] [PubMed]

44. Lavrent'ev, Y.G.; Usova, L.V. RMA89 Software Suit for Use with a CAMABAX Microprobe. *J. Anal. Chem. USSR* **1991**, *46*, 49–54.

45. Lavrent'ev, Y.G.; Karmanov, N.S.; Usova, L.V. Electron probe microanalysis of minerals: Microanalyzer or scanning electron microscope? *Russ. Geol. Geophys.* **2015**, *56*, 1154–1161. [CrossRef]

46. Nesterenko, G.V.; Zhmodik, S.M.; Airiyants, E.V.; Belyanin, D.K.; Kolpakov, V.V.; Bogush, A.A. Colloform high-purity platinum from placer deposit of Koura River (Gornaya Shoria, Russia). *Ore Geol. Rev.* **2017**, *91*, 236–245. [CrossRef]

47. Isaenko, M.P. *The Determinant of Textures and Structures of Ores*; Nedra: Moscow, Russia, 1983. (In Russian)

48. Krivenko, A.P.; Tolstykh, N.D.; Nesterenko, G.V.; Lazareva, E.V. Types of mineral associations of platinum metals in auriferous placers of the Altai-Sayan folded region. *Russ. Geol. Geophys.* **1994**, *35*, 70–78.

49. Kubaschewski, O. *Iron-Binary Phase Diagrams*; Springer: New York, NY, USA, 1982.

50. Okamoto, H. Fe-Pt (Iron-Platinum). *J. Phase Equilibria Diffus.* **2004**, *25*, 395. [CrossRef]

51. Dare, S.A.S.; Barnes, S.-J.; Prichard, H.M.; Fisher, P.C. Mineralogy and Geochemistry of Cu-Rich Ores from the McCreedy East Ni-Cu-PGE Deposit (Sudbury, Canada): Implications for the Behavior of Platinum Group and Chalcophile Elements at the End of Crystallization of a Sulfide Liquid. *Econ. Geol.* **2014**, *109*, 343–366. [CrossRef]

52. Godel, B.; Barnes, S.-J.; Maier, W.D. Platinum-Group Elements in Sulphide Minerals, Platinum-Group Minerals, and Whole-Rocks of the Merensky Reef (Bushveld Complex, South Africa): Implications for the Formation of the Reef. *J. Petrol.* **2007**, *48*, 1569–1604. [CrossRef]

53. Mungall, J.; Brenan, J. Partitioning of platinum-group elements and Au between sulfide liquid and basalt and the origins of mantle-crust fractionation of the chalcophile elements. *Geochim. Cosmochim. Acta* **2014**, *125*, 265–289. [CrossRef]

54. Baker, I.; Wu, M.; Wang, Z. Eutectic/eutectoid multi-principle component alloys: A review. *Mater. Charact.* **2019**, *147*, 545–557. [CrossRef]

55. Hansen, M.; Anderko, K. *Constitution of Binary Alloys*; McGraw-Hill: New York, NY, USA, 1958.

56. Maier, W.D.; Rasmussen, B.; Fetcher, I.R.; Godel, B.; Barnes, S.J.; Fisher, L.A.; Huhma, Y.S.H.; Lahaye, Y. Petrogenesis of the ~2·77 Ga Monts de Cristal Complex, Gabon: Evidence for Direct Pecipitation of Pt-arsenides from Basaltic Magma. *J. Petrol.* **2015**, *56*, 1285–1308. [CrossRef]

57. Sluzhenikin, S.F.; Mokhov, A.V. Gold and silver in PGE-Cu-Ni and PGE ores of the Noril'sk deposis, Russia. *Min. Depos.* **2015**, *50*, 465–492. [CrossRef]

58. Badanina, I.Y.; Malitch, K.N.; Lord, R.A.; Belousova, E.A.; Meisel, T.C. Closed-system behaviour of the Re–Os isotope system recorded in primary and secondary platinum-group mineral assemblages: Evidence from a mantle chromitite at Harold's Grave (Shetland Ophiolite Complex, Scotland). *Ore Geol. Rev.* **2016**, *75*, 174–185. [CrossRef]

59. Sinyakova, E.F.; Kosyakov, V.I. The behaviour of noble-metal admixtures during fractional crystallization of As- and Co-containing Cu-Fe-Ni sulfide melts. *Russ. Geol. Geophys.* **2012**, *53*, 1055–1076. [CrossRef]

minerals

MDPI

Article

Thalhammerite, Pd9Ag2Bi2S4, a New Mineral from the Talnakh and Oktyabrsk Deposits, Noril'sk Region, Russia

Anna Vymazalová [1,*], František Laufek [1], Sergey F. Sluzhenikin [2], Vladimir V. Kozlov [3], Chris J. Stanley [4], Jakub Plášil [5], Federica Zaccarini [6], Giorgio Garuti [6] and Ronald Bakker [6]

[1]   Czech Geological Survey, Geologická 6, 152 00 Prague 5, Czech Republic; frantisek.laufek@geology.cz
[2]   Institute of Geology of Ore Deposits, Mineralogy, Petrography and Geochemistry RAS,
      Staromonetnyi per. 12, Moscow 119017, Russia; sluzh@igem.ru
[3]   Oxford Instruments (Moscow Office), 26 Denisovskii Pereulok, Moscow 105005, Russia; v.kozlov@oxinst.ru
[4]   Department of Earth Sciences, Natural History Museum, London SW7 5BD, UK; c.stanley@nhm.ac.uk
[5]   Institute of Physics, AS CR v.v.i. Na Slovance 2, 182 21 Prague 8, Czech Republic; plasil@fzu.cz
[6]   Department of Applied Geosciences and Geophysics, University of Leoben, Peter Tunner Str. 5,
      A 8700 Leoben, Austria; federica.zaccarini@unileoben.ac.at (F.Z.); giorgio.garuti1945@gmail.com (G.G.);
      ronald.bakker@unileoben.ac.at (R.B.)
*   Correspondence: anna.vymazalova@geology.cz; Tel.: +420-251-085-228

Received: 19 July 2018; Accepted: 3 August 2018; Published: 8 August 2018

check for updates

Abstract: Thalhammerite, Pd9Ag2Bi2S4, is a new sulphide discovered in galena-pyrite-chalcopyrite and millerite-bornite-chalcopyrite vein-disseminated ores from the Komsomolsky mine of the Talnakh and Oktyabrsk deposits, Noril'sk region, Russia. It forms tiny inclusions (from a few µm up to about 40–50 µm) intergrown in galena, chalcopyrite, and also in bornite. Thalhammerite is brittle and has a metallic lustre. In plane-polarized light, thalhammerite is light yellow with weak bireflectance, weak pleochroism, in shades of slightly yellowish brown and weak anisotropy; it exhibits no internal reflections. Reflectance values of thalhammerite in air (R_1, R_2 in %) are: 41.9/43.0 at 470 nm, 43.9/45.1 at 546 nm, 44.9/46.1 at 589 nm, and 46.3/47.5 at 650 nm. Three spot analyses of thalhammerite give an average composition: Pd 52.61, Bi 22.21, Pb 3.92, Ag 14.37, S 7.69, and Se 0.10, total 100.90 wt %, corresponding to the empirical formula $Pd_{8.46}Ag_{2.28}(Bi_{1.82}Pb_{0.32})_{\Sigma2.14}(S_{4.10}Se_{0.02})_{\Sigma4.12}$ based on 17 atoms; the average of five analyses on synthetic thalhammerite is: Pd 55.10, Bi 24.99, Ag 12.75, and S 7.46, total 100.30 wt %, corresponding to $Pd_{8.91}Ag_{2.03}Bi_{2.06}S_{4.00}$. The density, calculated on the basis of the empirical formula, is 9.72 g/cm^3. The mineral is tetragonal, space group $I4/mmm$, with a 8.0266(2), c 9.1531(2) Å, V 589.70(2) Å^3 and Z = 2. The crystal structure was solved and refined from the single-crystal X-ray-diffraction data of synthetic Pd9Ag2Bi2S4. Thalhammerite has no exact structural analogues known in the mineral system; chemically, it is close to coldwellite (Pd_3Ag_2S) and kravtsovite ($PdAg_2S$). The strongest lines in the X-ray powder diffraction pattern of synthetic thalhammerite [d in Å (I) (hkl)] are: 3.3428(24)(211), 2.8393(46)(220), 2.5685(21)(301), 2.4122(100)(222), 2.3245(61)(123), 2.2873(48)(004), 2.2201(29)(132), 2.0072(40)(400), 1.7481(23)(332), and 1.5085(30)(404). The mineral honours Associate Professor Oskar Thalhammer of the University of Leoben, Austria.

Keywords: thalhammerite; platinum-group mineral; Pd9Ag2Bi2S4 phase; reflectance data; X-ray-diffraction data; crystal structure; Komsomolsky mine; Talnakh deposit; Noril'sk region; Russia

1. Introduction

Thalhammerite, ideally Pd9Ag2Bi2S4, was observed in the same holotype specimen as kravtsovite, PdAg2S [1], and vymazalováite, Pd3Bi2S2 [2]. The type sample (polished section) comes from vein-disseminated pyrite-chalcopyrite-galena ore from the Komsomolsky mine in the Talnakh deposit of the Noril'sk district, Russia. The sample was found at coordinates: 69°30′20″ N and 88°27′17″ E. The mineralization is characterized by lack of Ni minerals and high galena content and Pt-Pd-Ag bearing minerals in an association of pyrite and chalcopyrite. The host rocks of pyrite-chalcopyrite-galena ore are diopside-hydrogrosssular-serpentine metasomatites developed in diopside-monticellite skarns below the lower exocontact of the Talnakh intrusion (the eastern part of the Komsomolsky mine). Thalhammerite, in pyrite-chalcopyrite-galena ores, occurs in association with cooperite, braggite, vysotskite, stibiopalladinite, telargpalite, sobolevskite, kotulskite, sopcheite, insizwaite, kravtsovite, vymazalováite, Au-Ag alloys, and Ag-bearing sulphides, selenides, sulphoselenides, and tellurosulphoselenides. The mineral was also observed in vein-disseminated millerite-bornite-chalcopyrite ore from the Talnakh and Oktyabrsk deposits of the Noril'sk region [3]. The host rocks of millerite-bornite-chalcopyrite ore are pyroxene-hornfels at the lower exocontact of the Kharaelakh intrusion (the western part of the Komsomolsky mine). In millerite-bornite-chalcopyrite ore, thalhammerite occurs in association with kotulskite, telargpalite, laflammeite, and Au-Ag alloys.

The mineral likely formed under the same conditions as kravtsovite and vymazalováite, with decreasing temperature [3], most likely below 400 °C. Thalhammerite was also observed, in intergrowths with sobolevskite, in PGE ores from the Fedorov-Pana Layered Intrusive Complex, Russia (V.V. Subbotin—per. communication). Furthermore, the occurrence of unknown phases corresponding to Pb- and Tl-analogues of thalhammerite from the Fedorov-Pana Layered Intrusive Complex has been reported [4].

Both the mineral and name were approved by the Commission on New Minerals, Nomenclature and Classification of the International Mineralogical Association (IMA No 2017-111). The mineral name is for Dr. Oskar Thalhammer (b. 1956) Associate Professor at the University of Leoben, Austria for his contributions to the ore mineralogy and mineral deposits of platinum group elements. The type specimen is deposited at the Department of Earth Sciences of the Natural History Museum, London, UK, catalogue no. BM 2016, 150.

2. Appearance, and Physical and Optical Properties

Thalhammerite forms very small inclusions (from a few μm up to about 40–50 μm) in galena, chalcopyrite (Figure 1), and also in bornite.

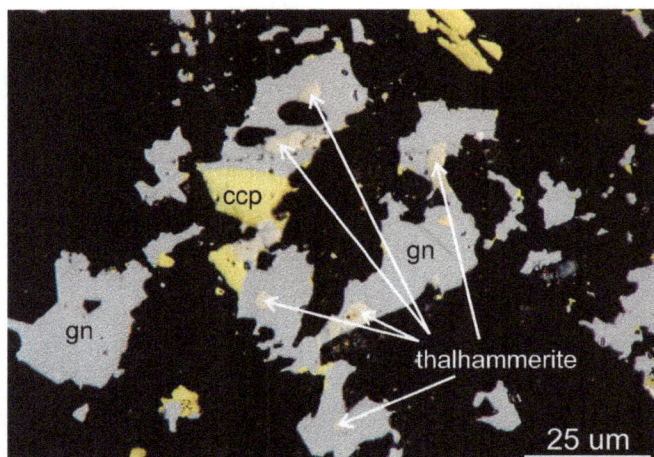

(a)

Figure 1. *Cont.*

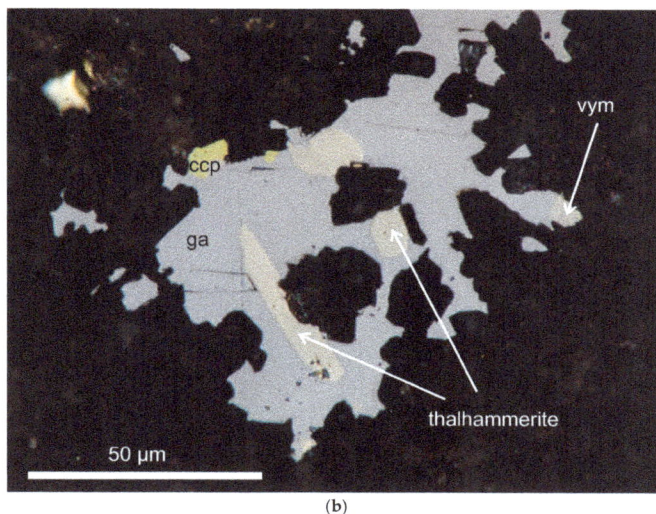

(b)

Figure 1. Digital image in reflected plane polarized light showing inclusions of thalhammerite in galena (gn) in association with (**a**) chalcopyrite (ccp) and (**b**) vymazalováite (vym).

The mineral occurs in aggregates (100–200 μm in size) formed by intergrowths of telargpalite, braggite, vysotskite, sopcheite, stibiopalladinite, sobolevskite, moncheite, kotulskite, malyshevite, insizwaite, acanthite, aurian silver, kravtsovite, and vymazalováite in association with galena, chalcopyrite, bornite, millerite, and pyrite.

Thalhammerite is opaque with a metallic lustre. The mineral is brittle. The density calculated on the basis of the empirical formula is 9.72 g/cm^3. In plane-polarized light, thalhammerite is light yellow with weak bireflectance, weak pleochroism, in shades of slightly yellowish brown and weak anisotropy. It exhibits no internal reflections.

Reflectance measurements were made in air relative to a WTiC standard on both natural and synthetic thalhammerite using a J and M TIDAS diode array spectrometer attached to a Zeiss Axiotron microscope. The results are tabulated (Table 1) and illustrated in Figure 2.

Table 1. Reflectance data for natural and synthetic thalhammerite.

λ (nm)	Natural		Synthetic	
	R$_1$ (%)	R$_2$ (%)	R$_1$ (%)	R$_2$ (%)
400	40.0	41.2	40.6	42.1
420	40.6	41.8	41.3	42.6
440	41.1	42.3	42.0	43.2
460	41.7	42.8	42.6	43.9
470	**41.9**	**43.0**	**42.9**	**44.3**
480	42.2	43.3	43.1	44.6
500	42.7	43.9	43.8	45.3
520	43.2	44.4	44.6	46.0
540	43.7	44.9	45.3	46.6
546	**43.9**	**45.1**	**45.6**	**46.9**
560	44.2	45.4	45.9	47.2
580	44.7	45.9	46.4	47.7
589	**44.9**	**46.1**	**46.7**	**47.9**
600	45.2	46.3	46.9	48.1
620	45.6	46.8	47.3	48.5
640	46.1	47.3	47.7	48.9

Table 1. *Cont.*

λ (nm)	Natural		Synthetic	
	R_1 (%)	R_2 (%)	R_1 (%)	R_2 (%)
650	**46.3**	**47.5**	**47.9**	**49.1**
660	46.5	47.8	48.0	49.2
680	47.0	48.3	48.3	49.5
700	47.4	48.9	48.6	49.8

Note. The values required by the Commission on Ore Mineralogy are given in bold.

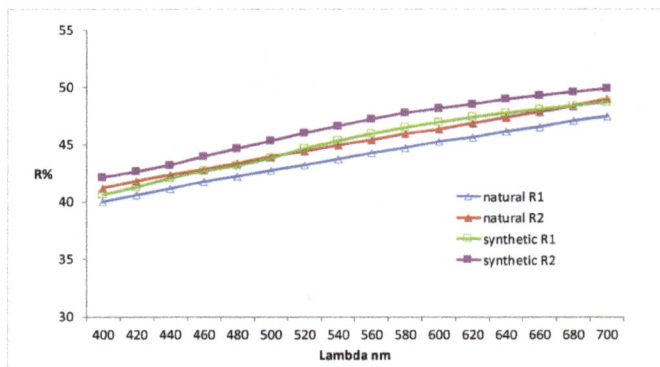

Figure 2. Reflectance data for thalhammerite compared to synthetic analogue, in air. The reflectance values (R%) are plotted versus the wavelength λ in nm.

3. Chemical Composition

Electron probe micro-analyses (EPMA) on grains of thalhammerite were obtained using a WDA Inca Wave 500 (Oxford Instruments NanoAnalysis, High Wycombe, UK) installed on an SEM Lyra 3GM (Tescan), with analytical conditions of 20 kV, 10 nA, and counting times of 30 s (on peak positions)/ 2×15 s (background on the left and right positions). The spectra were collected on PbM_α, BiM_α, PdL_α, AgL_α, SK_α, and SeL_α lines with standards of pure Se, Pd, Ag, Bi, synthetic PbTe, and natural FeS_2. Other elements were below the detection limit.

EPMA on synthetic thalhammerite were obtained using a CAMECA SX-100 electron probe microanalyzer in wavelength-dispersive mode with an electron beam focussed to 1–2 μm.

Pure elements and ZnS were used as standards and the radiations measured were BiM_α PdL_α, AgL_α, and SK_α, with an accelerating voltage of 15 kV, and a beam current of 10 nA measured on the Faraday cup.

EPMA compared with literature data are given in Table 2. The empirical formulae calculated on the basis of 17 *apfu* are $Pd_{8.46}Ag_{2.28}(Bi_{1.82}Pb_{0.32})_{\Sigma2.14}(S_{4.10}Se_{0.02})_{\Sigma4.12}$ for thalhammerite and $Pd_{8.91}Ag_{2.03}Bi_{2.06}S_{4.00}$ for its synthetic analogue, with the ideal formulae $Pd_9Ag_2Bi_2S_4$.

Table 2. Electron-microprobe analyses of natural and synthetic thalhammerite.

wt %	Pd	Ag	Pb	Bi	S	Se	Total
Thalhammerite							
	52.80	14.57	2.60	22.56	7.75	0.07	100.35
	53.40	14.29	3.05	22.09	7.62	0.03	100.47
	51.64	14.25	6.12	21.98	7.70	0.19	101.87
average	52.61	14.37	3.92	22.21	7.69	0.10	100.90
13/B-92 *	53.85	12.51		24.84	7.90		99.1
	52.77	12.27	1.77	24.29	7.45	0.57	99.12

Table 2. *Cont.*

wt %	Pd	Ag	Pb	Bi	S	Se	Total
Thalhammerite							
1/K-92 *	53.85	12.8		24.24	7.89		99.01
	52.77	11.83		25.73	7.99		99.53
	54.08	12.21		25.34	7.95		101.21
Synthetic Sample							
Exp37	54.18	13.69		25.02	7.59		100.48
	54.74	12.91		25.04	7.50		100.19
	54.66	12.46		25.90	7.39		100.42
	56.14	12.01		24.70	7.36		100.21
	55.78	12.67		24.27	7.44		100.16
	55.10	12.75		24.99	7.46		100.29
average	55.10	12.75		24.99	7.46		100.30

* Sluzhenikin and Mohkov [2].

4. Synthetic Analogue

The small size of thalhammerite embedded in galena (bornite) prevented its extraction and isolation in an amount sufficient for the relevant crystallographic and structural investigations. Therefore, these investigations were performed on the synthetic $Pd_9Ag_2Bi_2S_4$.

The synthetic phase of $Pd_9Ag_2Bi_2S_4$ was prepared in an evacuated and sealed silica-glass tube in a horizontal furnace in the Laboratory of Experimental Mineralogy of the Czech Geological Survey in Prague. To prevent loss of material to the vapour phase during the experiment, the free space in the tube was reduced by placing a closely-fitting silica glass rod against the charge.

The temperature was measured with Pt-PtRh thermocouples and is accurate to within ± 3 °C. A charge of about 300 mg was carefully weighed out from the native elements. We used, as starting chemicals, palladium (99.95%), silver (99.999%), bismuth (99.999%), and sulphur (99.999%). The starting mixture was sealed and annealed, quenched, and then ground in an agate mortar under acetone and reheated to 350 °C for 134 days. The sample was quenched by dropping the capsule in cold water.

5. X-ray Crystallography

5.1. Single-Crystal X-ray Diffraction

A small fragment of synthetic $Pd_9Ag_2Bi_2S_4$ was mounted on a glass fibre and examined using a Rigaku Super Nova single-crystal diffractometer with an Atlas S2 CCD detector utilizing MoKα radiation, provided by the microfocus X-ray tube and monochromatized by primary mirror optics. The ω rotational scans were used for collection of three-dimensional intensity data. From a total of 3659 reflections, 221 were classified as unique observed with $I > 3\rho(I)$. Corrections for background, Lorentz effects and polarization were applied during data reduction with the CrysAlis software. Empirical absorption correction was performed using the same software yielding R_{int} = 0.034. The crystal structure was solved with a charge-flipping method using the program Superflip [5] and subsequently refined by the full-matrix least-squares algorithm of JANA2006 program [6]. Because of the similarity of atomic number of Pd and Ag (46 and 47, respectively), it is nearly impossible to distinguish between these atoms from single-crystal (MoKα radiation) diffraction data. The refinement indicated five metallic positions, which one of them was assigned as Bi. The remaining metallic sites show multiplicities 2:8:8:4. Considering the empirical chemical composition $Pd_{8.91}Ag_{2.03}Bi_{2.06}S_{4.00}$ ($Z = 2$) and coordination environment of the $4e$ site, which was very different from the others (see structure description), the $4e$ site was refined as Ag position. Next, refinement cycles included all anisotropic displacement parameters, which revealed too large a value for Pd(2) position (U_{eq}(Pd2) = 0.0146 Å^2 cf. 0.0082 and 0.080 Å^2 for Pd(1) and Pd(3), respectively). Refinement of occupancy factors yielded 0.88 occupancy for the Pd(2) position; other positions were found to be fully occupied. Final refinement in the $I4/mmm$ space group for 21 parameters converged smoothly to the

$R = 0.0310$ and $wR = 0.0815$ for 221 observed reflections. Details of data collection, crystallographic data, and refinement are given in Table 3.

Table 3. Crystallographic data for the selected crystal of synthetic thalhammerite, $Pd_9Ag_2Bi_2S_4$.

Crystal Data	
Chemical formula (idealized)	$Pd_9Ag_2Bi_2S_4$
Space group	$I4/mmm$ (No. 139)
a [Å]	8.0266(2)
c [Å]	9.1531(2)
V [Å³]	589.70(2)
Z	2
Crystal size (mm)	$0.034 \times 0.027 \times 0.013$
Data Collection	
Diffractometer	SuperNova
Temperature (K)	293
Radiation	$MoK\alpha$ (0.7107 Å)
Theta range (°)	5.08–27.62
Reflections collected	3659
Independent reflections	226
Unique observed reflections $[I > 3(\sigma)]$	221
Index ranges	$-10 < h < 10$
	$-10 < k < 10$
	$-11 < l < 11$
Absorption correction method	Empirical
Structure Refinement	
Refinement method	Full matrix least-squares on F^2
Parameters/restrains/constrains	21/0/0
R, wR (obs)	0.0310/0.0815
R, wR (all)	0.0318/0.0817
Largest diff. peak and hole (e⁻/Å³)	1.20/−5.20

Atom coordinates and displacement parameters are listed in Table 4. Table 5 shows selected bond lengths.

Table 4. Fractional coordinates and anisotropic displacement parameters (Å²) for synthetic thalhammerite.

Atom	Pd(1)	Pd(2) *	Pd(3)	Ag	Bi	S
Wyckoff Position	2a	8f	8j	4e	4d	8h
x	1/2	1/4	1/2	1/2	1/2	0.2081(4)
y	1/2	1/4	0.2027(2)	1/2	0	0.2081(4)
z	1/2	1/4	0	0.1810(2)	1/4	0
U_{11}	0.0069(8)	0.0104(8)	0.0077(7)	0.0098(6)	0.0090(4)	0.0071(12)
U_{22}	0.0069(8)	0.0104(8)	0.0097(7)	0.0098(6)	0.0090(4)	0.0071(12)
U_{33}	0.0109(13)	0.0051(10)	0.0077(7)	0.0082(9)	0.0079(6)	0.012(2)
U_{12}	0	0.0017(6)	0	0	0	−0.0018(15)
U_{13}	0	0.0011(4)	0	0	0	0
U_{23}	0	0.0011(4)	0	0	0	0
U_{eq}	0.0083(6)	0.0086(5	0.0084(4)	0.0106(4)	0.0086(3)	0.0087(9)

* Refined with 0.88 occupancy.

Table 5. Selected bond distances (Å) in the thalhammerite crystal structure.

Pd(1)	4 × S	2.362(3)	Ag	4 × Pd(3)	2.905(1)
	2 × Ag	2.919(2)		4 × Pd(2)	2.9073(4)
Pd(2)	2 × S	2.3372(7)			
	2 × Bi	2.8378(1)	Bi1	4 × Pd(3)	2.808(1)
	2 × Ag	2.9073(4)		4 × Pd(2)	2.8378(1)
	4 × Pd(3)	3.0670(2)			

Table 5. *Cont.*

Pd(3)	2 × S	2.343(3)
	2 × Bi	2.8080(8)
	2 × Ag	2.905(2)
	4 × Pd(2)	3.0670(2)

It should be noted that the refined tetragonal structure model of thalhammerite is only a substructure. As was revealed by subsequent Rietveld refinement (see below), the powder X-ray diffraction pattern of synthetic thalhammerite shows at medium and high diffraction angles a few very weak unindexed peaks and very subtle peak splitting, which cannot be fitted using the tetragonal model. Attempts to refine the structure from single-crystal data in rhombic subgroups of *I4/mmm* (i.e., *Fmmm*, *Immm*) led to negligible lowering of *R*-factors (e.g., from 0.0313 to 0.0293) with a rapid increase of the refined parameters and correlations between them. Refinements in monoclinic subgroups failed. Additionally, neither of these low-symmetry models describe all peak splitting observed in powder diffraction patterns of synthetic thalhammerite. Therefore, we proposed only the tetragonal average substructure of thalhammerite, leaving some aspects of the structure unclear.

5.2. Powder X-ray Ddiffraction

The powder XRD pattern of synthetic thalhammerite was collected in the Bragg-Brentano geometry on a Bruker D8 Advance diffractometer equipped with the LynxEye XE detector and CuKα radiation. The data were collected in the range from $10°$ to $100°$ 2θ with a step size of $0.005°$ 2θ and 2 s counting time per step. The structure model obtained from a single-crystal XRD study of synthetic thalhammerite was used as a starting structural model in the subsequent Rietveld refinement. The FullProf program [7] was used and the pseudo-Voigt function was used to generate the shape of the diffraction peaks. The refined parameters include those describing peak shape and width, peak asymmetry, unit-cell parameters, the occupancy parameter of the Pd(2) position, and six isotropic displacement parameters.

In total, 17 parameters were refined. No fractional coordinates were refined. The final cycles of Rietveld refinement converged to the agreement factors $R_p = 0.077$ and $R_{wp} = 0.115$. The refinement indicated 7 wt % $Pd_3Bi_2S_2$ ($I2_13$) impurity in the investigated sample.

Figure 3 depicts two details of final Rietveld plot showing weak, however discernible, peak splitting at middle and high diffraction angles of 2θ (i.e., above $50°$). Attempts to index all observed diffractions in the powder pattern in the large and/or lower symmetry unit-cell remained unsuccessful and, therefore, the structure refinement was limited to the tetragonal substructure. Table 6 presents powder diffraction data for thalhammerite.

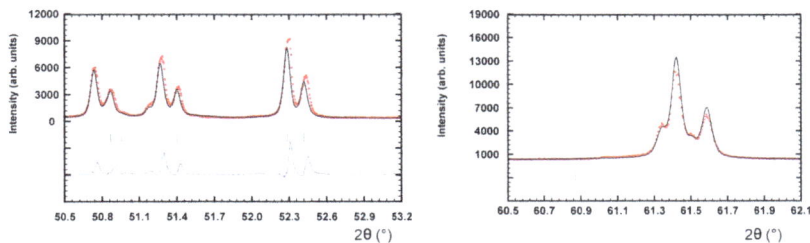

Figure 3. Details of the Rietveld profiles of synthetic thalhammerite showing the weak peak splitting, which cannot be fitted using the tetragonal cell. The observed (circles), calculated (solid), and difference profiles are shown. The vertical bars correspond to Bragg reflections.

6. Structure Description

The tetragonal substructure of thalhammerite contains three Pd, one Ag, Bi, and S sites, respectively. All sites, except the Pd(2) position, were found to be fully occupied. Its crystal structure is shown in Figure 4.

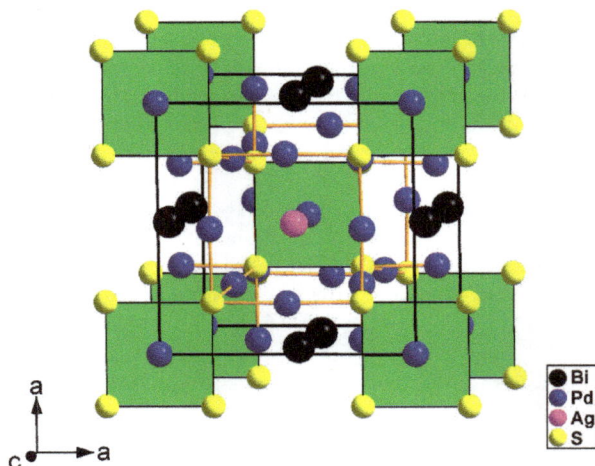

Figure 4. Crystal structure of thalhammerite showing the [PdS$_4$] squares and Pd–S bonds. Unit-cell edges are highlighted. Details show the Rietveld profiles of synthetic thalhammerite showing the weak peak splitting, which cannot be fitted using the tetragonal cell.

6.1. Coordination of Cations

The Pd(1) position is in the centre of regular square of S atoms with Pd(1)-S distances of 2.362(3) Å. The coordination is perfectly planar. Similar coordination was observed in vysotskite, PdS [8], which shows similar Pd–S separation of 2.34 Å. Such coordination geometry is typical for low-spin $4d^8$ Pd^{2+} cation in normal sulfides with M:S ratio equal to or smaller to one [9]. The Pd(1) coordination is further completed by two Ag atoms at 2.919(2) Å lying perpendicular to the [M(1)S$_4$] squares.

The Pd(2) (refined to 0.88 occupancy of Pd) and Pd(3) sites form complex polyhedron. Both Pd positions are coordinated by two S atoms at distances 2.3372(7) and 2.343(3) Å, a value very close to the Pd–S distance of 2.334(4) Å observed for the *zig-zag* chains in the structure of kravtsovite, PdAg$_2$S [1]. Whereas the S–Pd(2)–S group is perfectly linear, the S–Pd(3)–S shows a bonding angle of 177.9(1)°. Pd(2) is further coordinated by two Bi (2.8378(1) Å), two Ag (2.9073(4) Å), and two Pd(3) (3.0670(2) Å) atoms. Pd(3) also shows two Bi (2.8080(8) Å), two Ag (2.905(2) Å), and four Pd(3) (3.0670(2) Å) short contacts.

Ag site is surrounded by nine Pd atoms (Figure 5) forming a mono-capped tetragonal antiprismatic coordination. The Ag–Pd distances are in the range of 2.905(1) Å to 2.919(2) Å, comparable to those observed in lukkulaisvaaraite (Pd–Ag: 2.891(4)–3.037(4) Å; [10], where Ag atoms display tetragonal antiprismatic coordination.

As is shown in Figure 5, the Bi atom is coordinated by eight Pd atoms to form a bi-capped trigonal prism with Bi–Pd bond distances ranging from 2.808(1) to 2.8378(1) Å, values slightly shorter than those observed in structure of monoclinic PdBi (2.84–2.95) Å; [11]. There are no short (<3.5 Å) Bi–S contacts in the thalhammerite crystal structure. This contrasts with the environment of Bi in structure of chemically-related vymazalováite, Pd$_3$Bi$_2$S$_2$ [2,12], where Bi atoms show one additional S contact at 3.22(3) Å.

Table 6. X-ray powder diffraction data of thalhammerite (Cu*K*α radiation, Bruker D8 Advance, Bragg-Brentano geometry). Only reflections with $I_{(obs)} \geq 1$ are listed.

$I_{(obs)}$	h	k	l	$d_{(meas)}$	$d_{(calc)}$
11	1	0	1	6.0364	6.0338
11	1	1	0	5.6790	5.6767
13	0	0	2	4.5752	4.5736
8	2	0	0	4.0155	4.0140
18	1	1	2	3.5620	3.5615
24	2	1	1	3.3428	3.3420
2	2	0	2	3.0181	3.0169
9	1	0	3	2.8510	2.8504
46	2	2	0	2.8393	2.8383
21	3	0	1	2.5685	2.5684
100	2	2	2	2.4122	2.4117
61	1	2	3	2.3245	2.3241
48	0	0	4	2.2873	2.2868
29	1	3	2	2.2201	2.2197
2	2	3	1	2.1637	2.1634
17	1	1	4	2.1213	2.1212
40	4	0	0	2.0072	2.0070
3	3	3	0	1.8923	1.8922
8	4	0	2	1.8377	1.8378
15	2	3	3	1.7981	1.7982
18	2	2	4	1.7805	1.7807
23	3	3	2	1.7481	1.7485
4	1	3	4	1.6991	1.6991
2	4	2	2	1.6711	1.6710
5	1	4	3	1.6413	1.6410
1	2	1	5	1.6299	1.6300
4	4	3	1	1.5814	1.5814
1	5	1	0	1.5743	1.5744
8	0	3	5	1.5102	1.5103
30	4	0	4	1.5085	1.5085
9	1	1	6	1.4723	1.4724
13	4	4	0	1.4193	1.4192
7	4	4	2	1.3554	1.3554
12	2	2	6	1.3431	1.3431
9	2	5	3	1.3395	1.3393
1	3	5	2	1.3185	1.3184
19	3	1	6	1.3070	1.3070
7	1	5	4	1.2969	1.2968
9	6	2	0	1.2694	1.2693
3	1	2	7	1.2279	1.2279
18	6	2	2	1.2231	1.2231
1	1	6	3	1.2113	1.2112
10	4	4	4	1.2059	1.2058
11	3	3	6	1.1872	1.1871

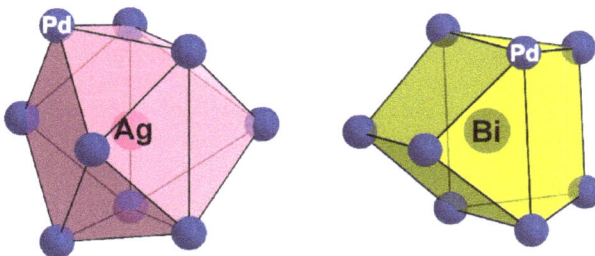

Figure 5. Coordination polyhedra of Ag (mono-capped tetragonal antiprism) and Bi (bi-capped trigonal prism) in the thalhammerite structure.

6.2. Modular Description

The thalhamerite crystal structure forms a three-dimensional framework. It contains features typical for intermetallic compounds (e.g., complex crystallochemical environment of metals) and, therefore, cannot be presented using a traditional cation-based coordination polyhedra approach.

Alternatively, the structure of thalhammerite can be conveniently described as an arrangement of two types of building blocks (cuboids) having common S atoms at the corners (Figure 6).

The first block (green in Figure 6) contains the [PdS$_4$] squares forming one face of the block and Ag atoms in its centre. Pd atoms are approximately located to the midpoints of the longer S-S edges. The second block (orange in Figure 6) contains Bi atoms in its centre. By analogy with the first block, the Pd atoms are located to the midpoints of the longer S–S edges. In the thalhammerite structure, two types of block alternate in a chess-boar fashion within the (001) plane and form chains along the c axis (Figure 6). It should be mentioned that, neglecting the Ag and Bi atoms, the packing of the green blocks automatically generates their duals, and the orange block, vice versa.

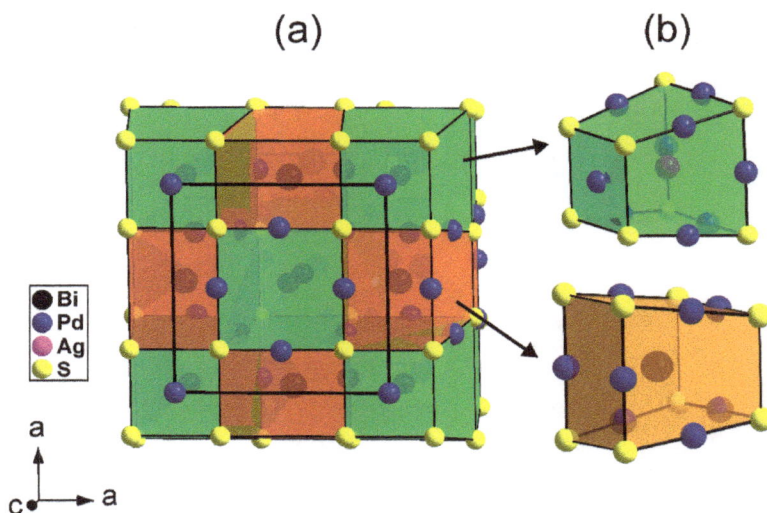

Figure 6. (**a**) Arrangement of two types of building blocks in the thalhammerite structure. (**b**) Detailed view showing the block containing Ag (green) and Bi (orange) atoms.

6.3. Relation to Other Minerals

The thalhammerite structure represents a unique structure type, and no exact structural analogue is hitherto known. It is worth noting that its structure merges structure motives typical for polar chalcogenides and intermetallic compounds. The [Pd(1)S$_4$] square-planar coordination is a hallmark of Pd-bearing sulfides with an M:S ratio equal to, or slightly smaller than, one. Contrary to that, (almost) linear coordination of Pd by two S atoms and number of further metal-metal contacts resulting in complex coordination geometry, can be observed in sulphides with intermetallic behaviour (e.g., kravtsovite PdAg$_2$S, Vymazalová et al., 2017 [1]).

Another chemically-related mineral, coldwellite, Pd$_3$Ag$_2$S (McDonald et al., 2015 [13]), adopts a cubic β-Mn-like structure and, hence, differs substantially from that of thalhammerite.

7. Proof of Identity of Natural and Synthetic Thalhammerite

The structural identity between the synthetic Pd$_9$Ag$_2$Bi$_2$S$_4$ and the natural material was confirmed by electron back-scattering diffraction (EBSD) and Raman spectroscopy.

7.1. Electron Back-Scattering Diffraction

The structural identity between the natural material and the synthetic $Pd_9Ag_2Bi_2S_4$ was confirmed by EBSD. A TESCAN Lyra 3GM field emission scanning electron microscope combined with EBSD system (Oxford Instruments AztecHKL system with NordlysNano EBSD camera) was used for the measurements. The surface of natural sample was prepared for investigation by broad beam argon ion milling using Gatan PECS II system operated at 1 kV. The solid angles calculated from the patterns were compared with our structural model for $Pd_9Ag_2Bi_2S_4$ synthetic phase match containing 12 reflectors to index the patters. The EBSD patterns (also known as Kikuchi patterns) obtained from the natural material (>50 measurements on different spots on natural thalhammerite grains) were found to match the patterns generated from our structural model for $Pd_9Ag_2Bi_2S_4$ synthetic phase, Figure 7.

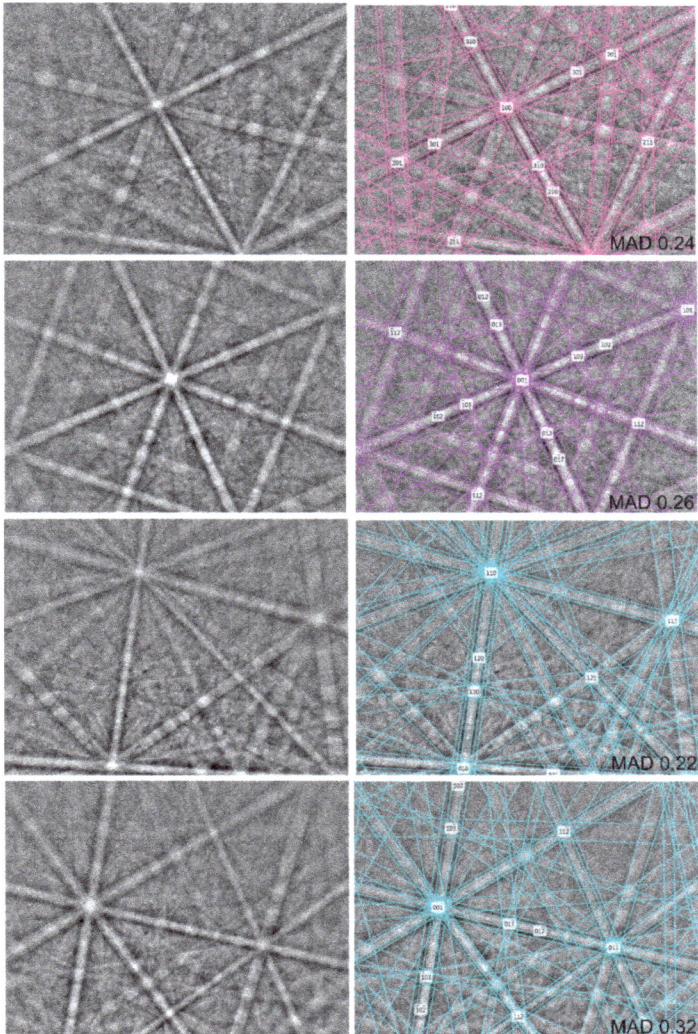

Figure 7. EBSD image of natural thalhammerite; in the right pane, the Kikuchi bands are indexed.

The values of the mean angular deviation (MAD, i.e., goodness of fit of the solution) between the calculated and measured Kikuchi bands range between 0.22° and 0.48°. These values reveal a very good match; as long as values of mean angular deviation are less than 1°, they are considered as indicators of an acceptable fit (HKL Technology, 2004).

7.2. Raman Spectroscopy

The Raman spectroscopy technique was applied to verify the structural identity between the synthetic $Pd_9Ag_2Bi_2S_4$ and the natural material (Figure 8).

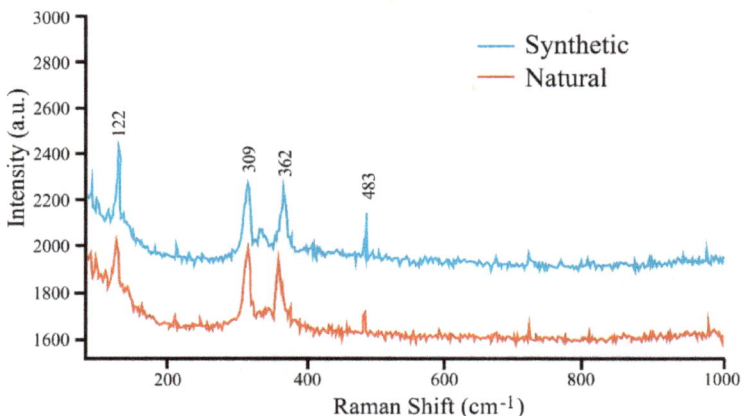

Figure 8. Comparison of Raman spectra in the synthetic $Pd_9Ag_2Bi_2S_4$ and in the natural material.

Raman spectra were obtained using a LABRAM (ISA Jobin Yvon) instrument installed at the University of Leoben, Austria. A frequency-doubled 100 mW Nd:YAG laser with an excitation of a wavelength of λ = 532.6 nm was used. The obtained Raman spectra of natural and synthetic $Pd_9Ag_2Bi_2S_4$ show four discernible absorption bands at the following values: 122, 309, 362, and 483 cm^{-1} (see Figure 8).

The EBSD study, Raman spectra, chemical identity and optical properties confirmed the identity of the natural and synthetic materials and thereby legitimise the use of the synthetic phase for the complete characterization of thalhammerite.

Author Contributions: All the authors (A.V., F.L., S.F.S., V.V.K., C.J.S., J.P., F.Z., G.G. and R.B.) discussed the obtained results, evaluated the data and wrote the article together. A.V. designed the article and conceived experiments; S.F.S. provided the samples and geological background; F.L., V.V.K., J.P. obtained the crystallographic data; C.J.S. provided optical properties; F.Z., G.G. and R.B. studied thalhammerite by Raman and evaluated the chemical data. All the authors revised and edited the manuscript.

Funding: The work was supported by the Grant Agency of the Czech Republic (project no. 18-15390S to A.V.), through an internal project 331400 from the Czech Geological Survey, the Russian Foundation for Basic Research (project RFBR 18-05-70073), and C.J.S. acknowledges Natural Environment Research Council, grant NE/M010848/1, Tellurium and Selenium Cycling and Supply.

Acknowledgments: The authors acknowledge Ulf Hålenius, Chairman of the CNMNC and its members for helpful comments on the submitted data. The authors are grateful to Zuzana Korbelová (Institute of Geology AS CR, v.v.i.) for carrying out the electron microprobe analyses. We thank two anonymous reviewers and the Editorial Board members, for their comments and improvements.

Conflicts of Interest: The authors declare no conflict of interest.

References

1. Vymazalová, A.; Laufek, F.; Sluzhenikin, S.F.; Stanley, C.J.; Kozlov, V.V.; Chareev, D.A.; Lukashova, M.L. Kravtsovite, $PdAg_2S$, a new mineral from Norilsk-Talnakh deposit, Russia. *Eur. J. Mineral.* **2017**, *29*, 597–602. [CrossRef]

2. Sluzhenikin, S.F.; Kozlov, V.V.; Stanley, C.J.; Lukashova, M.I.; Dicks, K. Vymazalováite, $Pd_3Bi_2S_2$, a new mineral from Norilsk -Talnakh deposit, Krasnoyarskiy region, Russia. *Mineral. Mag.* **2018**, *82*, 367–373. [CrossRef]

3. Sluzhenikin, S.F.; Mokhov, A.V. Gold and silver in PGE-Cu-Ni and PGE ores of the Norilsk deposit, Russia. *Mineral. Depos.* **2015**, *50*, 465–492. [CrossRef]

4. Subbotin, V.V.; Gabov, D.A.; Korchagin, A.U.; Savchenko, E.E. Gold and Silver in the Composition of PGE Ores of the Fedorov-Pana Layered Intrusive Complex. *Her. Kola Sci. Cent. RAS* **2017**, *1*, 53–65. (In Russian)

5. Palatinus, L.; Chapuis, G. SUPERFLIP—A computer program for the solution of crystal structures by charge flipping in arbitrary dimensions. *J. Appl. Crystallogr.* **2007**, *41*, 786–790. [CrossRef]

6. Petříček, V.; Dušek, M.; Palatinus, L. Crystallographic Computing System JANA2006: General features. *Z. Kristallogr.* **2014**, *229*, 345–352. [CrossRef]

7. Rodríguez-Carvajal, J. *Full Prof. 2k Rietveld Profile Matching & Integrated Intensities Refinement of X-ray and/or Neutron Data (Powder and/or Single-Crystal)*; Laboratoire Léon Brillouin, Centre d'Etudes de Saclay: Gif-sur-Yvette, France, 2006.

8. Brese, N.E.; Squattrito, P.J.; Ibers, J.A. Reinvestigation of the structure PdS. *Acta Crystallogr. Sect. C* **1985**, *C41*, 1829–1830. [CrossRef]

9. Dubost, V.; Balić-Žunić, T.; Makovicky, E. The crystal structure of $Ni_{9.54}Pd_{7.46}S_{15}$. *Can. Mineral.* **2007**, *45*, 847–855. [CrossRef]

10. Vymazalová, A.; Grokhovskaya, T.L.; Laufek, F.; Rassulov, V.A. Lukkulaisvaaraite, $Pd_{14}Ag_2Te_9$, a new mineral from Lukkulaisvaara intrusion, northern Russian Karelia, Russia. *Mineral. Mag.* **2014**, *78*, 1743–1754. [CrossRef]

11. Bhatt, Y.C.; Schubert, K. Kristallstruktur von PdBi.r. *J. Less Common Met.* **1979**, *64*, 17–24. [CrossRef]

12. Weihrich, R.; Matar, S.F.; Eyert, V.; Rau, F.; Zabel, M.; Andratschke, M.; Anusca, I.; Bernert, T. Structure, ordering and bonding of half antiperovskites $PbNi_{3/2}S$ and $BiNi_{3/2}S$. *Prog. Solid State Chem.* **2007**, *35*, 309–327. [CrossRef]

13. McDonald, A.M.; Cabri, L.J.; Stanley, C.J.; Good, D.J.; Redpath, J.; Spratt, J. Coldwellite, Pd_3Ag_2S, a new mineral species from the Marathon deposit, Coldwell Complex, Ontario, Canada. *Can. Mineral.* **2015**, *53*, 845–857. [CrossRef]

minerals

MDPI

Article

Platiniferous Tetra-Auricupride: A Case Study from the Bolshoy Khailyk Placer Deposit, Western Sayans, Russia

Andrei Y. Barkov [1,*]**, Nobumichi Tamura** [2]**, Gennadiy I. Shvedov** [3]**, Camelia V. Stan** [2]**, Chi Ma** [4]**, Björn Winkler** [5] **and Robert F. Martin** [6]

[1]   Research Laboratory of Industrial and Ore Mineralogy, Cherepovets State University,
      5 Lunacharsky Avenue, 162600 Cherepovets, Russia
[2]   Advanced Light Source, 1 Cyclotron Road, Lawrence Berkeley National Laboratory,
      Berkeley, CA 94720-8229, USA; ntamura@lbl.gov (N.T.); cstan@lbl.gov (C.V.S.)
[3]   Institute of Mining, Geology and Geotechnology, Siberian Federal University, 95 Avenue Prospekt im. gazety
      "Krasnoyarskiy Rabochiy", 660025 Krasnoyarsk, Russia; g.shvedov@mail.ru
[4]   Division of Geological and Planetary Sciences, California Institute of Technology, 1200 East California Blvd.,
      Caltech, 170-25 Pasadena, CA 91125, USA; chi@gps.caltech.edu
[5]   Institut für Geowissenschaften, Universität Frankfurt, Altenhöferallee 1, DE-60438 Frankfurt am Main,
      Germany; b.winkler@kristall.uni-frankfurt.de
[6]   Department of Earth and Planetary Sciences, McGill University, 3450 University Street,
      Montreal, QC H3A 0E8, Canada; robert.martin@mcgill.ca
*   Correspondence: ore-minerals@mail.ru; Tel.: +7-8202-51-78-27

Received: 12 February 2019; Accepted: 1 March 2019; Published: 7 March 2019

check for updates

Abstract: Tetra-auricupride, ideally AuCu, represents the only species showing the coexistence of Au with an elevated level of Pt, as in the case of a detrital grain studied structurally for the first time, from an ophiolite-associated placer at Bolshoy Khailyk, western Sayans, Russia. We infer that tetra-auricupride can incorporate as much as ~30 mol. % of a "PtCu" component, apparently without significant modification of the unit cell. The unit-cell parameters of platiniferous tetra-auricupride are: a 2.790(1) Å, c 3.641(4) Å, with c/a = 1.305, which are close to those reported for ordered AuCu(I) in the system Au–Cu, and close also to the cell parameters of tetraferroplatinum (PtFe), which both appear to crystallize in the same space group, $P4/mmm$. These intermetallic compounds and natural alloys are thus isostructural. The closeness of their structures presumably allows Pt to replace Au atoms so readily. The high extent of Cu + Au enrichment is considered to be a reflection of geochemical evolution and buildup in levels of the incompatible Cu and Au with subordinate Pt in a remaining volume of melt at low levels of fO_2 and fS_2 in the system.

Keywords: platiniferous tetra-auricupride; Pt-for-Au substitution; platinum; gold; ophiolite; Bolshoy Khailyk placer; western Sayans; Russia

---

1. Introduction

1.1. Sample Location and Objective

Tetra-auricupride, AuCu, an intermetallic mineral with a tetragonal symmetry, was first discovered in platiniferous ultrabasic rocks in the Sardala area, Xinjiang Autonomous Region, China [1]. It has since been reported from various localities worldwide (see below). In the majority of cases, the tetra-auricupride is platinum-free, and corresponds closely to the formula AuCu. However, several of them contain elevated levels of Pt; this is the case of the grain described here, originating in the Bolshoy Khailyk placer [2,3], western Sayans, Russia (Figure 1a,b).

The coexistence of Au and Pt is highly uncommon, and has not been observed in other species. Our main objective here was to characterize our specimen of platiniferous tetra-auricupride from Bolshoy Khailyk both compositionally and, for the first time, structurally.

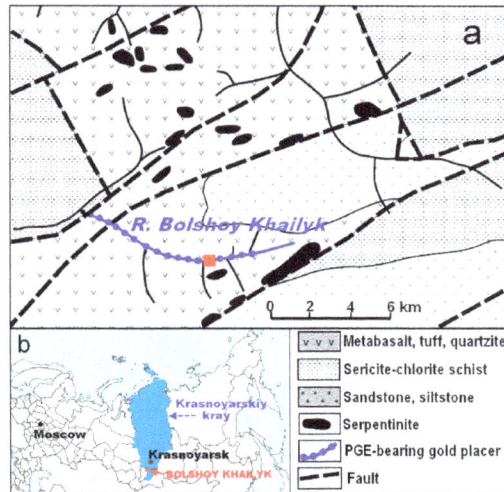

Figure 1. (**a**) Regional geology of the placer area at Bolshoy Khailyk (Figure 1a: simplified after [4]) and (**b**) a general map showing the location of the area in the Russian Federation (shown by red square symbol).

1.2. Worldwide Occurrences of Tetra-Auricupride

Tetra-auricupride has been documented in association with concentrically zoned Alaskan-Uralian-(Aldan)-type complexes at Tulameen*, BC, Canada; Nizhniy Tagil, Urals; and Kondyor*, Aldan Shield, Russia [5–7]. It also occurs in layered mafic–ultramafic complexes at Jijal, Pakistan [8]; Yoko-Dovyrenskiy, southern Siberia [9]; Burakovskiy, Karelia, Russia [10]; and Skaergaard, Greenland [11,12]; and in serpentinized ophiolitic rocks on the island of Skyros [13] and the Pindos complex, Greece [14]. It was also found in ore at the Kerr-Addison mine, ON, Canada [15]; in the Itabira district, Minas Gerais, Brazil [16]; the Bleida Far West mine, Morocco [17]; the Sieroszowice mine, Poland [18]; and in association with Cu–Ni sulfide deposits of the Noril'sk complex, Krasnoyarskiy kray, Russia [19,20] and in the alkaline Coldwell gabbro-syenite complex in Ontario, Canada [21]. In addition, tetra-auricupride occurs in various alluvial deposits in the Sotajoki* area, Finland [22]; Durance, France [23,24]; in placers of the rivers Zolotaya* and Bolshoy Khailyk*, western Sayans [2,3,25]; Olkhovaya-1, Kamchatka krai, Russia [26]; and placers of the southeastern Samar island, Philippines [27]. The tetra-auricupride is notably platiniferous in suites indicated by an asterisk.

2. Materials and Methods

Our materials involved data published in the literature sources and the original results of the present investigation obtained on a detrital grain of tetra-auricupride from the western Sayans. This grain hosts a great variety of minute inclusions, some of which were presently studied using single-crystal electron backscatter diffraction (EBSD). We also employed synchrotron micro-Laue diffraction, wavelength-dispersive analysis (WDS), and scanning-electron microscopy (SEM) combined with energy-dispersive analysis (EDS).

2.1. Occurrence and Associated Minerals at Bolshoy Khailyk

As noted, platiniferous tetra-auricupride was found as a detrital and composite grain associated with platinum-group minerals (PGMs) in a placer deposit at the River Bolshoy Khailyk, western Sayans, southern Krasnoyarskiy kray, Russia [2,3]. The river drains the Aktovrakskiy ophiolitic complex of dunite, harzburgite, and serpentinite, part of the Kurtushibinskiy belt (Figure 1).

An impressive enrichment in Ru is observed in the associated Os–Ir–Ru alloy minerals [3]. The minerals osmium, iridium, and ruthenium are the main PGMs in the Bolshoy Khailyk deposit. Isoferroplatinum-type alloys of Pt–Fe are subordinate. Alloys of the series $(Pt,Ir)(Ni,Fe,Cu)_{3-x}$–(Ir,Pt) $(Ni,Fe,Cu)_{3-x}$ are rare. The sulfide species observed in the placer represent members of the laurite–erlichmanite series, cooperite, bowieite (Cu-rich), a monosulfide-type phase $(Fe_{0.40}Ni_{0.39}Cu_{0.19})_{\Sigma0.98}S_{1.02}$, a bornite-like phase $(Cu_{4.06}Fe_{1.47})_{\Sigma5.5}S_{4.5}$, a godlevskite-like phase $Ni_{9.5}S_{7.5}$, and a thiospinel-like phase $Ni[Ir(Co,Cu,Fe)]_2S_4$. Less-common and rare minerals include sperrylite, a zoned oxide $Ru_6Fe^{3+}{}_2O_{15}$, and an uncommon variety of seleniferous and rhodiferous sperrylite $(Pt,Rh)(As,Se,S)_2$.

Inclusions of clinopyroxene (diopside: $Wo_{48.3-48.6}En_{48.4-48.5}Fs_{2.6}Ae_{0.4-0.7}$; Mg# 96.9–97.9), chromian spinel (magnesiochromite: Mg# up to 71), and serpentine are all rich in Mg, consistent with the ultramafic source-rocks. Actinolite, magnesio-hornblende, and barroisite also are present in inclusions.

The placer grain (Figure 2) hosts numerous inclusions: magnetite poor in Cr, Mg, (Rh,Co)-rich pentlandite, members of the tulameenite–ferronickelplatinum series, a tolovkite–irarsite–hollingworthite solid solution, and a Pt(Cu,Sn) phase [3]. In the present study, we reveal the presence of micrometric inclusions of geversite using electron backscatter diffraction (EBSD).

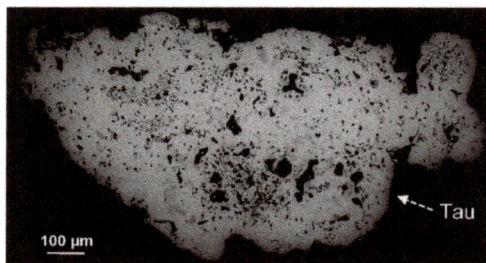

Figure 2. Back-scattered electron image showing the placer grain of platiniferous tetra-auricupride (Tau) from Bolshoy Khailyk. Black inclusions are filled with magnetite; gray phases are Rh–Co-bearing pentlandite and members of the tulameenite–ferronickelplatinum series.

2.2. WDS and SEM/EDS Analyses

Wavelength-dispersive analysis was done using a Camebax-micro electron microprobe at the Sobolev Institute of Geology and Mineralogy, Russian Academy of Sciences, Novosibirsk, Russia. The analytical conditions were 20 kV and 60 nA; the $L\alpha$ line was used for Ir, Rh, Ru, Pt, and Pd; the $M\alpha$ line was used for Os and Au; and the $K\alpha$ line was used for Fe, Ni, Cu, and Co. As standards, we used pure metals (for the platinum-group elements (PGEs) and Au), $CuFeS_2$, synthetic FeNiCo, and pure metals (for Fe, Cu, Ni, and Co). The minimum detection limit was ≤0.1 wt. % for results of the WDS analyses. The SEM/EDS analyses were carried out at 20 kV and 1.2 nA, using a Tescan Vega 3 SBH facility combined with an Oxford X-Act spectrometer at the Siberian Federal University, Krasnoyarsk, Russia. Pure elements (for PGEs, Fe, and Cu) were used as standards. The $L\alpha$ line was used for most of the PGEs except for Pt and Au ($M\alpha$ line); the $K\alpha$ line was used for Fe, Cu, and Ni.

2.3. Synchrotron Micro-Laue Diffraction Study

We have performed synchrotron X-ray scans of the grain of platiniferous tetra-auricupride from the western Sayans (Figure 2) on the basis of Laue microdiffraction measurements at beam line 12.3.2

of the Advanced Light Source (ALS). The Laue diffraction patterns were collected using a PILATUS 1M area detector in reflection geometry. The observed patterns were indexed and analyzed using XMAS (version 6) [28]. A monochromator energy scan was performed to determine the lattice parameters.

2.4. Single-Crystal Electron Backscatter Diffraction (EBSD)

Single-crystal electron backscatter diffraction (EBSD) analyses of micrometer-sized inclusions were performed using an HKL EBSD system on a ZEISS 1550VP Field-Emission SEM (Carl Zeiss Inc., Oberkochenc, Germany), operated at 20 kV and 6 nA in focused-beam mode with a 70° tilted stage and in a variable pressure mode (25 Pa). The focused electron beam is several nanometers in diameter. The spatial resolution for diffracted backscatter electrons is ~30 nm in size. The EBSD system was calibrated using a single-crystal silicon standard.

3. Results

3.1. Compositional Variations in Platiniferous Tetra-Auricupride

The formula of the Pt-bearing tetra-auricupride from Bolshoy Khailyk and four other occurrences (Table 1) can be written (Au,Pt)Cu.

Table 1. Compositions of platiniferous tetra-auricupride from Bolshoy Khailyk and other localities.

#	1	2	3	4	5	6	7	8	9
Locality	KHL	KHL	KHL	KHL	TUL	SOT	ZOL	KON	KON
Au	55.49–62.27	59.69	58.46–63.22	60.81	58.45	50.30	54.52	62.54–72.54	66.15
Pt	13.03–21.24	15.23	13.79–18.14	15.67	15.80	20.40	14.69	3.12–11.79	8.41
Pd	Bdl	bdl	bdl	bdl	bdl	5.49	1.64	bdl	bdl
Rh	bdl	bdl	bdl	bdl	bdl	bdl	0.51	bdl	bdl
Ir	bdl	bdl	bdl	bdl	bdl	bdl	1.96	bdl	bdl
Cu	23.68–26.07	25.43	24.10–25.25	24.52	23.51	22.00	22.68	23.52–24.69	24.15
Fe	bdl–0.72	bdl	bdl	bdl	bdl	0.66	bdl	bdl	bdl
Ni	Bdl	bdl	bdl	bdl	bdl	0.60	bdl	bdl	bdl
Atoms per formula unit (per a total of 2 apfu)									
Au	0.73–0.80	0.77	0.77–0.83	0.80	0.79	0.66	0.75	0.84–0.96	0.88
Pt	0.17–0.28	0.20	0.18–0.24	0.21	0.22	0.27	0.20	0.04–0.16	0.11
Au + Pt	0.96–1.01	0.98	0.99-1.02	1.00	1.01	0.92	0.95	0.98–1.02	1.00
Pd	0	0	0	0	0	0.13	0.04	0	0
Rh	0	0	0	0	0	0	0.01	0	0
Ir	0	0	0	0	0	0	0.03	0	0
Cu	0.96–1.04	1.02	0.98–1.01	1.00	0.99	0.89	0.97	0.98–1.02	1.00
Fe	0–0.03	0	0	0	0	0.03	0	0	0
Ni	0	0	0	0	0	0.03	0	0	0

Note: These results of EMP analyses expressed in wt. % were obtained using WDS, except for columns 1 and 2, which represent quantitative SEM/EDS data. KHL is Bolshoy Khailyk, western Sayans, Russia (This study); TUL is the Tulameen Alaskan-type complex, BC, Canada [5]; SOT is the Sotajoki area in Finland [22]; ZOL is the Zolotaya River placer, western Sayans, Russia [25]; KON is the Kondyor concentrically zoned complex, northern Khabarovskiy kray, Russia [7]. "bdl" means "not detected" or "not analyzed". Columns 1, 3, and 8 display the observed ranges, and nos. 2, 4, and 9 pertain to mean results of point analyses, which are based on a total of 22, 10, and 5 individual analyses, respectively.

The proportion of Pt and Au defines a linear correlation. Up to ~0.3 Pt atoms per formula unit (apfu) can substitute for Au in this series (Figure 3a,b). The contents of Fe and Ni are minor; comparatively large quantities of Pd (up to 5.5 wt. %) may also be present in solid solution in some of these examples [22].

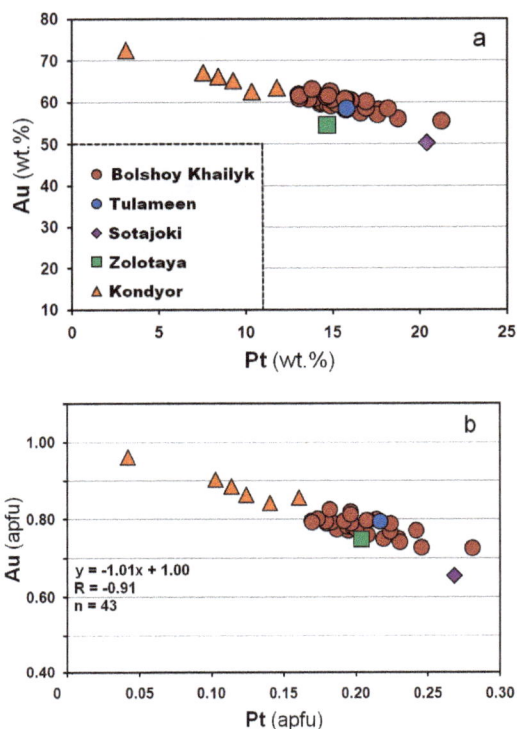

Figure 3. Plots of Pt vs. Au expressed in weight % (**a**) and values of atoms per formula unit (apfu), on the basis of a total of 2 apfu (**b**), showing compositional variations in grains of tetra-auricupride from Bolshoy Khailyk (this study) and various localities reported in the literature.

3.2. Synchrotron Micro-Laue Diffraction Study

The Laue microdiffraction patterns of the studied specimen of platiniferous tetra-auricupride from the western Sayans are shown in Figure 4a,b and Figure 5, and listed in Table 2. The inferred unit-cell parameters of this phase are: a 2.790(1) Å, c 3.641(4) Å, with c/a = 1.305; its space group is $P4/mmm$.

Table 2. X-ray diffraction pattern of platiniferous tetra-auricupride from Bolshoy Khailyk.

h	k	l	d(obs.)	I(rel.)
0	0	1	3.63819	9.5
1	0	0	2.79045	16.1
1	0	1	2.21541	100
1	1	0	1.97535	34.1
1	1	1	1.73554	28.8
1	0	2	1.52408	20.1
2	0	0	1.39583	16.1
1	1	2	1.33812	56.9
2	0	1	1.30304	32.1
2	1	0	1.24814	12.5
2	1	1	1.1808	46.5

Note: The powder-diffraction data were obtained by scans over a large area of the platiniferous tetra-auricupride phase (Figures 2 and 4a,b), d(obs.)—observed values; I(rel.)—intensity (I) is given in relative (rel.) units.

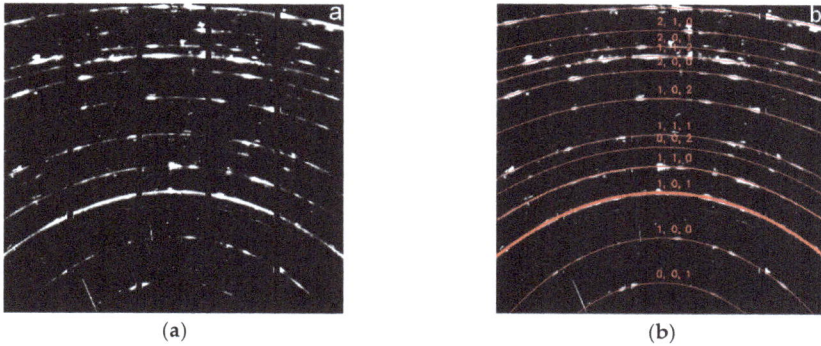

Figure 4. The observed and indexed powder-diffraction patterns (**a** and **b**, respectively) of a specimen of platiniferous tetra-auricupride at Bolshoy Khailyk, which were obtained by summing scans over a large area of the matrix phase in the detrital grain shown in Figure 2.

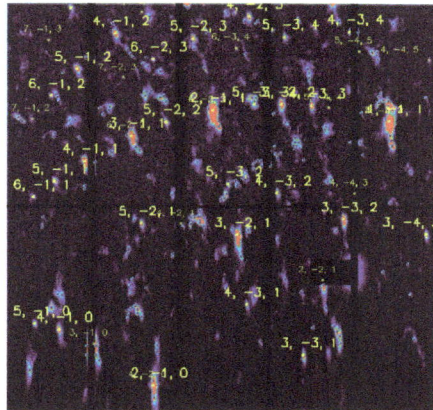

Figure 5. Indexed micro-Laue pattern of platiniferous tetra-auricupride from the western Sayans.

Our results are consistent with those obtained for the ordered AuCu(I) phase known in the Au–Cu system, which has the following parameters: *a* 2.785–2.810 Å, *c* 3.671–3.712 Å, space group *P*4/*mmm* [29]. Furthermore, they agree with the suggested revision [30]. Tetra-auricupride is quoted with *a* 2.800 Å, *c* 3.670 Å, space group *P*4/*mmm*, that is close to the parameters of tetraferroplatinum, PtFe: *a* 2.724 Å, *c* 3.720 Å, which were also revised by this author. We can conclude that the incorporation of up to 30 mol. % of a "PtCu" component does not notably change the unit cell of tetra-auricupride. Note that hongshiite, PtCu, is not an end member in this series. Hongshiite is trigonal, space group: *R*32, *R*3*m*, or *R*$\bar{3}$*m*, with the unit-cell parameters: *a* 10.713 Å, *c* 13.192 Å, *Z* = 48 [31,32]; synthetic PtCu also adopts the trigonal structure [33].

Note that the holotype tetra-auricupride, of composition $Au_{1.01}Cu_{0.99}$, is different. Chen et al. (1982) [1] reported it to have the cell parameters *a* 3.98 Å, *c* 3.72 Å, *Z* = 2, with *C*4/*mmm* as the probable space group. These values agree with those reported for the AuCu(II) phase documented in the system Au–Cu: *a* 3.96 Å, *b* 3.97 Å, *c* 3.68 Å, space group *Imma* [29].

The results of the EBSD analysis (Figure 6a,b) indicate that micrometric inclusions in the tetra-auricupride matrix consist of geversite, $PtSb_2$.

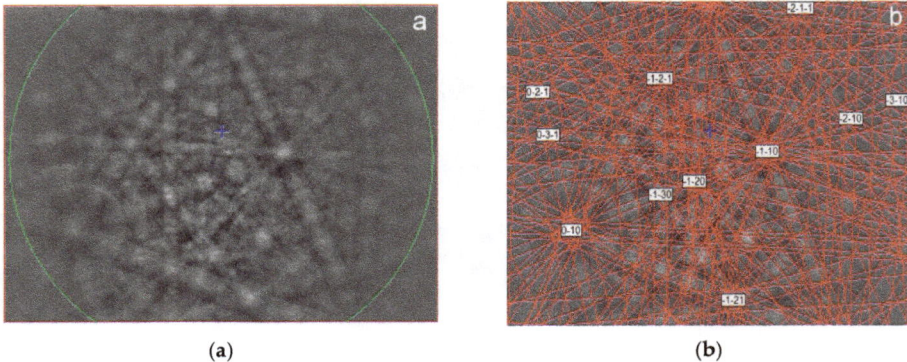

Figure 6. The observed and indexed EBSD patterns (**a** and **b**, respectively) obtained for micrometer-sized inclusion of geversite hosted by the placer grain of platiniferous tetra-auricupride at Bolshoy Khailyk.

4. Discussion and Concluding Comments

The strong dominance of Os–Ir–Ru alloy minerals, their patterns of Ru enrichment, and the regional association of detrital grains of PGMs and Pt-rich tetra-auricupride with exposed bodies of the Aktovrakskiy complex (Figure 1a) provide clear indications of an ophiolite origin of the PGE + Au mineralization at Bolshoy Khailyk. The platiniferous tetra-auricupride in the Zolotaya River placers is also spatially associated with serpentinites of this complex [25]. The high extent of Cu + Au enrichment expressed by the presence of tetra-auricupride is considered to be a reflection of geochemical evolution and the buildup in levels of the incompatible Cu and Au with subordinate Pt in a remaining volume of melt during the bulk crystallization of Os–Ir–Ru alloy phases, after the early crystallization of chromian spinel and olivine [3]. Low levels of O_2 and S_2 fugacities are inferred to have existed in the system.

The suggested crystallization of the Pt-rich tetra-auricupride from a fractionated melt at a moderately low temperature is consistent with compositions of magnetite inclusions, which are poor in Mg and Cr, in contrast to magnesiochromite inclusions in the associated grains of Os–Ir–Ru-rich PGM at Bolshoy Khailyk. It is known that tetra-auricupride (Pt-free) is a late phase deposited during serpentinization of ophiolitic rocks of Skyros Island, Greece [13]. Hydrothermal processes related to serpentinization are recognized to concentrate efficiently elevated levels of gold (1–10 ppm) in sulfide-rich varieties of serpentinites, or in carbonatized (listwaenite) varieties, or in silicified serpentinites associated with ophiolites [34,35].

We establish that in such a low-temperature environment, tetra-auricupride can incorporate as much as ~30 mol. % of a "PtCu" component, apparently without significant modification of the unit cell. This surprising level of incorporation of Pt for Au has not been documented in any other mineral. The unit-cell parameters of such platiniferous auricupride are close to those reported for ordered AuCu(I) in the Au–Cu system, and are also close to the cell parameters of tetraferroplatinum (PtFe). In addition, both seem to crystallize in the same space group, *P4/mmm* [30]. These intermetallic compounds and natural alloys are thus isostructural. The closeness of the two structures presumably allows platinum to displace gold atoms so readily, possibly at conditions of disequilibrium growth.

As noted, the tetra-auricupride grain enriched in Pt at Bolshoy Khailyk hosts a variety of inclusions: Cr–Mg–Mn-bearing magnetite, Co–(Rh)-rich pentlandite and various platinum-group minerals: the tulameenite–ferronickelplatinum series (Pt_2FeCu–Pt_2FeNi), a tolovkite (IrSbS)–irarsite (IrAsS)–hollingworthite (RhAsS) solid solution, geversite $PtSb_2$ (this study), and unnamed (Pt,Pd)(Cu,Sn), among others. On the basis of our experience, we note that EBSD analysis can be a useful tool to recognize micrometric inclusions of PGMs or other phases.

The ore-forming system thus involved at least 17 elements: Cu, Au, Pt, Rh, Pd, Ir, Fe, Co, Ni, S, Sb, As, Sn, Cr, Mn, Mg, and O, present as major or minor constituents in the mineral assemblage in

Minerals **2019**, *9*, 160

association with platiniferous tetra-auricupride. The observed diversity—the presence of compositions rich in Cu, Au, and Sn, along with PGE species having lower melting points (Rh and Pd) and metalloids (Sb and As)—is consistent with crystallization from a highly fractionated melt remaining after the deposition of primary phases such as highly magnesian olivine, magnesiochromite, and alloys of Os–Ir–Ru and $(Pt,Ir)_3Fe$. The upper limit of stability of a synthetic variant of ordered AuCu(II) is 410 °C, whereas the temperature of the phase transition of AuCu(II) to AuCu(I) varies under different conditions of synthesis and is generally close to 385 °C [29,36]. The observed incorporation of high amounts of Pt is expected to somewhat increase the crystallization temperature of the platiniferous AuCu(I) phase at Bolshoy Khailyk. Nevertheless, this phase presumably crystallized at a relatively low temperature, which is unlikely to have exceeded ~600–800 °C. This temperature is consistent with deposition from a droplet of residual melt.

Author Contributions: Authors wrote the article together. A.Y.B., R.F.M. and B.W. provided the interpretation of data and conclusions of the research, including results of synthesis of AuCu(I)-type compounds (B.W.). N.T., C.V.S. and C.M. contributed results of the micro-Laue and EBSD studies. G.I.S. collected placer samples and completed a field work on the Bolshoy Khailyk placer deposit.

Funding: A.Y.B. gratefully acknowledges a partial support of this investigation by the Russian Foundation for Basic Research (project # RFBR 19-05-00181).

Acknowledgments: We thank two reviewers and the Editorial board for their comments and suggestions. This research used beamline 12.3.2 at the Advanced Light Source, which is a DOE Office of Science User Facility under contract no. DE-AC02-05CH11231.

Conflicts of Interest: There is no conflict of interests.

References

1. Chen, K.; Yu, T.; Zhang, Y.; Peng, Z. Tetra-auricupride, CuAu discovered in China. *Sci. Geol. Sin.* **1982**, *17*, 111–116, (In Chinese with English abstract).

2. Shvedov, G.I.; Nekos, V.V. PGM of a placer at the R. Bolshoy Khailyk (western Sayans). *Geol. Resour. Krasnoyarskiy Kray* **2008**, *9*, 240–248. (In Russian)

3. Barkov, A.Y.; Shvedov, G.; Silyanov, S.; Martin, R.F. Mineralogy of platinum-group elements and gold in the ophiolite-related placer of the River Bolshoy Khailyk, western Sayans, Russia. *Minerals* **2018**, *8*, 247. [CrossRef]

4. Aleksandrov, G.P.; Gulyaev, Y.S. *The Geological Map of the USSR (scale 1:200 000; M-46-III)*; The Western Sayan Series; Nedra Press: Moscow, Russia, 1966.

5. Cabri, L.J.; Laflamme, J.H.G. Analyses of minerals containing platinum-group elements. In Platinum-group elements: Mineralogy, Geology, Recovery (L.J. Cabri, ed.). *Can. Inst. Min. Metall.* **1981**, *23*, 151–173.

6. Yushkin, N.P. (Ed.) *Mineralogy of the Urals: Native Elements, Carbides, Sulfides*; The Uralian branch of the USSR Academy of Sciences: Sverdlovsk, Russia, 1990; pp. 92–97. (In Russian)

7. Nekrasov, I.Y.; Lennikov, A.M.; Zalishchak, B.L.; Oktyabrsky, R.A.; Ivanov, V.V.; Sapin, V.I.; Taskaev, V.I. Compositional variations in platinum-group minerals and gold, Konder alkaline-ultrabasic massif, Aldan Shield, Russia. *Can. Miner.* **2005**, *43*, 637–654. [CrossRef]

8. Miller, D.; Loucks, R.R. Platinum-group Element Mineralization in the Jijal layered ultramafic-mafic complex, Pakistani Himalayas. *Econ. Geol.* **1991**, *86*, 1093–1102. [CrossRef]

9. Rudashevskiy, N.S.; Krecer, D.A.; Orsoev, D.A.; Kislov, E.V. Palladium-platinum mineralization in vein Cu–Ni-ores of Yoko-Dovyrenskiy layered massif. *Dokl. Akad. Nauk* **2003**, *391*, 519–522. (In Russian)

10. Grokhovskaya, T.L.; Lapina, M.I.; Ganin, V.A.; Grinevich, N.G. PGE mineralization in the Burakovskiy Layered Complex, Southern Karelia, Russia. *Geol. Rudn. Mestorozh.* **2005**, *47*, 315–341. (In Russian)

11. Holwell, D.A.; Keays, R.R.; McDonald, I.; Williams, M.R. Extreme enrichment of Se, Te, PGE and Au in Cu sulfide microdroplets: Evidence from LA-ICP-MS analysis of sulfides in the Skaergaard Intrusion, east Greenland. *Contrib. Mineral. Petrol.* **2015**, *170*, 1–26. [CrossRef]

12. Nielsen, T.F.; Andersen, J.Ø.; Holness, M.B.; Keiding, J.K.; Rudashevsky, N.S.; Rudashevsky, V.N.; Veksler, I.V. The Skaergaard PGE and gold deposit: The result of in situ fractionation, sulphide saturation, and magma chamber-scale precious metal redistribution by immiscible Fe-rich melt. *J. Petrol.* **2015**, *56*, 1643–1676. [CrossRef]

13. Tarkian, M.; Economou-Eliopoulos, M.; Eliopoulos, D.G. Platinum-group minerals and tetra-auricupride in ophiolitic rocks of Skyros island, Greece. *Miner. Petrol.* **1992**, *47*, 55–66. [CrossRef]

14. Kapsiotis, A.; Grammatikopoulos, T.A.; Tsikouras, B.; Hatzipanagiotou, K. Platinum-Group mineral characterization in concentrates from high-grade PGE Al-rich chromitites of Korydallos area in the Pindos ophiolite complex (NW Greece). *Resour. Geol.* **2010**, *60*, 178–191. [CrossRef]

15. Knipe, S.W.; Fleet, M.E. Copper-gold alloy minerals from the Kerr mine, Ontario. *Can. Miner.* **1997**, *35*, 573–586.

16. Kwitko, R.; Cabral, A.R.; Lehmann, B.; Laflamme, J.G.; Cabri, L.J.; Criddle, A.J.; Galbiatti, H.F. Hongshiite, PtCu, from itabirite-hosted Au-Pd-Pt mineralization (jacutinga), Itabira district, Minas Gerais, Brazil. *Can. Miner.* **2002**, *40*, 711–723. [CrossRef]

17. El Ghorfi, M.; Oberthür, T.; Melcher, F.; Lüders, V.; El Boukhari, A.; Maacha, L.; Ziadi, R.; Baoutoul, H. Gold-palladium mineralization at Bleïda Far West, Bou Azzer-El Graara Inlier, Anti-Atlas, Morocco. *Miner. Depos.* **2006**, *41*, 549–564. [CrossRef]

18. Pieczonka, J.; Piestrzyński, A.; Mucha, J.; Głuszek, A.; Kotarba, M.; Więcław, D. The red-bed-type precious metal deposit in the Sieroszowice-Polkowice copper mining district, SW Poland. *Ann. Soc. Geologorum Pol.* **2008**, *78*, 151–280.

19. Spiridonov, E.M.; Kulagov, E.A.; Serova, A.A.; Kulikova, I.M.; Korotaeva, N.N.; Sereda, E.V.; Tushentsova, I.N.; Belyakov, S.N.; Zhukov, N.N. Genetic Pd, Pt, Au, Ag, and Rh mineralogy in Noril'sk sulfide ores. *Geol. Ore Depos.* **2015**, *57*, 402–432. [CrossRef]

20. Vymazalová, A.; Laufek, F.; Sluzhenikin, S.F.; Stanley, C.J. Norilskite, (Pd,Ag)$_7$Pb$_4$, a new mineral from Noril'sk -Talnakh deposit, Russia. *Miner. Mag.* **2017**, *81*, 531–541. [CrossRef]

21. McDonald, A.M.; Cabri, L.J.; Stanley, C.J.; Good, D.J.; Redpath, J.; Lane, G.; Spratt, J.; Ames, D.E. Coldwellite, Pd$_3$Ag$_2$S, a new mineral species from the Marathon deposit, Coldwell complex, Ontario, Canada. *Can. Miner.* **2015**, *53*, 1–13. [CrossRef]

22. Törnroos, R.; Vuorelainen, Y. Platinum-group metals and their alloys in nuggets from alluvial deposits in Finnish Lapland. *Lithos* **1987**, *20*, 491–500.

23. Ohnenstetter, M.; Johan, Z.; Fisher, W.; Amosse, J. *Platinum-Group Minerals from the Durance River Alluvia, France*; Papunen, H., Ed.; Bulletin of the Geological Society of Finland: Espoo, Finland, 1989; Volume 61, p. 46.

24. Johan, Z.; Ohnenstetter, M.; Fischer, M.; Amossé, J. Platinum-group minerals from the Durance River alluvium, France. *Miner. Petrol.* **1990**, *42*, 287–306. [CrossRef]

25. Tolstykh, N.D.; Krivenko, A.P.; Pospelova, L.N. Unusual compounds of iridium, osmium and ruthenium with selenium, tellurium and arsenic from placers of the Zolotaya River (western Sayans). *Zap. Vsesoyuzn. Miner. Obshch.* **1997**, *126*, 23–34. (In Russian)

26. Sandimirova, E.I.; Sidorov, E.G.; Chubarov, V.M.; Ibragimova, E.K.; Antonov, A.V. Native metals and intermetallic compounds in heavy concentrate halos of the Ol'khovaya-1 River, Kamchatsky Mys Peninsula, eastern Kamchatka. *Geol. Ore Depos.* **2014**, *56*, 657–664. [CrossRef]

27. Oberthür, T.; Melcher, F.; Weiser, T.W. Detrital platinum-group minerals and gold in placers of southeastern Samar Island, Philippines. *Can. Miner.* **2017**, *55*, 45–62. [CrossRef]

28. Tamura, N. XMAS: A Versatile Tool for Analyzing Synchrotron X-ray Microdiffraction Data. In *Strain and Dislocation Gradients from Diffraction*; Barabash, R., Ice, G., Eds.; Imperial College Press: London, UK, 2014; pp. 125–155.

29. Okamoto, H.; Chakrabarti, D.J.; Laughlin, D.E.; Massalski, T.B. The Au-Cu (Gold-Copper) system. *J. Phase Equilib.* **1987**, *8*, 454–474. [CrossRef]

30. Bayliss, P. Revised unit-cell dimensions, space group, and chemical formula of some metallic minerals. *Can. Miner.* **1990**, *28*, 751–755.

31. Yu, Z. New data on hongshiite. *Bull. Inst. Geol.* **1982**, *1982*, 75–81, (In Chinese with English abstract). *Am. Miner.* **1984**, *69*, 411–412.

32. Yu, Z. New data for hongshiite. *Acta Geol. Sin.* **2001**, *75*, 400–403, (In Chinese with English Abstract).

33. Wyckoff, R.W.G. *The Second Edition of Structure of Crystals by Wyckoff Published by The Chemical Catalog Company*; INC: New York, NY, USA, 1931; p. 225.
34. Buisson, G.; Leblanc, M. Gold in mantle peridotites from upper Proterozoic ophiolites in Arabia, Mali, and Morocco. *Econ. Geol.* **1987**, *82*, 2091–2097. [CrossRef]
35. Leblanc, M. Platinum-Group Elements and Gold in Ophiolitic Complexes: Distribution and Fractionation from Mantle to Oceanic Floor. *Petrol. Struct. Geol.* **1991**, 231–260. [CrossRef]
36. Lamiri, I.; Martínez-Blanco, D.; Abdelbaky, M.S.M.; Mari, D.; Hamana, D.; García-Granda, S. Investigation of the order–disorder phase transition series in AuCu by in-situ temperature XRD and mechanical spectroscopy. *J. Alloys Compd.* **2019**, *770*, 748–754. [CrossRef]

minerals

MDPI

Review

The Fate of Platinum-Group Minerals in the Exogenic Environment—From Sulfide Ores via Oxidized Ores into Placers: Case Studies Bushveld Complex, South Africa, and Great Dyke, Zimbabwe

Thomas Oberthür

Bundesanstalt für Geowissenschaften und Rohstoffe (BGR), Stilleweg 2, D-30655 Hannover, Germany;
mut.oberthuer@yahoo.de

Received: 14 August 2018; Accepted: 4 December 2018; Published: 9 December 2018

check for updates

Abstract: Diverse studies were performed in order to investigate the behavior of the platinum-group minerals (PGM) in the weathering cycle in the Bushveld Complex of South Africa and the Great Dyke of Zimbabwe. Samples were obtained underground, from core, in surface outcrops, and from alluvial sediments in rivers draining the intrusions. The investigations applied conventional mineralogical methods (reflected light microscopy) complemented by modern techniques (scanning electron microscopy (SEM), mineral liberation analysis (MLA), electron-probe microanalysis (EPMA), and LA-ICPMS analysis). This review aims at combining the findings to a coherent model also with respect to the debate regarding allogenic versus authigenic origin of placer PGM. In the pristine sulfide ores, the PGE are present as discrete PGM, dominantly PGE-bismuthotellurides, -sulfides, -arsenides, -sulfarsenides, and -alloys, and substantial though variable proportions of Pd and Rh are hosted in pentlandite. Pt–Fe alloys, sperrylite, and most PGE-sulfides survive the weathering of the ores, whereas the base metal sulfides and the (Pt,Pd)-bismuthotellurides are destroyed, and ill-defined (Pt,Pd)-oxides or -hydroxides develop. In addition, elevated contents of Pt and Pd are located in Fe/Mn/Co-oxides/hydroxides and smectites. In the placers, the PGE-sulfides experience further modification, whereas sperrylite largely remains a stable phase, and grains of Pt–Fe alloys and native Pt increase in relative proportion. In the Bushveld/Great Dyke case, the main impact of weathering on the PGM assemblages is destruction of the unstable PGM and PGE-carriers of the pristine ores and of the intermediate products of the oxidized ores. Dissolution and redistribution of PGE is taking place, however, the newly-formed products are thin films, nano-sized particles, small crystallites, or rarely μm-sized grains primarily on substrates of precursor detrital/allogenic PGM grains, and they are of subordinate significance. In the Bushveld/Great Dyke scenario, and in all probability universally, authigenic growth and formation of discrete, larger PGM crystals or nuggets in the supergene environment plays no substantial role, and any proof of PGM "neoformation" in a grand style is missing. The final PGM suite which survived the weathering process en route from sulfide ores via oxidized ores into placers results from the continuous elimination of unstable PGM and the dispersion of soluble PGE. Therefore, the alluvial PGM assemblage represents a PGM rest spectrum of residual, detrital grains.

Keywords: Bushveld Complex; South Africa; Great Dyke; Zimbabwe; platinum-group minerals; primary ores; oxide ores; placers; allogenic; authigenic

---

Contributing to the Debate Allogenic versus Authigenic Origin of Placer PGM

1. Introduction

The world's prime sources of platinum-group elements (PGE) are layered intrusions of Proterozoic and Archean age (e.g., the Bushveld Complex, South Africa; the Great Dyke, Zimbabwe; the Stillwater Complex, USA). Placer deposits in Russia and in Colombia were the world's only providers of platinum-group elements up to early in the 20th century, when the rich primary deposits of the Bushveld Complex in South Africa, of the Sudbury district in Canada, and later Norilsk in Russia came into production [1–3]. Today, only about 1–2% of the world's PGE production originates from placers mainly in the Russian Far East, the Ural Mountains, and Colombia [4]. Remarkably, the identification of detrital platinum-group minerals (PGM) in alluvial gravels on the farm Maandagshoek, then eastern Transvaal, South Africa, led to the discovery of the largest PGE deposits on Earth in the Bushveld Complex in 1924 [5–8]. Today, South Africa is the dominant country in mining and supplying PGE to the world, entirely from the Bushveld Complex, followed by Russia's Norilsk-Talnakh field and modest contributions from Zimbabwe, Canada, and the USA.

The 2054.4 ± 1.3-Ma-old [9] Bushveld Complex in north-eastern South Africa is the largest layered intrusion on Earth, covering an area of at least 65,000 km^2 (Figure 1). The PGE ores are principally related to sulfide mineralization. Three major, laterally extensive ore bodies are mined, namely the Merensky and the UG-2 chromitite reef in the western and eastern Bushveld, and the Platreef in the northern Bushveld [10,11]. In addition, increasing amounts of PGE are extracted as by-products from the chromitite seams of the Bushveld Complex [12].

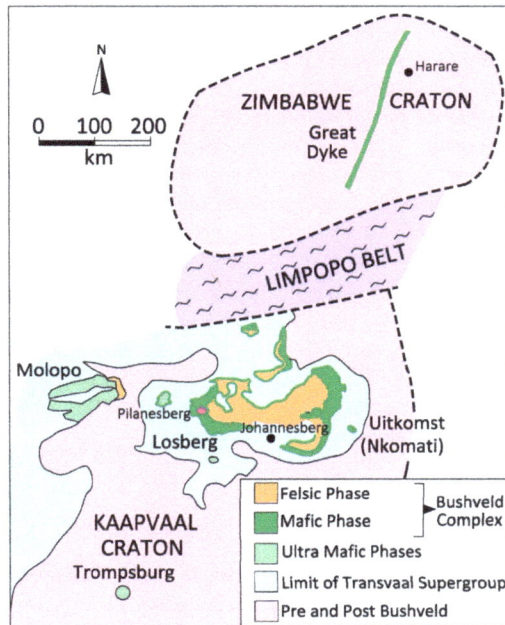

Figure 1. Location of the Bushveld Complex and the Great Dyke in southern Africa (Reproduced with permission from M. Viljoen, *Episodes* [10]).

The 2575.4 ± 0.7-Ma-old [13] Great Dyke of Zimbabwe (Figure 1) is a 550 km long mafic/ultramafic layered intrusion with a NNE strike and a maximum width of about 11 km that transects the Zimbabwe Craton [14]. Within the Great Dyke, economic PGE mineralization is restricted to sulfide disseminations in pyroxenites of the 1–5 m thick Main Sulfide Zone (MSZ), sited some meters below the transition of the Ultramafic and the Mafic Sequence of the Great Dyke [15,16]. The MSZ displays a regular geochemical and mineralogical fine structure which is regarded to reflect primary magmatic features of consecutive batches of sulfide accumulation, concomitant scavenging of PGE, and fractionation [16,17].

In both intrusions, the pristine PGE mineralization is related to sulfide accumulations (mainly pyrrhotite, pentlandite, chalcopyrite, and sporadic pyrite), in a manner typical of Ni–Cu-sulfide mineralization worldwide [18]. The PGE are bimodally distributed; they occur at variable proportions in the form of discrete PGM, and also in variable proportions substituting in the crystal lattices of sulfides (mainly Pd and Rh in pentlandite) [16,19–23].

Under the subtropical conditions encountered in the Bushveld Complex and along the Great Dyke, the outcropping PGE-rich sulfide ores are typically weathered to depths of ca. 20–30 m as observed in drill core and in open pit workings [16,24–26]. The most important features of the oxidation zone are the removal of sulfides and a significant loss of palladium concomitant with the extensive destruction of most species of the primary PGM assemblage. Large proportions of the PGE are dispersed, and either lost from the system or re-concentrated in secondary iron/manganese oxides/hydroxides. Currently, oxidized PGE ores are not being processed, neither in the Bushveld Complex nor in the Great Dyke nor in any other PGE-producing mine worldwide, as all previous attempts to recover PGE from this type of ore by conventional methods failed due to insufficient recovery rates [24].

In completion of the studies, the final step of changes of the PGM assemblage, the detrital PGM in stream sediments of rivers draining the Bushveld and the Great Dyke was investigated. This placer PGM assemblage differs starkly from that in pristine and oxidized ores; it is dominated by stable compounds, foremost Pt–Fe alloys, followed by sperrylite and cooperite/braggite [27–30].

During the past years, our working group at the BGR had access to underground exposures, cores, open pits, surface exposures, and stream sediments of rivers draining the Bushveld and the Great Dyke, and the research results were published in internal reports and in international journals. This contribution summarizes the findings of the individual studies and intends to combine these with the intention of following the fate of the platinum-group minerals in the exogenic environment. The work concentrates on the PGM and aims at advancing our understanding of the processes during weathering of the primary ores via oxidized equivalents and into placers mainly with respect to the redistribution of PGE and concurrent mineralogical changes.

2. Materials and Methods

Complete profiles of pristine ores were obtained from various mines and prospects of Impala Platinum in the western limb (Karee Mine; Merensky Reef and UG-2 chromitite), Anglo Platinum and Nkwe Platinum mainly in the eastern limb (Merensky Reef and UG-2; platinum pipes), and the northern limb (Platreef) of the Bushveld Complex in South Africa. Equivalent samples of the MSZ of the Great Dyke were taken underground and from drill core at the Hartley, Ngezi, Unki, and Mimosa mines as well as at the Mhondoro Concession in Zimbabwe.

Bushveld oxide ores comprised Merensky Reef from the eastern limb (farms Twickenham and Richmond) and Platreef material from surface exposures and drill core taken at Anglo Platinum projects and mines. MSZ oxide ores were taken on surface, in open pits, and from drill core at the Hartley, Ngezi, Unki, and Mimosa mines.

The placer sample campaigns in rivers draining the Bushveld Complex in South Africa yielded close to 6500 PGM grains in the samples from ten panning localities (one rich placer site with ca. 6000 PGM grains!), and 390 grains of detrital PGM were obtained by panning 28 (seven with output) samples in rivers draining the Great Dyke in Zimbabwe.

Some hundred polished sections of sulfide and oxide ores were prepared and investigated by reflected light and scanning electron (SEM) microscopy, as well as electron-probe microanalysis (EPMA). In selected sections of pristine ore samples, sulfides, and secondary phases (Fe–Mn-oxides/hydroxides, PGE-oxides) were analyzed in situ for their PGE contents by either EPMA or Laser Ablation Inductively Coupled Plasma Mass Spectrometry (LA-ICPMS). Finally, some samples were treated by combined Electric Pulse Disaggregation (EPD) and Hydroseparation (HS).

More details on samples, sample localities, analytical procedures, instruments used, and analytical parameters of the studies are provided in the respective publications [12,17,20–22].

3. Results

3.1. Pristine Sulfide Ores

The pristine Merensky Reef, Platreef, and MSZ ores generally contain between about 0.1 and 10 vol % sulfides. Interstitial sulfides and sulfide aggregates, up to several mm and locally some cm across, consist of pyrrhotite, pentlandite, chalcopyrite, and subordinate pyrite. As a rule, the PGM are associated with the sulfides, commonly intergrown with or sited at sulfide–sulfide or sulfide–silicate grain boundaries, at the peripheries or at contacts of the sulfide grains. Most PGM occur as single grains, however, inter-PGM intergrowths are also present. Grain sizes of the PGM usually range from <5 to 50 μm but may reach up to about 400 μm in the longest dimension.

In the chromitite reefs (LG, MG, UG-2) of the Bushveld Complex, sulfide contents are usually below 0.1 vol %, however, a primary sulfide–PGE relation is indicated as elevated PGE contents are only encountered up-sequence from a point (LG-5) when sulfur saturation was achieved during differentiation [10,11,18,31]. The low sulfide contents are interpreted to result from sulfide–chromite reactions whereby sulfide iron was taken up by chromite and sulfur was expelled from the system [18].

The number of platinum-group minerals (PGM) accepted by the International Mineralogical Association (IMA) was 138 in June 2014, whereby Pd-dominated PGM prevail (62) followed by Pt-rich (31), Ir-rich (17), and Rh-rich (14) species; for Ru and Os, 8 and 6 PGM are on record [32]. However, in the primary sulfide ores studied, only a small number of PGM species is of greater significance, and these are mainly compounds with pnictogens (As, Sb, Bi) and chalcogens (S, Se, Te), or PGE alloys.

The following PGM groups are of relevance, and examples of the PGM and their associations are depicted in Figure 2:

- Bismuthides and tellurides mainly of Pt and Pd, called (Pt,Pd)-bismuthotellurides in the following, which show high degrees of Pt ↔ Pd and Te ↔ Bi substitution. A first group [MeX$_2$] comprises moncheite [PtTe$_2$], maslovite [PtBiTe], merenskyite [PdTe$_2$], and michenerite [PdBiTe], and a second, rarer group [MeX] consists of kotulskite [PdTe]-sobolevskite [PdBi] (Figure 2A,B,G).
- Sulfides, encompassing the Pt–Pd monosulfides cooperate-braggite-vysotskite [PtS–(Pt,Pd,Ni)S–PdS] and called (Pt,Pd)-sulfides in the following (Figure 2B,C); malanite-cuprorhodsite solid solution [CuPt$_2$S$_4$–CuRh$_2$S$_4$] (Figure 2C); and laurite [RuS$_2$], which often carries the bulk IPGE budget of the ores.
- Arsenides, the most ubiquitous representative being sperrylite [PtAs$_2$] (Figure 2D,E).
- Sulfarsenides with limited substitution comprise the common Rh-mineral hollingworthite [RhAsS], rarer platarsite [PtAsS], irarsite [IrAsS], and ruarsite [RuAsS].
- Antimonides like stibiopalladinite [Pd$_5$Sb$_2$].
- Zvyagintsevite [Pd$_3$Pb] and members of the solid solution rustenburgite-atokite [Pt$_3$Sn–Pd$_3$Sn] are locally abundant in the Bushveld ores.
- Alloys mainly of Pt with Fe, Cu, or Ni (Figure 2D,F), and inter-PGE alloys (e.g., Pt–Pd, Os–Ir–Ru).

The PGM assemblages of the pristine Bushveld ores vary within wide limits, and the compilation given here largely reflects the data presented in previous publications [20,21,25,33–51].

The Merensky Reef consistently has high contents of (Pt,Pd)-bismuthotellurides (~10–40%), (Pt,Pd)-sulfides (~5–70%) and sperrylite (~1–10%); sulfarsenides (up to 50%) and Pt–Fe alloys (up to 90%!) dominate locally [19–21,33–36].

Platinum-group minerals in the Platreef are mainly (Pt,Pd)-bismuthotellurides (20–40%), (Pt,Pd)-sulfides, and sperrylite (~10% each), zvyagintsevite (~10%), and PGE-sulfarsenides (~10%). Malanite, laurite, and Pt–Fe alloys are also present in smaller amounts [25,41–46].

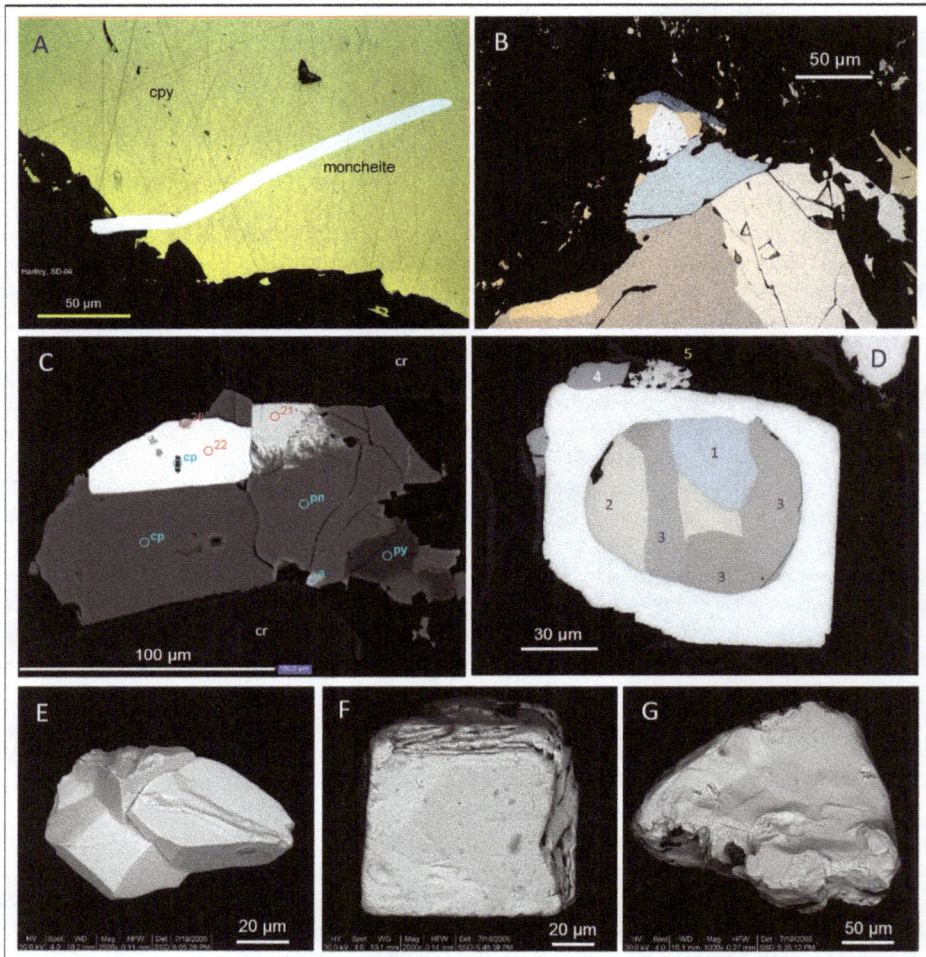

Figure 2. Platinum-group minerals (PGM), pristine ores, in reflected light, oil immersion (**A,B,D**), Backscatter electron (BSE) image (**C**), and BSE images of PGM grains from pristine ores of the Main Sulfide Zone, Great Dyke, extracted by electric pulse disaggregation, EPD (**E–G**). (**A**) Lath of moncheite in chalcopyrite (cpy). MSZ, Hartley Mine. (**B**) Aggregate of pyrrhotite (brownish) and pentlandite (cream-white) at bottom part, overgrown by braggite (bluish), grain of moncheite (white, soft) partly surrounded by chalcopyrite (yellow), and molybdenite (gray, crocodile-shaped). MSZ, Hartley Mine. (**C**) Sulfide aggregate between two chromite grains (cr) consisting of chalcopyrite (cp), pentlandite (pn), and pyrite (py). White grain on top is cooperite/braggite (22) intergrown with malanite (light grey, 21). Merensky Reef, Garatau, eastern Bushveld. (**D**) Composite grain of Pt–Fe alloy (white rim; $Pt_{65}Fe_{30}$), geversite [$PtSb_2$] (1, bluish), stumpflite [Pt(Sb,Bi)] (2, yellowish), genkinite [$(Pt,Pd)_4Sb_3$] (3, light brown), sperrylite (4), and tulameenite [Pt_2CuFe] (5). Driekop Pipe, eastern Bushveld. (**E**) Sperrylite crystal. (**F**) Pt–Fe alloy. (**G**) Moncheite.

The UG-2 is characterized by the dominance of (Pt,Pd)-sulfides and lesser amounts of malanite/cuprorhodsite, followed by Pt-Fe alloys. Laurite is common, and (Pt.Pd)-bismuthotellurides are generally rare though locally present.

Similarly, within the LG/MG chromitites, (Pt,Pd)-sulfides, namely cooperite-braggite, and also malanite-cuprorhodsite are abundant, as are laurite, followed by PGE-sulfarsenides, sperrylite, and Pt–Fe alloys. Notably, all chromitite PGE ores are characterized by elevated contents of zvyagintsevite [12,37,40].

The ultramafic Pt-pipes of the eastern Bushveld need to be mentioned here as we investigated eluvial and alluvial PGM in near-by placers [28,30]. Concentrates from the best-studied Driekop pipe contain about 50% Pt–Fe alloys (Figure 2D), 15% each of sperrylite and geversite [PtSb$_2$], 15% PGE-sulfarsenides (hollingworthite and irarsite), and 5% of other PGM. (Pt,Pd)-bismuthotellurides and (Pt,Pd)-sulfides are virtually absent [47–50].

Platinum-group minerals of the Main Sulfide Zone of the Great Dyke [16,17,51] comprise (Pt,Pd)-bismuthotellurides, sperrylite, the (Pt,Pd)-sulfides cooperite and braggite, and some rarer phases (Figure 2A,B). Mineral proportions by number show the predominance of (Pt,Pd)-bismuthotellurides (50.1%), followed by sperrylite (19%), cooperite/braggite (8.5%), the PGE-sulfarsenides hollingworthite, platarsite, irarsite, and ruarsite (11.9%), laurite (5.0%), Pt–Fe alloys (2.4%), and some rarer PGM [16]. Interestingly, the MSZ samples of the Great Dyke's North chamber (Hartley, Mhondoro, Ngezi) have higher proportions of (Pt,Pd)-bismuthotellurides (64%) and cooperite/braggite (13%), and lower proportions of sperrylite (11%), PGE sulfarsenides (2.4%), and laurite (1.7%) compared to MSZ samples of the South chamber (Unki and Mimosa) with (Pt,Pd)-bismuthotellurides (35%), cooperite/braggite (4%), sperrylite (28%), PGE sulfarsenides (22%), and laurite (8.6%). The elevated proportions of PGE arsenides and sulfarsenides in the South chamber (together 50%) most likely indicate higher fugacities of arsenic in the South chamber magmas compared to the North chamber during MSZ formation. Further, the PGM spectrum of the Great Dyke is Pt dominated, as about two out of three PGM are Pt-rich (67.7%), followed by Pd- (19.8%) and Rh-rich (6.6%) compounds. Ru- and Ir-rich minerals make up 3.7 and 2 percent, respectively.

Notably, in the pristine sulfide ores of both the Bushveld and the Great Dyke, variable amounts of the PGE are hosted in sulfides. Apparently, Pd and Rh readily substitute for Ni and Fe in the crystal lattice of pentlandite, and accordingly, extensive proportions of the Pd and Rh budget of the ores are hosted in pentlandite [16,17,19–25,44,46,52,53]. Maximum concentrations of Pd and Rh in pentlandite are 757 ppm Pd and 649 ppm Rh for the Merensky Reef, 29,975 ppm Pd and 23,817 ppm Rh for the UG-2, and 746 ppm Pd and 405 ppm Rh for the Platreef. Pentlandite in the LG-6 and MG-1/2 chromitites constantly has maximum concentration levels above 1000 ppm of both Pd and Rh. Maximum contents are 7731 ppm Pd (LG-6) and 11,366 ppm Rh (MG-1/2). For the MSZ, maximum contents of 2506 ppm Pd and 562 ppm Rh were reported from the Hartley Mine. In all cases, PGE contents in chalcopyrite, pyrrhotite, and pyrite are in general insignificant, and the silicates and chromite usually contain no detectable amounts of PGE [19,20].

Mass balance calculations revealed that in the Merensky Reef, between 13 and 100% of the Pd and from 3 to nearly 100% of the Rh hosted by pentlandite [19,20]. In the UG-2, pentlandite consistently hosts elevated proportions of the whole-rock Pd (up to 55%) and Rh (up to 46%) budget, whereas Pt is almost absent in the base metal sulfides (BMS) [21]. In the Platreef, almost all Pd and Rh may be present in solid solution in pentlandite [54].

In summary, the source material of our study, the PGM assemblages of the pristine ores, have pronounced mineralogical similarities but the proportions of the various PGM vary within wide limits. The PGE are bimodally distributed; whereas Pt is dominantly present in the form of discrete PGM like (Pt,Pd)-bismuthotellurides, PGE-sulfides (cooperite/braggite and malanite/cuprorhodsite), sperrylite, and Pt–Fe alloys, large though variable proportions of the Pd and Rh are hosted in pentlandite. Part of the IPGE (Ru, Os, Ir) may be hosted in sulfides and sulfarsenides, however, a great deal of the IPGE budget of the ores is present in laurite.

This generalized inventory above is the basis of our examination of the fate of the abovementioned PGM and PGE-bearing sulfides during the weathering of the ores.

3.2. Oxidized Ores

The behaviour of the PGE in the exogenic cycle was examined in several profiles of oxidized Merensky Reef of the eastern limb and Platreef ores of the northern limb of the Bushveld and on oxidized MSZ ores. Geochemically, in the oxidized ores of both the Bushveld and the Great Dyke, the general metal distribution patterns of the pristine ores are grossly preserved. However, at similar Pt grades, significant proportions of Pd have been lost from the system [27,55–57]. This indicates that Pd is more mobile than Pt and is dispersed in the supergene environment as already shown for the Merensky Reef in the seminal book of Wagner [7]; Pt/Pd values of 2.7 for sulfide ore and 5.1 for oxidized ore were reported [58]. For the UG-2 from the Union section, Pt/Pd values of 2.2 for primary and up to 3.2 for oxidized ore were provided [56]. Pt/Pd ratios of 0.75 and 1.15 were reported for pristine and oxidized Platreef, respectively [25]. Similar relationships between pristine and oxidized PGE mineralization, namely average Pt/Pd values of 1.28 and 2.43, respectively, are on record from the MSZ of the Great Dyke [16,17,24,27].

Mineralogically, in the oxidized ores of both the Great Dyke and the Bushveld, rare relict sulfides, mainly pyrrhotite, are surrounded by rims of iron hydroxides (Figure 3A,B). In case of pervasive oxidation, weathering leads to the complete destruction of all base metal sulfides (BMS) and concurrent replacement by iron hydroxides, which may carry up to 5 wt % Ni and/or Cu. However, the iron hydroxides are not considered major carriers of Ni and Cu in the oxidized ores as, according to microprobe analyses, a large proportion of the Ni and Cu is hosted in chlorites and smectites [57].

In general, sperrylite and cooperite/braggite grains show no distinct features of alteration (Figure 3A). Pt–Fe alloy grains, both compact and porous ones, were found in samples from Hartley and Ngezi. The porous grains of Pt–Fe alloy (close to Pt_3Fe in composition; Figure 3C,D) probably represent replacements of other precursor PGM of unknown chemical composition, or they are relics of spongy to emulsion-like Pt–Fe alloys which were intergrown with other PGM or sulfides, as testified from the pristine MSZ [16] and the Merensky Reef ores [T. Oberthür, unpublished data]. As early as 1939, texturally similar porous grains of "native Pt" from oxidized Merensky Reef were reported and the early authors proposed that these grains represent relicts of sperrylite or cooperite grains [59].

The (Pt,Pd)-bismuthotellurides, common in the pristine ores, show features of alteration and decomposition at various degrees. At incipient oxidation, islands of (Pt,Pd)-bismuthotellurides are surrounded by somewhat porous secondary (Pt,Pd)-oxides/hydroxides (Figure 3E,F), occasionally with shrinkage cracks. In pervasively oxidized ores, all (Pt,Pd)-bismuthotellurides have disappeared and locally some colloform, banded grains of PGE-oxides/hydroxides may represent complete replacements of former PGM (Figure 3G). Further, only sparse relict PGE-sulfarsenides were encountered in the oxidized ores. "Pt-alloys with low sums and weak reflectivity" were first reported by Evans et al. [60] from the Great Dyke, and a first description of oxidation products of PGM from the Bushveld (UG-2) was presented by Hey [56]. Today, the existence of secondary (Pt,Pd)-oxides/hydroxides ("PGE-oxides") has been established worldwide by a number of authors, however, their physico-chemical nature remains ill-defined [56,61–68]. The only PGE-oxide approved (but assigned "questionable") by the IMA is palladinite [PdO], occurring as an ocherous coating on palladian gold (porpezite) in Itabira, Brazil [69].

The polished section studies of oxidized ores revealed the presence of some PGM grains of a seemingly secondary nature. They comprise small (<5 μm in size) grains of poorly defined phases. Some of them may actually be oxides or hydroxides, or alteration products of primary PGM, or neoformations (Figure 4A,C). Compounds of Pt–S, Pd–S, Pt–Pd–As–Cu, Pd–Cu–Fe, Pt–Fe, and Pt were also identified, either hosted by Fe hydroxides or by hydrous silicates, or commonly by amphibole, chlorite-, or smectite-like phases. None of the PGM grains had stoichiometric composition; concentrations of sulfur and arsenic, if present, are much lower than in their presumed precursor phases [61,70].

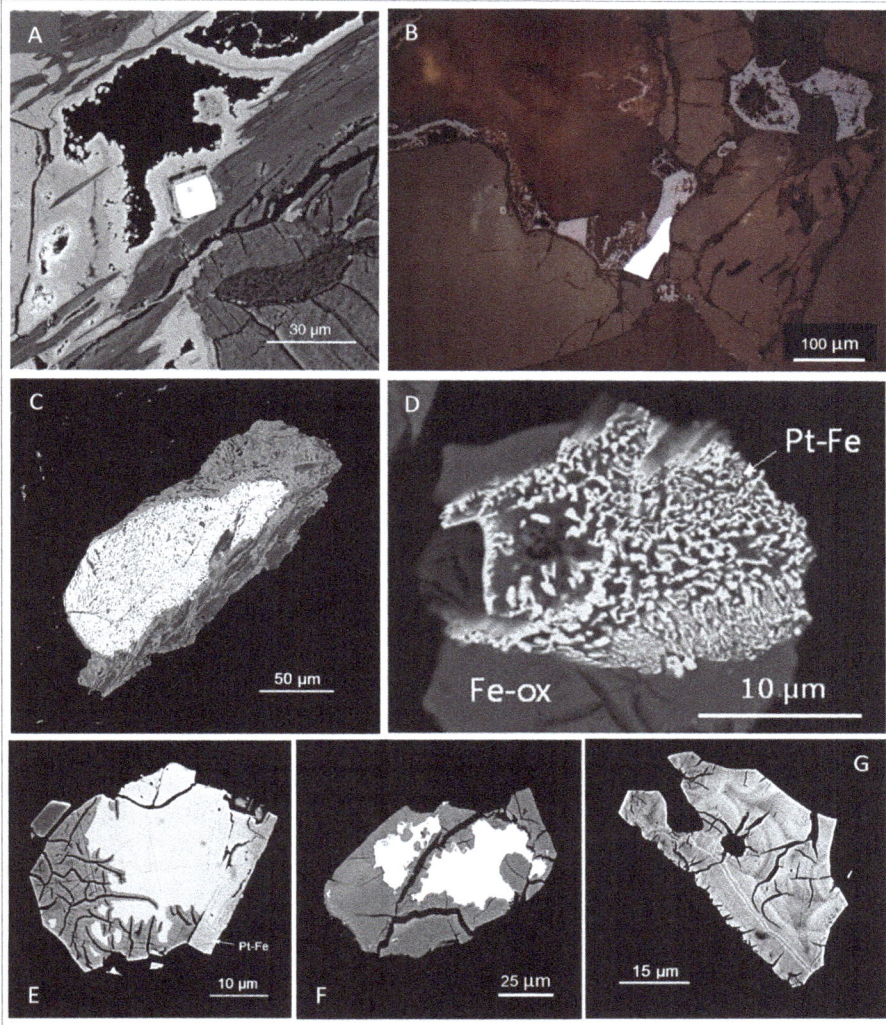

Figure 3. Relict platinum-group minerals, replacements, secondary formations, in oxidized ores. BSE images of polished sections, except (**B**) which is in reflected light. (**A**) Euhedral, relict sperrylite grain (white) in weathered MSZ ore consisting of iron-hydroxides (medium gray) and smectites (dark gray). HOP-05, ps 5663a, Hartley Mine. (**B**) Relict cooperite/braggite (white) surrounded by goethite (gray-bluish) and silicates (darkest phases). Merensky Reef, Richmond, eastern Bushveld [26]. (**C**) Porous grain of Pt–Fe alloy surrounded by a mixture of secondary oxides/hydroxides and silicates. HOP-206a, ps 5910a, Hartley Mine. (**D**) Pt–Fe phase of spongy texture intergrown with Fe hydroxide. Concentrate after hydroseparation, Hartley Mine. (**E**) Pt-bearing sobolevskite (light grey) being replaced by PGE-oxides/hydroxides (left and bottom). Note attached grain of Pt–Fe alloy (Pt–Fe). Adit A, Ngezi concession. (**F**) Grain of michenerite (white, center) in disintegration. Alteration rim (gray) of (Pt,Pd)-oxide/hydroxide phases shows Pd, Cu, and Fe as major elements. HOP-206a, ps 5910a, Hartley Mine. (**G**) Colloform, banded grain of PGE-oxide/hydroxide phase with shrinkage cracks. NGZ 1C, ps 5711b, from Adit A, Ngezi concession.

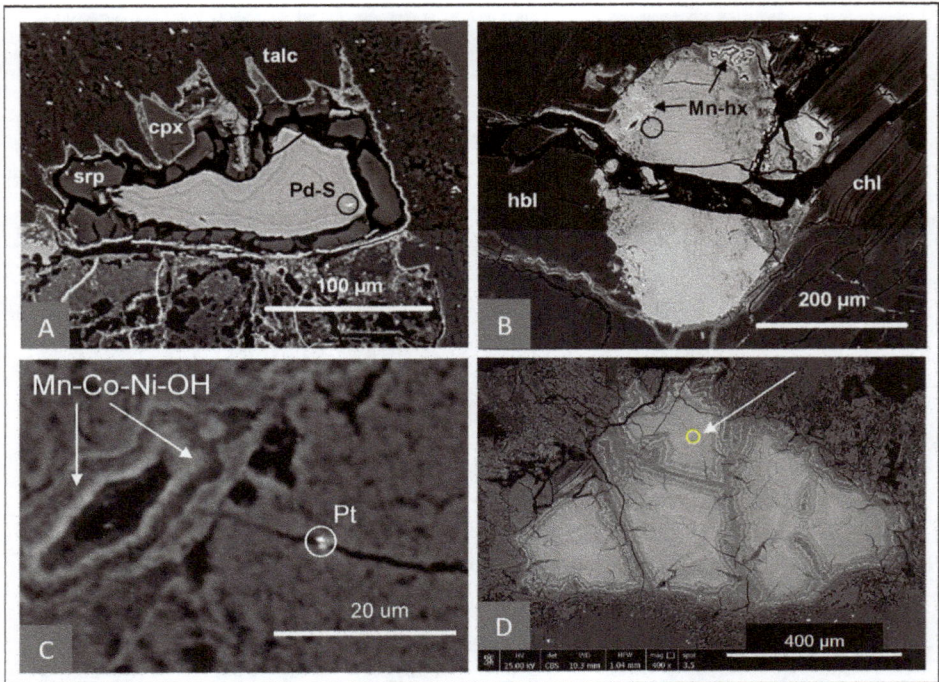

Figure 4. BSE images of secondary minerals in oxidized ores. (**A**) Iron-hydroxide with colloform texture and inclusion of Pd–S compound (circle:) in a matrix of Fe-rich serpentine (srp), talc, and clinopyroxene (cpx). The Fe-hydroxide carries 30 ppm Pt. AS 6828, Mimosa Mine. (**B**) Iron-hydroxide aggregate enclosed by chlorite (chl) and "hornblende" (hbl). Within the Fe-hydroxide, small areas and veinlets of Mn–Co-hydroxide (Mn–hx) carry 150–400 ppm Pt. Small PGM (Pt) in circle. The chlorite is rich in Ni and Cu (up to 4 and 6 wt %, resp.). AS 5320a, Mimosa Mine. (**C**) Porous Fe-hydroxide (matrix) with areas of Mn–Co–Ni-hydroxides (Mn–Co–Ni–OH) which contain ~200 ppm Pt. Small spec of PGM (Pt) in fracture. Ngezi Mine. (**D**) Interstitial iron-hydroxide with colloform textures. Analyzed point (circle, arrow) has 214 ppm Pt and 192 ppm Pd. Platreef, Overysel, Mogalakwena Mine. Reproduced with permission from Malte Junge [25].

Within the oxidized ores, substantial proportions of the PGE are hosted in Fe- and Mn-oxides/hydroxides. They also occur as vein-like structures that crosscut the silicates. Both types of Fe-/Mn-oxides/hydroxides reveal characteristic layered and zoned internal textures (Figure 4A). At Mimosa mine [61,70], Fe-hydroxides pseudomorphous after sulfide droplets may carry small grains of secondary PGM (mainly Pt, but also relict PGM), whereas the vein-like hydroxides are barren of PGM.

The unusual correlation of presumably silicate-bound elements and metals corroborates that mixtures of Fe hydroxide with Si-rich material are present, probably amorphous or very fine-grained clayey substances. In the Great Dyke oxide ores, the Fe hydroxide aggregates may carry up to 230 ppm Pt and 150 ppm Pd (EPMA trace analysis). In some cases, Fe-hydroxides, pseudomorphous after sulfides are veined by bluish/grey Mn–Co–Ni–Cu-hydroxides (17–47 wt % Mn, 7–18 wt % Co, 8–13 wt % Ni, 4–23 wt % Cu) that have elevated concentrations (40–400 ppm) of Pt, however, Pd contents are below the EPMA detection limit of 25 ppm [61,66]. In the Hartley mine ores, iron-oxides/hydroxides revealed highly variable concentrations of Pt (up to 3600 ppm) and Pd (up to 3100 ppm), and Mn-oxides/hydroxides are significant carriers of Pt (up to 1.6 wt %) and Pd (up to

157 ppm) [57]. From the Platreef ores (Figure 4D), up to 416 ppm Pd, 335 ppm Rh, and 803 ppm Pt are on record from secondary minerals [25].

Secondary silicates comprise serpentine minerals, smectite (nontronite), and chlorite-like phases. Smectite and chlorite carry up to several wt % Ni and Cu and occasionally PGE; trace analyses gave high contents of Pt (up to 1800 ppm) and Pd (up to 1600 ppm).

Summary—PGM and PGE Carriers in Oxidized Ores

Weathering of the pristine, sulfide-rich ores has a pervasive impact on the ores. Within the oxidized ores, the sulfides are destroyed, and the PGM assemblage suffers destruction especially of the (Pt,Pd)-bismuthotelluride and the PGE-sulfarsenide species. Pt and Pd, liberated from either PGM or sulfides, are now found in PGE oxides/hydroxides or in appreciable amounts in secondary iron and iron/manganese oxides/hydroxides, or in secondary phyllosilicates. A large proportion of the mobile Pd is carried away in solution. Neoformation of small, secondary PGM plays a subordinate role. In brief, the PGE are polymodally distributed in the oxidized ores and they occur in different modes:

(1) As relict primary PGM (mainly sperrylite, cooperite/braggite, and Pt–Fe alloys),
(2) in solid solution in relict sulfides (dominantly Pd and Rh in pentlandite),
(3) as secondary PGM neoformations (rare, mainly small grains, e.g., native Pt),
(4) as PGE oxides/hydroxides that either replace primary PGM or represent neoformations,
(5) in iron oxides/hydroxides (up to some thousand ppm Pt and Pd),
(6) in iron/manganese oxides/hydroxides (up to 1.6% Pt and 1150 ppm Pd) [57], and
(7) in secondary phyllosilicates (up to a few hundred ppm Pt and Pd).

The proportions of the various PGE-bearing phases vary considerably from mine to mine and between samples. This probably reflects both variations in primary ore mineralogy and depth within the weathering profile (incipient to pervasive oxidation). Semiquantitative mass balance calculations and findings related to the metallurgical treatment of oxides ores indicate survival of ca. 10–30% of the primary PGM assemblage, mainly Pt-rich PGM like sperrylite, cooperite/braggite, and Pt–Fe alloys. All (Pt,Pd)-bismuthotellurides and the PGE-sulfarsenides were destroyed and at least part of their PGE contents are located in secondary phases. Up to 50% of the original Pd is lost from the system, probably transported away by acid surface waters, and the remaining Pd is largely hosted in PGE oxides/hydroxides, iron and iron/manganese oxides/hydroxides, and secondary phyllosilicates [57,61]. The fate of Rh remains open, as only rare carriers of Rh (PGM or secondary phases) were detected in the oxidized ores.

3.3. PGM Grains in Alluvial Sediments of Rivers Draining the Bushveld Complex and the Great Dyke

3.3.1. SEM Observations of Single Grains—Morphology, Intergrowths, Alteration, Modification

The assemblages of detrital PGM found in rivers draining the Bushveld Complex and the Great Dyke indicate further mineralogical changes [27–30,71].

Within the Bushveld Complex, our first study [28] concentrated on the placers of the farm Maandagshoek in the eastern Bushveld, i.e., the locality where Dr. Hans Merensky, based on placer PGM, identified platiniferous pipes first and then the Merensky Reef in 1924 [5–8]. Our sampling localities were, at maximum, 1–2 km away from the probable sources (Merensky Reef, UG-2 and other chromitites, platiniferous dunite pipes). Accordingly, many grains of the detrital PGM assemblage are still rather unaffected (Figure 5a,b,d,f) whereas others have experienced some physical modification or chemical corrosion [28,30,71].

Figure 5. Detrital PGM, scanning electron microscope (SEM) images. (**a**) Pt–Fe alloy grain (1.6 mm in diameter) with smooth surface. Moopetsi river, Maandagshoek, eastern Bushveld. (**b**) Idiomorphic Pt–Fe alloy grain. Note slightly contorted edges. Locality as Figure 5a. (**c**) Well-rounded grain of Pt–Fe alloy. Makwiro river near Hartley Mine, Great Dyke, Zimbabwe. (**d**) Grain of Pt–Fe alloy with cubic crystal faces and smooth surface polish, intergrown with platelets of laurite (on top). Locality as Figure 5a. (**e**) Well-rounded grain of native Pt. Dithokeng river, northern Bushveld. (**f**) Splinter of large grain of cooperite. Locality as Figure 5a. (**g**) Sperrylite crystal showing little attrition. Locality as Figure 5a. (**h**) Well crystallized sperrylite grain with crystallographically oriented etch pits. Makwiro River, Hartley Mine. (**i**) Sperrylite with a thin surface coating of native platinum. Dwars river, eastern Bushveld. (**j,k**) Sperrylite surface overgrown by platelets of native platinum, (**k**) is surface magnification. Der Brocken, eastern Bushveld. (**l,m**) Tiny crystals of native Pt (light grey) on grain of pentlandite (dark grey). (**m**) Magnification from (**l**) showing crystals of native Pt (light grey). Brakspruit, western Bushveld.

SEM investigations of 1425 PGM grains provided an excellent overview and showed that the PGM assemblage is mainly composed of grains of native Pt, Pt–Fe and Pt–Fe–Cu–Ni alloys (together 73.2% by number of grains), cooperite/braggite (14.2%), and sperrylite (10.2%). The remainder (2.4%) consists of a variety of rarer PGM. Grain sizes range from 40 μm to 1.6 mm in diameter, and the highest numbers of PGM grains were found in the fraction <125 μm. Pt–Fe alloy grains are up to 1.6 mm in diameter (mostly between 100 and 200 μm) and have various surface morphologies. Grains with well-rounded shapes are most frequent, although cubic crystals are also present. Intergrowths with laurite embedded in or attached to Pt–Fe alloy grains are common (Figure 5d). Sperrylite grains are generally multifaceted single crystals without any signs of corrosion or mechanical wear (Figure 5g). Cooperite/braggite grains are mainly present as splintered, broken grains. Besides monomineralic PGM grains, many grains are intergrowths of between two and six different PGM, the most common association being Pt–Fe alloy intergrown with laurite (Figure 5d).

The other localities of the Bushveld studied [30] revealed similar gross PGM assemblages with minor variation (Figure 5e,i) also due to the smaller amount of PGM grains (*n* = 54) recovered. PGM grain sizes are mostly in the range from ~50 to 150 μm (maximum diameter 600 μm). The overall PGM proportions are: native Pt and Pt–Fe alloys (54%), sperrylite (33%), cooperite/braggite (11%), and stibiopalladinite (one grain). Accordingly, nearly 98% of the detrital PGM are Pt minerals.

In the river sediments along the Great Dyke [29], PGM mineral proportions by number of grains (*n* = 390) obtained by SEM are: sperrylite (45.6%; Figure 5h), native Pt and Pt–Fe alloys (together 42.1%; Figure 5c), cooperite and braggite (6.7%), and a number of rarer Pd-bearing PGM (3.4%). Notably, Os–Ir–Ru alloy grains (2.0%) were only found in samples from the Umtebekwe River south of Unki mine and probably originate from the economically important chromitite mineralization in the near-by Archean Shurugwi greenstone belt. Most PGM grains are monomineralic, however, several intergrowths, mainly between Pt–Fe alloy and laurite, were observed, and also various inclusions of different minerals in the detrital PGM grains. Grain sizes of the PGM range from 90 to 300 μm (mostly between 100–200 μm); maximum grain sizes determined are 480 μm for sperrylite, and 285 μm for a Pt–Fe alloy grain.

In summary and generalized, the Pt–Fe alloy grains span a large compositional range from [Pt_3Fe] to [$Pt_{1.5}Fe$] and have various surface morphologies. Grains with well-rounded shapes (Figure 5a,c) are common, although cubic crystals are also present (Figure 5b), as are occasional intergrowths mainly with laurite (Figure 5d). Grains of native Pt may contain elevated contents of Pd (%-range), are usually well-rounded and show mechanical wear on their surfaces (Figure 5e). Sperrylite grains are commonly multifaceted single crystals. Although some grains show etch pits, channels, or fractures, the majority of the sperrylite grains are without obvious signs of corrosion or mechanical wear (Figure 5g,h). Cooperite/braggite grains are mainly present as splintered, broken grains (Figure 5f). PGM neoformation is restricted to sporadic, thin coatings of native Pt on sperrylite and cooperite/braggite (Figure 5i–k). Deposition of tiny Pt crystals on a pentlandite grain was observed once (Figure 5l,m).

The above SEM studies provided an insight into grain morphologies and also allowed to obtain semi-quantitative analytical data of the mineral grains. However, SEM analyses are performed on the surfaces of grains only and therefore, thin crusts on the grains or overgrowths may show compositions that differ from the internal composition of the grains. Indeed, polished section studies demonstrate that some PGM grains assigned to native Pt or Pt–Fe alloy have cores of cooperite, braggite, or in rarer cases, sperrylite, as will be shown in the following.

3.3.2. Detrital PGM: Intergrowths, Inclusions, Alteration, Modification

Features of magmatic-hydrothermal alteration of Bushveld PGM were described in detail elsewhere [71]. Based on light microscopic and EPMA studies, this section will concentrate on the characteristics of secondary alteration of the detrital PGM in the weathering environment, from oxidized ores to final placer deposition. Further details on alteration, intergrowths, and inclusions will be disclosed, and the question of PGM neoformation will be illuminated.

As mentioned, in the Maandagshoek placer, many grains of the detrital PGM assemblage are still rather unaffected whereas others have experienced some physical and chemical modifications [28,71]. Figure 6a is an example showing well-crystallized grains of Pt–Fe alloy intergrown with other PGM, and a porous grain of native Pt. The PGM grain depicted in Figure 6b shows an unusual, complex intergrowth of some rather exotic PGM. This author is convinced that these textures cannot survive extended alluvial transport, corroborating the earlier notion of short transport distance.

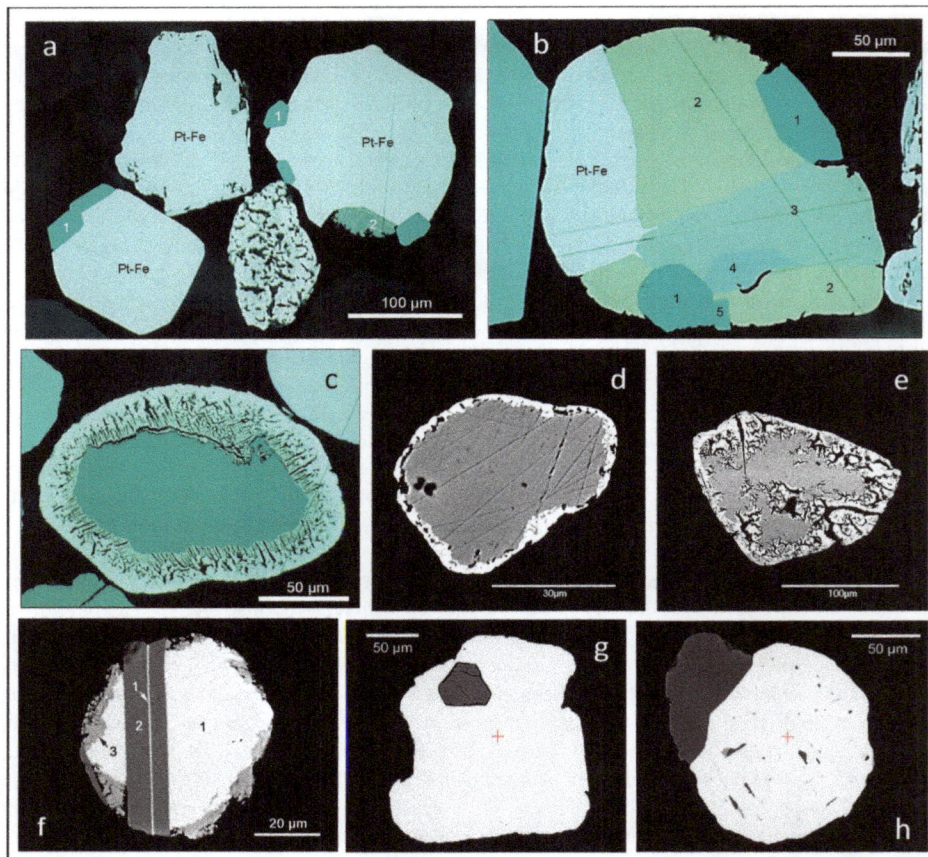

Figure 6. Detrital PGM in polished sections. Reflected light, oil immersion (**a**–**c**), and BSE images (**d**–**h**). (**a**) Three grains of Pt–Fe alloy (Pt–Fe), with attached laurite (1) and RhFeNi sulfide (2). Porous grain (lower half, center) is native platinum. Maandagshoek. (**b**) Composite grain consisting of Pt–Fe alloy, laurite (1), törnroosite [$Pd_{11}As_2Te_2$] (2), palladoarsenide [Pd_2As] (3), sperrylite (4), and irarsite (5). Maandagshoek. (**c**) Alteration of cooperate-braggite, replaced by platinum. Note intruding channels and outer, compact rim. Maandagshoek. (**d**) Sperrylite grain with rim of pure Pt. Umtshingwe River, Great Dyke. (**e**) Strongly corroded cooperite grain being replaced by pure Pt (white). Umtebekwe River, Great Dyke. (**f**) Polyphase grain of Pt–Fe alloy (1) with a lamellar Ru_{70} alloy (2), overgrown by a rim of [Ni_2PtFe] (3). Maandagshoek. (**g**) Grain of Pt–Fe alloy (white) with idiomorphic inclusion of laurite. Umsweswe River, Great Dyke. (**h**) Well-rounded grain of Pt–Fe alloy (white) with numerous inclusions of various PGM (palladodymite (Pd,Rh)$_2$As, unnamed RhS, unnamed Pd$_3$Te), intergrown with unnamed Rh(Ni,Cu)$_2$S$_3$ (dark grey; top left). Umsweswe River, Great Dyke.

In both the Bushveld and the Great Dyke samples, sperrylite is usually unaltered. However, a few grains have narrow (<10 μm), discontinuous to complete rims of pure platinum (Figure 6d) as also observed in the SEM studies (Figure 5i–k), or corrosion channels on their surface (Figure 5h).

Relics of primary Pt–Pd sulfides (cooperite/braggite) are rimmed by porous native platinum with abundant root-like channels that occasionally form a dense network penetrating into the Pt–Pd sulfide (Figure 6c,e). The platinum rims may have elevated concentrations of Ni (up to 5.2 at %) and Pd (up to 30 at %) which also point to the precursor phases. Therefore, it is assumed that many porous to compact grains of native platinum are the end-products of Pt–Pd–Ni sulfide alteration [71]. Accordingly, most of the native platinum grains detected in the SEM studies probably originate from the alteration of Pt–Pd–Ni sulfides.

Further modifications observed are local alteration of Pt–Fe alloy grains by inhomogeneous Pt–Fe oxide phases and replacement of laurite by Ru-rich oxides or hydroxides. Minor to trace concentrations of PGE in secondary phases such as Fe oxide and hydroxide, chlorite, and smectite forming a continuous rim around PGM grains are considered to have formed in situ within the placer deposits.

Intergrowths of the PGM mainly consist of Pt–Fe alloys intergrown with other PGM (Figure 6f,h), and inclusions detected mainly in Pt–Fe alloy grains comprise a wide spectrum of sulfides and PGM (Figure 6g,h). In all probability, the observed intergrowths and inclusion types point to early formation in the history of the grains, at the magmatic stage.

Unambiguous neoformations are the coatings of native Pt on cooperite/braggite, and more sporadic on sperrylite or pentlandite (Figures 5i–m and 6c–e). Visibly, these thin coatings source their metal contents from the enclosed grains (PtS, PtAs$_2$), following the simplified reaction (example: cooperite): PtS + 3/2O$_2$ + H$_2$O = Pt^0 + SO$_4$$^{2-}$ + 2H$^+$ [71]

Dissolution takes place under oxidizing conditions, and reprecipitation of Pt may be assisted by electrochemical processes [72]. Note that a large proportion of native platinum grains is regarded to be the end-product of Pt–Pd–Ni sulfide alteration [71]. Chemically, the neoformations observed are native Pt, not Pt–Fe alloys. Further, small overgrowths and crusts of zvyagintsevite [Pd$_3$Pb] record the mobilization of Pd and reaction with stray Pb, however, the general trend of the placer PGM assemblage is a further increase of the Pt/Pd ratio, and now >95% of the PGM are Pt-minerals. The newly grown PGM occur as small platelets or crystals (Figure 5j–m), and any growth to larger crystals or nuggets was not observed. The common Pt–Fe alloy grains occasionally carry inclusions of sulfides and other PGM which most probably were included at the magmatic stage. Inclusions trapped in an oxidizing, secondary environment like iron/manganese oxides or hydroxides or silicates are absent.

4. Discussion

The paper reports on the PGM assemblages identified in the Bushveld and the Great Dyke, specifically in:

(1) the primary, pristine ores,
(2) the near-surface oxidized/weathered ores, and,
(3) associated placers.

One important aspect to be discussed on a wider base below aims at contributing to the debate allogenic versus authigenic origin of placer PGM, i.e., whether the placer PGM:

(1) originate from the primary ores (residual/allogenic/detrital grains), or,
(2) developed during weathering/oxidation of the primary ores (authigenic grains), or,
(3) were newly formed (authigenic grains/"neoformation") in the alluvial placer environment.

4.1. Development of the Bushveld/Great Dyke PGM Assemblages in the Course of Weathering

The processes of PGE redistribution and the behavior of the PGM in the weathering cycle are much debated and cases of both dispersion and concentration, destruction and neoformation have been proposed by various authors (for details see Sections 4.3.1 and 4.3.2 below). In contributing to the discussion, it is fortunate that near-continuous underground and surface exposures of the Bushveld and Great Dyke reefs allowed us to investigate some aspects of the fate of the PGE and PGM in the exogenic environment in detail. The basic results of our studies were reported above and are depicted schematically in Figure 7.

In the pristine sulfide ores of both the Bushveld and the Great Dyke, the PGE are bimodally distributed: (i) Pt, the IPGE (Ru,Os,Ir) and variable proportions of Pd and Rh are present in the form of discrete PGM, dominantly PGE-bismuthotellurides, -sulfides, -arsenides, -sulfarsenides, and -alloys; (ii) Substantial though variable proportions of Pd and Rh are hosted in pentlandite.

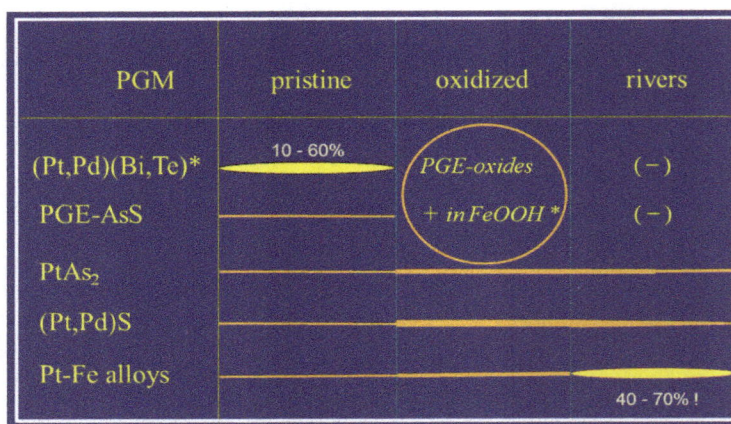

Figure 7. Schematic graph showing the principal changes of the PGM assemblages from pristine, sulfide ores via oxidized ores into placer accumulations. Abbreviations: (Pt,Pd)(Bi,Te)* = (Pt,Pd)-bismuthotellurides; FeOOH* stands for Fe,Mn,Co-oxides/hydroxides. Note that Pt–Fe alloys include native Pt.

In the weathered/oxidized ores, the PGE are polymodally distributed: PGE-sulfides, -arsenides, and Pt–Fe alloys remain the only relics of the pristine PGM assemblage. During weathering of the ores, all base metal sulfides, the (Pt,Pd)-bismuthotellurides and most PGE-sulfarsenides are destroyed. Part of their PGE contents are found in the form of ill-defined (Pt,Pd)-oxides or -hydroxides, and in secondary Fe/Mn/Co-oxides/hydroxides as well as smectites which carry elevated though variable contents of Pt and Pd. Possible PGM neoformations comprise rare and tiny (<3 µm) specks of native Pt or Pt oxide in Fe-hydroxides, and small (1–5 µm) grains of zvyagintsevite. Notably, a large proportion of the mobile Pd (up to 50%) is lost from the system. The weathering event is mainly characterized by destruction of primary sulfides and unstable PGM, and dispersion of the PGE.

Finally, the PGE are unimodally distributed in the alluvial sediments as only discrete PGM are present. Obviously, the often porous (Pt,Pd)-oxides or -hydroxides and secondary Fe/Mn/Co-oxides/hydroxides, important carriers of Pt and Pd in the oxidized ores, do not survive the mechanical transport into the rivers and their metal contents are either transported away as detritus in the fine sediment fraction or in solution. The remaining PGM assemblage is characterized by continued partial alteration or destruction of most PGE-sulfide grains. In contrast, sperrylite largely remains a stable mineral, and ubiquitous Pt–Fe alloy and native Pt grains have gained in importance. Notably, >95% of the placer PGM are Pt-rich, indicating a further loss of the mobile Pd.

Native Pt as crusts and overgrowths and forming porous or compact grains is a newcomer that mainly formed more or less in situ on and replaces pre-existing PGE-sulfide (cooperite/braggite) and to a lesser extent sperrylite grains. Rare, newly grown coatings of native Pt on pentlandite indicate a low-temperature solution state of Pt, however, the precipitates occur as small platelets or crystals only and any indications of growth to larger crystals or nuggets (mm size) are absent.

4.2. Placer PGM Worldwide—An Overview

Placers in Colombia and in Russia (Ural Mountains) were the major source of PGE for many years before the discovery of the rich deposits of the Bushveld Complex in South Africa. The initial major detection of platinum in the Bushveld Complex, which subsequently led to the discovery of the Merensky Reef, was made in 1924 by panning in a river bed on the farm Maandagshoek in the eastern Bushveld [5–8,10]. Early reports describe several alluvial diggings in the Bushveld Complex that produced some platinum [7]. However, as mining commenced on the rich dunite pipes and reef-type deposits of the Bushveld, alluvial PGM soon became forgotten because prospecting work in the 1920s did not reveal any profitable placers of significance [7].

Indeed, economic accumulations of PGM in stream sediments draining PGE-bearing layered intrusions such as the Bushveld Complex or the Great Dyke are unknown [2,3,7,28–30]. Their PGM assemblages are dominated by Pt–Fe alloys, sperrylite, and cooperite/braggite (Table 1), distinctly contrasting to the spectra of placer PGM derived from Alaskan/Uralian type complexes, and also from PGM assemblages originating from ophiolitic, generally uneconomic PGE mineralization (Table 1). Therefore, the "layered intrusion" assemblage can be regarded to represent a useful proximity indicator of near-by platinum mineralization.

Table 1. PGM proportions (in %) of placers originating from layered intrusions (1 + 2), Alaskan-type intrusions (3–5), ophiolite-related localities (6–8), the Rhine River (9), and the Witwatersrand gold paleoplacer (10). At some localities, no exact numbers are available, and x means = present. Abbreviations: $PtAs_2$ = sperrylite; RuS_2–OsS_2 = laurite-erlichmanite; PGE-sulfides = various PGE-sulfides; PGE–AsS = various PGE-sulfarsenides.

	PGM Locality	Ru–Os–Ir Alloys	Pt–Fe Alloys	$PtAs_2$	RuS_2–OsS_2	PGE Sulfides	PGE–AsS	Others	Ref.
1	*Great Dyke Rivers*		42	46	x	7		5	[29]
2	*Bushveld Rivers*		54	33		11		2	[30]
3	*Tulameen, Canada*	2	98						[73]
4	*Choco, Columbia*	1	97		1	1	x	x	[2]
5	*Urals, Russia*	2	97		1				[74]
6	*Chindwin, Burma*	96	x	x	x	x	4		[75]
7	*Aikora River, PNG*	88	12						[76]
8	*Samar, Philippines*	41	40		17			2	[77]
9	*Rhine River*	70	15	10	x			5	[78]
10	*Witwatersrand*	80	10	10		x	x		[79]

The most productive PGE placer deposits are associated with "Alaskan" or "Uralian"-type ultramafic-mafic complexes in tectonic belts, the best-known examples being many locations in the Ural Mountains in Russia, the Tulameen Complex in Canada, and alluvials in the Chocó department of Colombia [1–3,73]. Their PGM assemblages are characterized by the dominance of Pt–Fe alloys (>95% of the PGM assemblage), followed by inter-IPGE (Ru–Os–Ir) alloys (Table 1).

Numerous PGM placer occurrences linked to "Alpine-type" intrusions (ophiolite complexes) are on record worldwide and are well-studied, however, they are generally uneconomic [2,3,78,80]. Their PGM assemblages mainly comprise Pt–Fe alloys and Ru–Os–Ir alloys in about equal though variable proportions (Table 1).

The comparison provided in Table 1 includes the enigmatic PGM assemblages of the Witwatersrand gold-uranium paleoplacers (sources are probably Archean komatiites), and of the Rhine river in Germany. The assessment demonstrates that different PGM assemblages prevail in relation to the various source rock types and between different localities with similar source areas. Therefore, care must be exercised in comparing PGM from these different PGM sources, and this will be kept in mind in the discussion following.

4.3. Placer PGM—Residual Grains (Allogenic/Detrital) or Authigenic ("Neoformation")?

4.3.1. PGM in Laterites

It is felt that topic Section 4.3 deserves some deeper going and extensive discussion, which will follow now. Those readers less attracted by this theme may proceed to Section 4.3.3.

In the context of the discussion on PGM "neoformation", several studies have shown that lateritization can lead to solution, transport, and concentration of Pt and Pd [64,65,81]. In lateritized ophiolitic pyroxenites from the Pirogues River area of New Caledonia, up to 2 ppm (although typically 400 ppb) Pt were analyzed and this enrichment has been attributed to the lateritization process [64,81]. Pt mobility in the lateritic environment was also investigated at Andriamena in Madagascar [81], where a variety of primary, magmatic PGE minerals were observed including arsenides, sulfides, and stibiopalladinite, however, only sperrylite appears to have been resistant and occurs in the weathered zone, together with Pt–Fe alloy which is suggested to be newly-formed from a pre-existing grain of Pt, Fe (Pd, Ni) sulfide [81]—grossly being reminiscent of replacements of cooperite-braggite grains in the Bushveld alluvials (Figure 6c–e). Equally, an almost complete removal of Pd is reported [81].

More recent work on PGM from Ni-laterites (with ophiolite parent rocks) of the Dominican Republic [82] led the authors conclude that PGE are mobile on a local scale leading to in situ growth of PGM within limonite, probably by bio-reduction and/or electrochemical metal accretion. Besides primary PGM inclusions in fresh Cr-spinel (laurite, bowieite), supposedly secondary PGM (e.g., Ru–Fe–Os–Ir compounds) from weathering of pre-existing PGM were described, and also PGM precipitated after PGE mobilization within the laterite ("neoformation"). One elongated Pt–Fe–Ni grain (20 μm) is characterized by delicate botryoidal textures interpreted as in situ growth at surface conditions. No larger or well-crystallized PGM were observed.

Follow-up research [83] revealed the presence of so-called multistage PGM grains, namely porous Os–Ru–Fe–(Ir) grains overgrown by Ni–Fe–Ir and Ir–Fe–Ni–(Pt) compounds which in turn are overgrown by Pt–Ir–Fe–Ni phases. A model for multistage PGM grain formation is proposed: (i) hypogene PGM are transformed to secondary PGM by desulphurization during serpentinization; (ii) at the stages of serpentinization and/or at the early stages of lateritization, Ir is mobilized and recrystallizes on porous surfaces of secondary PGM; and (iii) at the late stages of lateritization, biogenically mediated "neoformation" (and accumulation) of Pt–Ir–Fe–Ni nanoparticles occurs. It is suggested [83] that in situ growth of Pt–Ir–Fe–Ni alloy nuggets of isometric symmetry may be possible within Ni laterites from the Dominican Republic.

The observed multistage IPGE-rich PGM grains are porous, corroded, locally oxidized, and overgrown by PGM nanoparticles. Clearly, the PGM assemblage has experienced a series of overprints including serpentinization (strongly reducing) and lateritization (oxidizing), and the observations can also be interpreted as corrosion and dissolution of primary PGM (intergrowth of various inter-IPGE alloys, here Ru–Os–Ir ± Pt alloys, laurite—typical PGM from ophiolites) and direct re-precipitation of dissolved PGE. Further, the result of an XRD analysis of one Pt–Ir–Fe–Ni alloy grain, namely an X-ray pattern identical to awaruite [83], is an unequivocal indication that this grain formed under strongly reducing conditions during serpentinization [84,85], and not through supergene (oxidizing) processes. Notably, Pt-rich awaruite grains have been described from both the Bushveld and Great Dyke placers [29,71], and Ir-rich awaruite grains were reported from ophiolitic chromitites in Pakistan [86].

In conclusion, any indisputable and convincing mineralogical evidence of in situ growth of larger and chemically well-defined PGM grains in the lateritic environment and at low temperatures is still lacking. No evidence of well-crystallized newly-formed PGM is on hand; in fact, all products of "neoformation" are thin films, nanoparticles and/or crystallites on substrates of earlier, generally larger, pre-existing PGM.

4.3.2. PGM in Placers

As indicated above, the origin of certain alluvial PGM grains is a matter of controversial debate. Two major models are in vogue:

(i)   Primary PGM formation within mafic/ultramafic intrusions, followed by weathering with no or only minor further alteration prior to alluvial concentration.
(ii) Solution of the PGE during weathering of the source rocks, probably aided by organic reactions, followed by the supergene growth of the macroscopic grains within the weathering zone and final alluvial concentration.

In the alluvial samples from the Bushveld and the Great Dyke, the largest proportion of the PGM assemblage (~40–70%) consists of Pt–Fe alloy and native Pt grains, compared to the scarcity (mostly $\pm$10%) of these PGM in the primary ores (Figure 7 and Table 1). Indeed, explanations are needed for this discrepancy.

Based on textural and geochemical arguments such as the presence of inclusions of a variety of sulfides, spinels, silicates, and other PGM as well as Os isotope compositions, the origin of Pt–Fe alloy, the most abundant PGM in most placers worldwide [2,3,80], is considered by most researchers to reflect an origin from high-temperature (i.e., magmatic) processes [2,3,87–90].

In contrast, a secondary origin of PGM in placers and soils is anticipated by several authors [81,91–98]. As size, shape, composition and micro texture of many eluvial and alluvial PGM differ from those observed in bedrocks and ores, it was proposed that secondary PGM come into being in a simplified process: (1) Serpentinization or weathering leads to the decomposition of base-metal sulfides carrying PGE in solid solution; (2) PGE are removed and transported as colloidal particles; (3) The colloids may coalesce or accrete to form larger particles and aggregates of PGE alloys.

An examination of gold and PGM (Pt–Fe and Os–Ir alloys) in offshore placers near Goodnews Bay, Alaska [99], revealed textures related to both derivation of PGM grains from mechanically weathered primary ore (i.e., typical assemblages of inclusions of PGE-arsenides, -sulfides and -tellurides; exsolution phenomena), and subsequent accretion (i.e., micro-crystalline assemblages of PGM in grain-rim cavities, suggesting leaching and crystallization).

Similar features were also observed in and on PGM grains of the Bushveld/Great Dyke placer assemblage [27–30]. Indeed, the above observations [99] pertain to many occurrences of placer PGM. Surface corrosion features are obvious in many localities, and a certain degree of PGE mobility has been documented by newly-formed nm- to μm-sized crystalline PGM [27–30,100–102] or PGE-biofilms [103] on and in cavities of pre-existing residual PGM grains. However, relative to the total budget of the alluvial PGM assemblages, these PGM neoformations play an insignificant role.

Bowles and coworkers [93–98] studied alluvial PGM grains (mostly Pt–Fe alloys, laurite-erlichmanite, Os–Ir alloys) from Sierra Leone and reported contrasting features of the primary and placer PGM [97], a main reason for the proposal that the latter developed as a result of breakdown of the primary PGM during weathering, movement of the PGE in solution, and growth of new PGM in placers with a different mineral assemblage, mineralogy and mineral chemistry [97]. The processes postulated for PGM "neoformation" include [97]: (1) long term weathering in an area of warm climate and high rainfall; (2) destruction of the primary PGM and transport of the PGE in solution; (3) organic compounds, such as humic or fulvic acids that are abundant in tropical rain forest soils appear likely to be involved in solution and transport of the PGE; (4) differential movement of the PGE in solution; (5) concentration of the least mobile PGE in soils close to the rivers, the more mobile PGE (especially

Pd) being carried away in solution; (6) accretion of the remaining PGE to form a new PGM assemblage, possibly where there is a change in Eh and pH conditions favourable for deposition; and (7) bacterial action may assist, or be responsible for, the PGM growth.

The main evidence claimed that PGM grow in situ as "neoformations" is that they comprise a different mineral assemblage from the PGM in the source rocks. The differences are [97]: (1) PGE arsenides, tellurides, and sulfides present in the host rocks become less abundant in or disappear completely from the alluvial suite, and Pt–Fe alloys including tulameenite, Os–Ir alloys, and laurite—erlichmanite become more abundant; (2) oxidized PGM appear in weathered or oxidized rocks, Cloudy, porous, or filamentous altered PGM may also be present; (3) there is a loss of the more soluble Pd which is present in the host rocks, but is much less abundant in the alluvial mineral suite; (4) a considerable difference in size (typically three orders of magnitude) exists between micrometre-sized PGM in the host rocks and millimetre-sized PGM in the alluvial suite; (5) some PGM in an alluvial suite show delicate crystal features that would not survive mechanical transport, these features include dendritic growth, perfect crystal faces, and perfect edges and corners between crystal faces. The corners are the most susceptible to mechanical damage and the presence of undamaged corners provides evidence for the absence of abrasion. The dendritic PGM are not known to occur in the host rocks. (6) There is over plating of mineral faces, colloform, cyclic, and fibrous textures not characteristic of PGM in the host rocks.

The observations of points (1) to (3) mirror those of the Bushveld/Great Dyke case. Less stable PGM are partly altered or destroyed during weathering, combined with a relative increase in Pt–Fe alloys; PGE-oxides appear and a large proportion of the bulk Pd is lost.

Point (4), the observed discrepancy in size distributions of detrital PGM grains (mostly between 100–200 μm and larger) compared to PGM in pristine ores (<5 to 50 μm) has been resolved for the Great Dyke case by treatment of a pristine MSZ sample (ca. 1 kg) by electric pulse disintegration. Altogether, 75 PGM grains larger than 50 μm were hand-picked from the treated material. The observed maximum true diameters were 480 μm for (Pt,Pd)-bismuthotellurides, 85 μm for sperrylite, 195 μm for cooperite/braggite, and 300 μm for Pt–Fe alloy grains [16,17,27], emphasizing that coarser PGM grains are present in the pristine MSZ ores. Notably, the presence of larger (~100 μm) and well-crystallized PGM grains extracted by careful crushing of primary Bushveld ores was reported recently [104,105], and wherever placer PGM were followed to their source rocks (e.g., Great Dyke, Bushveld, Urals, Colombia), chemically and size-equivalent precursor PGM were detected [16,17,27,87,106,107]. Evidently, larger PGM grains are present in the pristine ores and invalidate the argument of point (4).

It is claimed in point (5) that some alluvial PGM show delicate crystal features that would not survive mechanical transport. However, this generalized statement is not adequate as distinct differences exist in the chemical and physical stabilities of different PGM. Accordingly, detrital grains of the comparably soft native Pt and Pt–Fe alloys (e.g., Vickers hardness VHN_{50} of 303–321 for Pt_3Fe; [108]) often show evidence of physical abrasion such as bent and rounded corners and smooth, occasionally scratched or polished grain surfaces. In contrast, Os–Ir–Ru grains display little signs of corrosion or mechanical wear due to the great hardness (Os: VHN_{100} = 689–734; Ir: VHN_{100} = 841–900; Ru: VHN_{100} = 841–907; [108]) and chemical resistivity of this mineral group, and the same applies to the hardest PGM known, laurite (VHN_{100} = 1650 and 2012) and erlichmanite (VHN_{50} = 1358) [108]. Consequently, undamaged and perfect crystals of Os–Ir–Ru and laurite-erlichmanite [RuS_2–OsS_2] are on record from placers worldwide [2,3,77,78]. For example, the perfect hexagonal Os grains found in the Rhine river were probably transported up to some hundred kilometres [78]. Furthermore, well-crystallized or delicate alluvial grains could have well been transported within a protecting rock matrix, or alternatively not as bedload, but in suspension [2,3,77,78]. In conclusion, the argumentation of point (5) is faint.

Point (6), over plating of mineral faces, colloform, cyclic, and fibrous internal textures were described from a well-crystallized eluvial erlichmanite grain from Serra Leone and regarded as not characteristic of PGM in the host rocks, and "neoformation" by late-stage hydrothermal or secondary processes was proposed [96]. However, exactly this grain, which also carries bornite and chalcopyrite inclusions and displays a heterogenous internal distribution of Ru and Os, was analyzed for its Os isotope composition, and the authors conclude [109]: "that the nuggets were not formed during lateritization nor were they formed in-situ in sediments. The delicate morphology of PGM nuggets simply manifests very short transportation distances from the eroded site, as expected in eluvial (residual) deposits. The $^{187}Os/^{186}Os$ data are consistent with the formation of PGM nuggets in the melt and their detrital origin."

Recently, a Pt–Fe alloy grain from the Freetown Layered Complex, Sierra Leone, containing numerous and diverse inclusions of laurite, irarsite-hollingworthite (IrAsS–RhAsS), Pd–Te–Bi–Sb phases, Ir-alloy, Os-alloy, Pd-bearing Au, a Rh–Te phase, Pd–Au alloy, and Pd–Pt–Cu alloy was presented [98]. This grain was interpreted to be "neoform" growths in the organic- and bacterial-rich soils of the tropical rain forest cover of the Freetown intrusion [98]. However, such an assemblage of PGM, typical of magmatic, high temperature ores, is highly improbable to form from dispersed PGE, As, S, Te, Bi, etc. at ambient surficial conditions. In evidence to the contrary, previous work [27,57] and the present review (e.g., Figure 3E,F) have shown that the (Pt,Pd)-bismuthotellurides and -tellurides are the PGM species most susceptible to oxidation/weathering and are unstable in the supergene environment. PGM of this mineral group and the equally unstable sulfarsenide inclusions have only survived oxidation as they are well-shielded by the enclosing Pt–Fe alloy grain, and the described multi-PGM assemblage [98] as well as their Pt–Fe alloy host are definitely not products of supergene formation. Concentration of rare and stray elements should follow a chemical gradient which is not in evidence in this environment. Accordingly, an origin from the magmatic realm of this remarkable detrital Pt–Fe alloy grain is suggested. Its excellent state of preservation points to a near-by source. In fact, the external surface of the Pt–Fe alloy grain with intergrown PGM sticking out closely resembles eluvial grains (for example, similar grains from Onverwacht in South Africa [30]).

An investigation of placer PGM surfaces [103] demonstrated the existence of biological PGE cycling in some environments, whereby microbial biofilms play a key role in transforming PGM into more mobile forms (for example, nanoparticles). The authors [103] suggest that the microbial biofilms may enable the aggregation of PGE nanoparticles to form secondary PGM grains. Again, the products are small and distributed on the surface layers of pre-existing PGM substrates, and any proof of PGM growth to larger grains or nuggets still remains to be furnished.

Opponents of the school of supergene PGM "neoformation" admitted only two likely exceptions where "Pt nuggets" may have formed in a surficial environment, namely palladian gold, potarite, and native platinum in alluvial sediments from Devon, England, and botryoidal, zoned Pt–Pd nuggets from the Bom Sucesso stream in Brazil [2,3,87]. However, contrasting views on the Brazilian occurrences still exist: An evaluation of the Brazilian example [110,111] led to the conclusion [112] that the alluvial PGM most likely are detrital grains from mineralization formed by low-temperature hydrothermal fluids. In contrast, it is postulated [113] that they have a bio-organic origin, linking high levels of iodine in these grains to their formation from organic matter-rich waters. In conclusion, the origin of these Pt–Pd nuggets still remains enigmatic.

4.3.3. PGE Mobility and PGM Neoformation—Concluding Remarks

The above discussion demonstrates that the debate on the origin of PGM in placers—primary, detrital/allogenic grains versus authigenic "neoformation"—is not resolved. However, this is also partly due to a mix-up of terms and definitions, and dubious comparisons. For instance, the terms "PGM" or "Pt nuggets" embrace Pt–Fe alloys ("ferroan Pt") and native Pt (which may contain some % Pd) as in the Bushveld/Great Dyke case; the Brazilian example involves dendritic, mamillary or botryoidal grains of zoned Pt–Pd alloys of variable composition and partly containing potarite in the core [2,87,110–113], the PGM described from the Dominican Republic are Ru–Os–Ir ± Pt alloys [82,83], and in the Sierra Leone occurrences Pt–Fe alloys and erlichmanite-laurite [OsS_2-RuS_2] are the PGM studied [96–98]. The PGM dealt with need to be named unambiguously because, as shown above, different PGM will behave differently during weathering, Further, the definite relation of the placer PGM species to their source rocks—layered intrusions, Uralian/Alaskan complexes, or ophiolites (Table 1)—needs to be considered.

No doubt, the six PGE are in general though variably mobile in the supergene environment. First signs are detected in the weathering zone of the pristine ores, where oxidation in a hydrous environment leads to the destruction of the sulfides, releasing contained PGE (mainly Pd, Rh) and producing sulfuric acid, concomitant with the destruction of specific PGM, and "neoformation" more or less in situ of PGE-oxides/-hydroxides. PGE released from the destruction of sulfides and unstable PGM follow two paths: (i) Pt and part of the Pd are fixed (either adsorbed or in the crystal lattices) in iron- and iron-manganese oxides/hydroxides and smectites, and (ii) a great proportion of the dissolved Pd (up to 50%) is lost from the ores, taken away in solution. Rare newly-formed PGM in the Bushveld/Great Dyke oxidized ores consist of tiny (~1–5 µm) specks of Pt or Pt-oxides and zvyagintsevite usually on pre-existing residual PGM. Accretion to larger PGM crystals is not noted. The main effect of oxidation and weathering on the pristine PGM assemblage is continuous destruction of unstable compounds.

In the Bushveld and Great Dyke alluvial samples, breakdown and dissolution of the remaining PGM assemblage persists. Evidently, the often porous (Pt,Pd)-oxides/-hydroxides and the secondary Fe/Mn/Co-oxides/hydroxides, important carriers of Pt and Pd in the oxidized ores, do not survive the mechanical transport into the rivers and their metal contents were either transported away in the fine sediment fraction or in solution. The remaining PGM assemblage is characterized by continued partial alteration or destruction of most PGE-sulfide grains, as best seen in the immature Maandagshoek placer PGM assemblage [28,30,49,71], where some polyphase grains consisting of diverse PGM (various alloys, sulfides, sulfarsenides, tellurides) have survived, and simultaneously cooperite/braggite and rare Ru- and Rh-sulfides are still in disintegration [71]. Native Pt as crusts and overgrowths and forming porous or compact grains largely is a "neoformation" that mainly formed in situ by alteration of pre-existing cooperite/braggite grains. Sperrylite largely remains stable, and ubiquitous Pt–Fe alloy and native Pt grains have gained in prominence. In fact, ~40–70% of the PGM assemblage consists of Pt–Fe alloy and native Pt grains, compared to their scarcity (mostly ±10%) in the primary ores. Rare Pd-dominated minerals (Pd–Hg ± As and Pd–Sb ± As) probably are newly-formed compounds which may have formed during supergene processes.

Some Pt–Fe alloy grains are intergrown with or host a variety of inclusions of other PGM or sulfides, equivalent to grains in the pristine ores, and therefore indicate a consanguineous, magmatic origin. Based on comparable observations, similar deductions were put forward for placers of the Urals [87], Colombia [106,107], and worldwide [2,3,87].

The replacement of PGM-sulfides by native Pt and the newly grown coatings of native Pt on relict sulfides in the Bushveld alluvials indicate that low-temperature solution and redeposition of Pt is possible to a certain degree. The resulting precipitates occur as small (<1–5 µm) microcrystalline platelets or crystallites, or colloform, or porous grains, mostly deposited on or in cavities of pre-existing detrital PGM.

In the Bushveld/Great Dyke case, the predominance (~40–70%) of Pt–Fe alloy and native Pt grains of the PGM assemblage is explained in a satisfactory manner by the assumption that they are direct descendants from the pristine, primary ores (Merensky Reef, UG-2, Platreef, MSZ). Some Pt–Fe alloy grains host a variety of inclusions of sulfides and other PGM indicating a consanguineous, magmatic origin. Pt–Fe alloy and native Pt are the PGM most stable in the weathering environment. Therefore, the placer PGM assemblages represent a mineral rest spectrum that has survived physical and chemical attack. Demonstrably, the order of decreasing stability in the supergene environment is: (1) Pt–Fe alloys (very stable) → (2) sperrylite (stable) → (3) cooperite/braggite (variably stable/"meta-stable") → (4) PGE-bismuthotellurides and PGE-sulfarsenides (unstable).

Both in the oxidized ores and in the alluvial environment, the main impact of weathering on the PGM assemblages is destruction. Redistribution of PGE (PGM "neoformation") is taking place, however, on a very limited scale only. As a rule, the newly-formed products are nano-sized particles, small crystallites or rarely lowest range µm-sized grains, predominantly sited on precursor detrital/allogenic PGM grains, and they are of subordinate significance. Any growth to larger crystals or nuggets (mm size) was not observed, and in the Bushveld/Great Dyke case, any proof of PGM "neoformation" in a grand style is missing.

This finding can safely be extended to the common Pt–Fe alloys in placer occurrences worldwide. The missing link of the "neoformationists", a look into the cradle of PGM creation in the supergene environment, is still outstanding, as is any evidence of authigenic PGM formation and growth to larger crystals or nuggets.

5. Conclusions

The behavior of the platinum-group minerals (PGM) in the weathering cycle was studied in the Bushveld Complex, South Africa (Merensky Reef, UG-2 chromitite, Platreef) and the Great Dyke, Zimbabwe (Main Sulfide Zone). The main findings comprise:

- In the pristine sulfide ores of both the Bushveld and the Great Dyke, the PGE are bimodally distributed: Pt and variable proportions of Pd and Rh are present in the form of discrete PGM, dominantly PGE-bismuthotellurides, -sulfides, -arsenides, -sulfarsenides, and -alloys, and substantial though variable proportions of Pd and Rh are hosted in pentlandite.
- Near surface, in the oxidized ores, the PGE become polymodally distributed: In the course of weathering of the ores, PGE-sulfides, -arsenides and Pt–Fe alloys remain the only relics of the pristine PGM assemblage. The base metal sulfides and the (Pt,Pd)-bismuthotellurides are destroyed, and ill-defined (Pt,Pd)-oxides or -hydroxides develop. Further, elevated contents of Pt and Pd are found in Fe/Mn/Co-oxides/hydroxides and smectites.
- In the alluvial sediments, the PGE are unimodally contained in discrete PGM. The assemblages of detrital PGM are characterized by partial alteration or destruction of most remaining PGE-sulfide grains, whereas sperrylite largely survives as a stable phase, and Pt–Fe alloy grains predominate.
- Accordingly, the order of decreasing stability in the supergene environment is as follows: (1) Pt–Fe and Os–Ir–Ru alloys (very stable) → (2) sperrylite (stable) → (3) cooperite/braggite (variably stable/"meta-stable") → (4) PGE-bismuthotellurides, PGE-sulfarsenides, and PGE-oxides (unstable).
- In the Bushveld/Great Dyke case, and in all probability also worldwide, "neoformation", i.e., authigenic growth, of discrete, larger PGM in both oxidized ores and placers plays no substantial role. Dissolution and redistribution of PGE is taking place, however, the newly-formed products are nano-sized particles, small crystallites, or rarely µm-sized grains primarily sited on substrates of precursor detrital/allogenic PGM grains, and they are of subordinate significance. Any growth to larger crystals or nuggets (mm size) was not observed, and in the Bushveld/Great Dyke case, any proof of PGM "neoformation" in a grand style is missing.
- The final PGM suite which survived the weathering process on route from sulfide ores via oxidized ores into placers results from the continuous elimination of unstable PGM and the dispersion of soluble PGE. Therefore, the alluvial PGM assemblage represents a PGM rest spectrum of residual, detrital grains.

Funding: This research received no external funding.

Acknowledgments: I am grateful to all my helpful and productive colleagues at the BGR in Hannover who accompanied me during nearly 20 years of "platinum research". Joint field work was fantastic with Thorolf Weiser, Frank Melcher, Patrick Herb, Ulrich Schwarz-Schampera, Lothar Gast, Christian Wöhrl (nee Wittich), Malte Junge as well as Maximilian Korges and Gregor Borg (Halle), Inga Osbahr and Reiner Klemd (Erlangen), as well as

Dennis Krämer (nee Mohwinkel) and Michael Bau (Bremen). Up-to date laboratory work was ably performed by the colleagues and friends above as well as Peter Müller, Patrick Herb, Sarah Gregor, Marek Locmelis, Gregor Kuhlmann, Maria Sitnikova, a number of fine students from German and international universities, and the highly skilled technicians at the BGR labs, namely Donald Henry, Detlef Klosa, Frank Korte, Jerzy and Julian Lodziak, Thaddaeus Malarski and Dieter Weck.

Great thanks go to the mining companies Anglo Platinum, Impala Platinum, Nkwe Platinum, Sylvania Platinum, BHP Zimbabwe, Zimplats and Mimosa Mines, and especially their geologists and consultants Greg Holland, Ray Brown, Andrew Du Toit, Martine and Harry Wilhelmij, Humphrey O'Keeffe, Martin Prendergast, Peter Vanderspuy and Klaus Kappenschneider for their continuous support of our fieldwork in Zimbabwe. The hospitality and support of members of the Zimbabwe Geological Survey and the University of Zimbabwe is highly appreciated.

Our partners and friends in South Africa were Ray Brown (again), Grant Cawthorn, Gordon Chunnett, Paul Gutter, Dennis Hoffmann, Tawanda Manyeruke, Jan van der Merwe, Eveline de Meyer, Wally Mugabi, Trust Muzondo, Jacques Roberts, Robert Schouwstra, Martin Slabbert, and Rolf Frankenhauser. Finally, it is hoped that I have not forgotten anyone—if so my apologies! And thanks also to the valuable input of reviewer Giorgio Garuti, two anonymous colleagues and the Academic Editor of MINERALS which helped to straighten the golden thread of the paper.

Conflicts of Interest: The author declares no conflict of interest.

References

1. Mertie, J.B. Economic Geology of the Platinum Metals. *US Geol. Surv. Prof. Pap.* **1969**, *63*, 120.

2. Cabri, L.J.; Harris, D.C.; Weiser, T.W. The mineralogy and distribution of Platinum Group Mineral (PGM) placer deposits of the world. *Explor. Min. Geol.* **1996**, *5*, 73–167.

3. Weiser, T.W. Platinum-group minerals (PGM) in placer deposits. In *The Geology, Geochemistry, Mineralogy and Mineral Beneficiation of Platinum-Group Elements*; Cabri, L.J., Ed.; Canadian Institute of Mining, Metallurgy and Petroleum (CIM): Westmount, QC, Canada, 2002; CIM Special Volume 54, pp. 721–756.

4. Bundesanstalt für Geowissenschaften und Rohstoffe (BGR), Hannover, Germany. BGR Data Bank, PGE Production 2017. Available online: www.bgr.bund.de (accessed on 12 August 2018).

5. Merensky, H. *The Various Platinum Occurrences on the Farm Maandagshoek No. 148*; Unpublished Memorandum to Lydenburg Platinum Syndicate; Archives of the Merensky Trust: Duivelskloof, South Africa, 1924.

6. Merensky, H. Die neuentdeckten Platinfelder im mittleren Transvaal und ihre wirtschaftliche Bedeutung. *Zeitschrift der Deutschen Geologischen Gesellschaft* **1926**, *78*, 296–314.

7. Wagner, P.A. *The Platinum Deposits and Mines of South Africa*; Oliver & Boyd: London, UK, 1929; 326p.

8. Cawthorn, R.G. The discovery of the platiniferous Merensky Reef in 1924. *S. Afr. J. Geol.* **1999**, *102*, 178–183.

9. Scoates, J.S.; Friedman, R.M. Precise Age of the platiniferous Merensky Reef, Bushveld Complex. South Africa, by the U-Pb zircon chemical abrasion ID-TIMS technique. *Econ. Geol.* **2008**, *103*, 465–471. [CrossRef]

10. Viljoen, M. The Bushveld Complex—Host to the world's largest platinum, chromium and vanadium resources. *Episodes* **2016**, *39*, 239–268. [CrossRef]

11. Naldrett, T.; Kinnaird, J.; Wilson, A.; Chunnett, G. Concentration of PGE in the Earth's Crust with Special Reference to the Bushveld Complex. *Earth Sci. Front.* **2008**, *15*, 264–297. [CrossRef]

12. Oberthür, T.; Junge, M.; Rudashevsky, N.; de Meyer, E.; Gutter, P. Platinum-group minerals in the LG and MG chromitites of the Bushveld Complex, South Africa. *Miner. Depos.* **2016**, *51*, 71–87. [CrossRef]

13. Oberthür, T.; Davis, D.W.; Blenkinsop, T.G.; Höhndorf, A. Precise U-Pb mineral ages, Rb-Sr and Sm-Nd systematics for the Great Dyke, Zimbabwe—Constraints on late Archean events in the Zimbabwe Craton and Limpopo Belt. *Precambrian Res.* **2002**, *113*, 293–305. [CrossRef]

14. Worst, B.G. *The Great Dyke of Southern Rhodesia*; Bulletin 47; Southern Rhodesian Geological Survey: Salisbury, Zimbabwe, 1960.

15. Prendergast, M.D.; Wilson, A.H. The Great Dyke of Zimbabwe—II: Mineralisation and mineral deposits. In *Magmatic Sulphides—The Zimbabwe Volume*; Prendergast, M.D., Jones, M.J., Eds.; Institution Mining and Metallurgy: London, UK, 1989; pp. 21–42.

16. Oberthür, T. Platinum-Group Element Mineralization of the Main Sulfide Zone, Great Dyke, Zimbabwe. *Rev. Econ. Geol.* **2011**, *17*, 329–349.

17. Oberthür, T.; Weiser, T.W.; Gast, L.; Kojonen, K. Geochemistry and mineralogy of the platinum-group elements at Hartley Platinum Mine, Zimbabwe. Part 1: Primary distribution patterns in pristine ores of the Main Sulfide Zone of the Great Dyke. *Miner. Depos.* **2003**, *38*, 327–343. [CrossRef]

18. Naldrett, A. *Magmatic Sulphide Deposits—Geology, Geochemistry and Exploration*; Springer: Berlin, Germany, 2004; p. 728.

19. Godel, B.; Barnes, S.-J.; Maier, W.D. Platinum-Group Elements in Sulphide Minerals, Platinum-Group Minerals, and Whole-Rocks of the Merensky Reef (Bushveld Complex, South Africa): Implications for the Formation of the Reef. *J. Petrol.* **2007**, *48*, 1569–1604. [CrossRef]

20. Osbahr, I.; Klemd, R.; Oberthür, T.; Brätz, H.; Schouwstra, R. Platinum-group element distribution in base-metal sulfides of the Merensky Reef from the eastern and western Bushveld Complex, South Africa. *Miner. Depos.* **2013**, *48*, 211–232. [CrossRef]

21. Osbahr, I.; Oberthür, T.; Klemd, R.; Josties, A. Platinum-group element distribution in base-metal sulfides of the UG2, Bushveld Complex, South Africa—A reconnaissance study. *Miner. Depos.* **2014**, *49*, 655–665. [CrossRef]

22. Junge, M.; Wirth, R.; Oberthür, T.; Melcher, F.; Schreiber, A. Mineralogical siting of platinum-group elements in pentlandite from the Bushveld Complex, South Africa. *Miner. Depos.* **2015**, *50*, 41–54. [CrossRef]

23. Weiser, T.; Oberthür, T.; Kojonen, K.; Johanson, B. Distribution of trace PGE in pentlandite and of PGM in the Main Sulfide Zone (MSZ) at Mimosa Mine, Great Dyke, Zimbabwe. In *8th Internat Platinum Symposium*; Symposium Series S 18; South African Institute Mining Metallurgy: Johannesburg, South Africa, 1998; pp. 443–445.

24. Oberthür, T.; Melcher, F.; Buchholz, P.; Locmelis, M. The oxidized ores of the Main Sulfide Zone, Great Dyke, Zimbabwe: Turning resources into minable reserves—Mineralogy is the key. *J. South. Afr. Inst. Min. Metall.* **2013**, *133*, 191–201.

25. Junge, M. The Fate of Platinum-Group Elements during Weathering Processes—With Special Focus on Pristine and Weathered Platreef Ores at the Mogalakwena Mine in the Bushveld Complex. Ph.D. Thesis, Leibniz-Universität Hannover, Hannover, Germany, 2017; 265p.

26. Korges, M. Supergene Mobilization and Redistribution of Platinum-Group Elements in the Merensky Reef, Eastern Bushveld Complex, South Africa. Master's Thesis, Martin-Luther-Universität Halle-Wittenberg, Halle, Germany, 2015; 116p.

27. Oberthür, T.; Weiser, T.W.; Gast, L. Geochemistry and mineralogy of the platinum-group elements at Hartley Platinum Mine, Zimbabwe. Part 2: Supergene redistribution in the oxidized Main Sulfide Zone of the Great Dyke, and alluvial platinum-group minerals. *Miner. Depos.* **2003**, *38*, 344–355. [CrossRef]

28. Oberthür, T.; Melcher, F.; Gast, L.; Wöhrl, C.; Lodziak, J. Detrital platinum-group minerals in rivers draining the eastern Bushveld Complex, South Africa. *Can. Mineral.* **2004**, *42*, 563–582. [CrossRef]

29. Oberthür, T.; Weiser, T.W.; Melcher, F.; Gast, L.; Wöhrl, C. Detrital platinum-group minerals in rivers draining the Great Dyke, Zimbabwe. *Can. Mineral.* **2013**, *51*, 197–222. [CrossRef]

30. Oberthür, T.; Weiser, T.W.; Melcher, F. Alluvial and Eluvial Platinum-Group Minerals (PGM) from the Bushveld Complex, South Africa. *S. Afr. J. Geol.* **2014**, *117*, 255–274. [CrossRef]

31. Naldrett, A.J.; Kinnaird, J.; Wilson, A.; Yudovskaya, M.; McQuade, S.; Chunnett, G.; Stanley, C. Chromite composition and PGE content of Bushveld chromitites: Part 1—the Lower and Middle Groups. *Appl. Earth Sci. (Trans. Inst. Min. Metall. B)* **2009**, *118*, 131–161. [CrossRef]

32. Hazen, R.M.; Grew, E.S.; Downs, R.T.; Golden, J.; Hystad, G. Mineral ecology: Chance and necessity in the mineral diversity of terrestrial planets. *Can. Mineral.* **2015**, *53*, 295–324. [CrossRef]

33. Schouwstra, R.; Kinloch, E.D.; Lee, C. A short geological review of the Bushveld Complex. *Platin. Met. Rev.* **2000**, *44*, 33–39.

34. Kinloch, E.D. Regional Trends in the Platinum-Group Mineralogy of the Critical Zone of the Bushveld Complex, South Africa. *Econ. Geol.* **1982**, *77*, 1328–1347. [CrossRef]

35. Kinloch, E.D.; Peyerl, W. Platinum-group minerals in various rocks of the Merensky Reef: Genetic implications. *Econ. Geol.* **1990**, *85*, 537–555. [CrossRef]

36. Rose, D.; Viljoen, F.; Knoper, M.; Rajesh, H. Detailed assessment of platinum-group minerals associated with chromitite stringers in the Merensky Reef of the eastern Bushveld Complex, South Africa. *Can. Miner.* **2011**, *49*, 1385–1396. [CrossRef]

37. Voordouw, R.J.; Gutzmer, J.; Beukes, N.J. Zoning of platinum group mineral assemblages in the UG2 chromitite determined through in situ SEM-EDS-based image analysis. *Miner. Depos.* **2010**, *45*, 47–159. [CrossRef]

38. Junge, M.; Oberthür, T.; Melcher, F. Cryptic variation of chromite chemistry, platinum-group-element and -mineral distribution in the UG-2 chromitite—An example from the Karee Mine, western Bushveld Complex, South Africa. *Econ. Geol.* **2014**, *109*, 795–810. [CrossRef]

39. Kottke-Levin, J.; Tredoux, M.; Gräbe, P.-J. An investigation of the geochemistry of the Middle Group of the eastern Bushveld Complex, South Africa Part 1—The chromitite layers. *Appl. Earth Sci. Trans. Inst. Min. Metall. B* **2009**, *118*, 111–130. [CrossRef]

40. Junge, M.; Oberthür, T.; Osbahr, I.; Gutter, P. Platinum-group elements and minerals in the Lower and Middle Group chromitites of the western Bushveld Complex, South Africa. *Miner. Depos.* **2016**, *51*, 841–852. [CrossRef]

41. McDonald, I.; Holwell, D.A.; Armitage, P.E.B. Geochemistry and mineralogy of the Platreef and 'Critical Zone' of the northern lobe of the Bushveld Complex, South Africa: Implications for Bushveld stratigraphy and the development of PGE mineralisation. *Miner. Depos.* **2005**, *40*, 526–549. [CrossRef]

42. Kinnaird, J.; Hutchinson, D.; Schurmann, L.; Nex, P.A.M.; de Lange, R. Petrology and mineralization of the southern Platreef, northern limb of the Bushveld Complex, South Africa. *Miner. Depos.* **2005**, *40*, 576–597. [CrossRef]

43. Holwell, D.A.; McDonald, I.; Armitage, P.E.B. Platinum-group mineral assemblages in the Platreef at the South-Central Pit, Sandsloot Mine, northern Bushveld Complex, South Africa. *Mineral. Mag.* **2006**, *70*, 83–101. [CrossRef]

44. Manyeruke, T. Compositional and Lithological Variation of the Platreef on the Farm Nonnenwerth, Northern Lobe of the Bushveld Complex: Implications for the Origin of the Platinum-Group Element Mineralization. Ph.D. Thesis, University of Pretoria, Pretoria, South Africa, 2007; 248p.

45. van der Merwe, F.; Viljoen, F.; Knoper, M. The mineralogy and mineral associations of platinum group elements and gold in the Platreef at Zwartfontein, Akanani Project, Northern Bushveld Complex, South Africa. *Mineral. Petrol.* **2012**, *106*, 25–38. [CrossRef]

46. Klemd, R.; Junge, M.; Oberthür, T.; Herderich, T.; Schouwstra, R.; Roberts, J. Platinum-group element concentrations in base-metal sulphides from the Platreef, Mogalakwena Platinum Mine, Bushveld Complex, South Africa. *S. Afr. J. Geol.* **2016**, *119*, 623–638. [CrossRef]

47. Tarkian, M.; Stumpfl, E. Platinum mineralogy of the Driekop mine, South Africa. *Miner. Depos.* **1975**, *10*, 71–85. [CrossRef]

48. Stumpfl, E.; Rucklidge, J.C. The platiniferous dunite pipes of the eastern Bushveld. *Econ. Geol.* **1982**, *77*, 1419–1431.

49. Melcher, F.; Lodziak, J. Platinum-group minerals of concentrates from the Driekop platinum pipe, Eastern Bushveld Complex—Tribute to Eugen F. Stumpfl. *Neues Jahrbuch für Mineralogie Abhandlungen* **2007**, *183*, 173–195. [CrossRef]

50. Oberthür, T.; Melcher, F.; Sitnikova, M.; Rudashevsky, N.S.; Rudashevsky, V.N.; Cabri, L.J.; Lodziak, J.; Klosa, D.; Gast, L. Combination of Novel Mineralogical Methods in the Study of Noble Metal Ores—Focus on Pristine (Bushveld, Great Dyke) and Placer Platinum Mineralisation. In Proceedings of the Ninth International Congress for Applied Mineralogy, Brisbane, QLD, Australia, 8–10 September 2008; pp. 187–193.

51. Coghill, B.M.; Wilson, A.H. Platinum-group minerals in the Selukwe subchamber, Great Dyke, Zimbabwe: Implications for PGE collection mechanisms and post-formational redistribution. *Mineral. Mag.* **1993**, *57*, 613–633. [CrossRef]

52. Oberthür, T.; Cabri, L.J.; Weiser, T.; McMahon, G.; Müller, P. Pt, Pd and other trace elements in sulfides of the Main Sulfide Zone, Great Dyke, Zimbabwe—A reconnaissance study. *Can. Mineral.* **1997**, *35*, 597–609.

53. Kuhlmann, G.; Oberthür, T.; Melcher, F.; Lodziak, J. *Bushveld Komplex, Südafrika: UG2-Chromitit-Horizont, Mineralogisch—Geochemische Feinstratigraphie—Schwerpunkt Platinmetall-Verteilung*; Unpublished Internal Report Tgb.-Nr. 11327/06; Bundesanstalt für Geowissenschaften und Rohstoffe (BGR): Hannover, Germany, 2006; 166p.

54. Holwell, D.A.; McDonald, I. Distribution of platinum-group elements in the Platreef at Overysel, northern Bushveld complex: A combined PGM and LA-ICP-MS study. *Contrib. Mineral. Petrol.* **2007**, *154*, 171–190. [CrossRef]

55. Fuchs, W.A.; Rose, A.W. The geochemical behaviour of platinum and palladium in the weathering cycle in the Stillwater Complex, Montana. *Econ. Geol.* **1974**, *69*, 332–346. [CrossRef]

56. Hey, P. The effects of weathering on the UG-2 chromitite reef with special reference to the platinum-group minerals. *S. Afr. J. Geol.* **1999**, *102*, 251–260.

57. Locmelis, M.; Melcher, F.; Oberthür, T. Platinum-Group Element Distribution in the Oxidized Main Sulfide Zone, Great Dyke, Zimbabwe. *Miner. Depos.* **2010**, *45*, 93–109. [CrossRef]

58. Cousins, C.A.; Kinloch, E.D. Some observations on textures and inclusions in alluvial platinoids. *Econ. Geol.* **1976**, *71*, 1377–1398. [CrossRef]

59. Schneiderhöhn, H.; Moritz, H. Die Oxydationszone im platinführenden Sulfidpyroxenit (Merensky-Reef) des Bushvelds in Transvaal. *Zentralblatt Mineralogie Geologie Paláontologie* **1939**, *Abteilung A*, 1–12.

60. Evans, D.M.; Buchanan, D.L.; Hall, G.E.M. Dispersion of platinum, palladium and gold from the Main Sulphide Zone, Great Dyke, Zimbabwe. *Trans. Inst. Min. Metall.* **1994**, *103*, B57–B67.

61. Oberthür, T.; Melcher, F. PGE and PGM in the supergene environment: The Great Dyke, Zimbabwe. In *Exploration for Platinum-Group Element Deposits*; Mungall, J.E., Ed.; Short Course Series; Mineral Association of Canada: Quebec City, QC, Canada, 2005; Volume 35, pp. 97–112.

62. Weiser, T. The quantitative proof of the existence of PGE-oxides. In Proceedings of the 6th Internat Platinum Symposium, Perth, Australia, 8–11 July1991; Abstract Volume 52.

63. Jedwab, J. Oxygenated platinum-group element and transition metal (Ti, Cr, Mn, Co, Ni) compounds in the supergene domain. *Chronique de la Recherche Minière* **1995**, *520*, 47–53.

64. Augé, T.; Legendre, O. Platinum-group element oxides from the Pirogues ophiolitic mineralization, New Caledonia: Origin and significance. *Econ. Geol.* **1994**, *89*, 1454–1468. [CrossRef]

65. Augé, T.; Maurizot, P.; Breton, J.; Eberlé, J.-M.; Gilles, C.; Jézéquel, P.; Mézière, J.; Robert, M. Magmatic and supergene platinum-group minerals in the New Caledonia ophiolite. *Chronique de la Recherche Miniere* **1995**, *520*, 3–26.

66. Evans, D.M.; Spratt, J. Platinum and palladium oxides/hydroxides from the Great Dyke, Zimbabwe, and thoughts on their stability and possible extraction. In *Applied Mineralogy*; AA Balkema: Rotterdam, The Netherlands, 2000; pp. 289–292.

67. McDonald, I.; Ohnenstetter, D.; Ohnenstetter, M.; Vaughan, D.J. Palladium Oxides in Ultramafic Complexes Near Lavatrafo, Western Andriamena, Madagascar. *Mineral. Mag.* **1999**, *63*, 345–352. [CrossRef]

68. Cabral, A.R.; Lehmann, B.; Kwitko, R.; Jones, R.; Pires, F.; Rocha Filho, O.; Innocentini, M. Palladium-oxygenated compounds of the Gongo Soco mine, Quadrilátero Ferrífero, central Minas Gerais, Brazil. *Mineral. Mag.* **2001**, *65*, 169–179. [CrossRef]

69. Jedwab, J.; Criddle, A.J.; du Ry, P.; Piret, P.; Stanley, C.J. Rediscovery of palladinite (PdO tetrag.) from Itabira (Minas Gerais, Brazil) and from Ruwe (Shaba, Zaire). In Proceedings of the IAGOD Meeting, Orleans, France, 6 September 1994.

70. Gregor, S. *Evaluation of PGE Mineralization in the Oxidized Main Sulfide Zone at Mimosa Mine, Great Dyke, Zimbabwe*; Unpublished Internal File 10748/04; Bundesanstalt für Geowissenschaften und Rohstoffe (BGR): Hannover, Germany, 2004; 55p.

71. Melcher, F.; Oberthür, T.; Lodziak, J. Modification and alteration of detrital platinum-group minerals from the Eastern Bushveld Complex, South Africa. *Can. Mineral.* **2005**, *43*, 1711–1734. [CrossRef]

72. Möller, P.; Kersten, G. Electrochemical accumulation of visible gold on pyrite and arsenopyrite surfaces. *Miner. Depos.* **1994**, *29*, 404–413. [CrossRef]

73. Nixon, G.T.; Cabri, L.J.; Laflamme, J.H.G. Platinum-group element mineralization in lode and placer deposits associated with the Tulameen Alaskan-type complex, British Columbia. *Can. Mineral.* **1990**, *28*, 503–535.

74. Malitch, K.N.; (Institute of Geology and Geochemistry, UB RAS, Ekaterinburg, Russia). Personal communication, 2016.

75. Hagen, D.; Weiser, T.; Than, H. Platinum-group minerals in Quaternary gold placers in the upper Chindwin area of northern Burma. *Mineral. Petrol.* **1990**, *42*, 265–286. [CrossRef]

76. Weiser, T.W.; Bachmann, H.-G. Platinum-group minerals from the Aikora River area, Papua New Guinea. *Can. Mineral.* **1999**, *37*, 1131–1145.

77. Oberthür, T.; Melcher, F.; Weiser, T. Detrital platinum-group minerals and gold in placers of south-eastern Samar Island, Philippines. *Can. Mineral.* **2017**, *54*, 45–62. [CrossRef]

78. Oberthür, T.; Melcher, F.; Goldmann, S.; Wotruba, H.; Dijkstra, A.; Gerdes, A.; Dale, C. Mineralogy and mineral chemistry of detrital heavy minerals from the Rhine River in Germany as evidence of their provenance, sedimentary and depositional history: Focus on platinum-group minerals and remarks on cassiterite, columbite-group minerals, and uraninite. *Int. J. Earth Sci.* **2016**, *105*, 637–657.

79. Feather, C.E. Mineralogy of Platinum-Group Minerals in the Witwatersrand, South Africa. *Econ. Geol.* **1976**, *71*, 1399–1428. [CrossRef]

80. O'Driscoll, B.; González-Jiménez, J.M. Petrogenesis of the Platinum-Group Minerals. *Rev. Mineral. Geochem.* **2016**, *81*, 489–578. [CrossRef]

81. Salpeteur, I.; Martel-Jantin, B.; Rakotomanana, D. Pt and Pd mobility in ferralitic soils of the West Andriamina Area (Madagascar). Evidence of a supergene origin of some Pt and Pd Minerals. *Chronique de la Recherches Minière* **1995**, *520*, 27–45.

82. Aiglsperger, T.; Proenza, J.A.; Zaccarini, F.; Lewis, J.F.; Garuti, G.; Labrador, M.; Longo, F. Platinum group minerals (PGM) in the Falcondo Ni laterite deposit, Loma Caribe peridotite (Dominican Republic). *Miner. Depos.* **2015**, *50*, 105–123. [CrossRef]

83. Aiglsperger, T.; Proenza, J.A.; Font-Bardia, M.; Baurier-Aymat, S.; Galí, S.; Lewis, J.F.; Longo, F. Supergene neoformation of Pt-Ir-Fe-Ni alloys: Multistage grains explain nugget formation in Ni-laterites. *Miner. Depos.* **2017**, *52*, 1069–1083. [CrossRef]

84. Ramdohr, P. A widespread mineral association, connected with serpentinization. *Neues Jahrbuch Mineralogie Abhandlungen* **1967**, *107*, 241–265.

85. Eckstrand, O.R. The Dumont serpentinite: A model for control of nickeliferous opaque mineral assemblages by alteration reactions in ultramafic rocks. *Econ. Geol.* **1975**, *70*, 183–201. [CrossRef]

86. Ahmed, Z.; Bevan, J.C. Awaruite, iridian awaruite and a new Ru-Os-Ir-Ni-Fe alloy from the Sakhakot-Qila complex, Malakand agency, Pakistan. *Mineral. Mag.* **1981**, *44*, 225–230. [CrossRef]

87. Weiser, T.W. Platinum-Group Minerals (PGM) from placer deposits in the mineral collection of the Museum of Natural History, Vienna, Austria. *Annalen Naturhist. Museum Wien* **2004**, *105A*, 1–28.

88. Malitch, K.N.; Thalhammer, O.A.R. Pt-Fe nuggets from clinopyroxenite–dunite massifs, Russia: A structural, compositional and osmium-isotope study. *Can. Mineral.* **2002**, *40*, 395–417. [CrossRef]

89. Daouda Traoré, D.; Beauvais, A.; Augé, T.; Parisot, J.C.; Colin, F.; Cathelineau, M. Chemical and physical transfers in an ultramafic rock weathering profile: Part 2. Dissolution vs. accumulation of platinum group minerals. *Am. Mineral.* **2008**, *93*, 31–38. [CrossRef]

90. Okrugin, A.V. Origin of platinum-group minerals in mafic and ultramafic rocks: From dispersed elements to nuggets. *Can. Mineral.* **2011**, *49*, 1397–1412. [CrossRef]

91. Augusthitis, S.S. Mineralogical and geochemical studies of the platiniferous dunite-birbirite-pyroxenite complex of Yubdo, Birbir, W. Ethiopia. *Chemie der Erde* **1965**, *24*, 159–165.

92. Ottemann, J.; Augusthitis, S.S. Geochemistry and origin of "platinum-nuggets" in laterite covers from ultrabasic rocks and birbirites of W. Ethiopia. *Miner. Depos.* **1967**, *1*, 269–277. [CrossRef]

93. Stumpfl, E.F. The genesis of platinum deposits: Further thoughts. *Miner. Sci. Eng.* **1974**, *6*, 120–141.

94. Bowles, J.F.W. The development of platinum-group minerals in laterites. *Econ. Geol.* **1986**, *81*, 1278–1285. [CrossRef]

95. Bowles, J.F.W. Platinum–iron alloys, their structural and magnetic characteristics in relation to hydrothermal and low-temperature genesis. *Mineral. Petrol.* **1990**, *43*, 37–47. [CrossRef]

96. Bowles, J.F.W. The development of platinum-group minerals (PGM) in laterites: Mineral morphology. *Chronique de la Recherche Minière* **1995**, *520*, 55–63.

97. Bowles, J.F.W.; Suárez, S.; Prichard, H.M.; Fisher, P.C. The mineralogy, geochemistry and genesis of the alluvial platinum-group minerals of the Freetown Layered Complex, Sierra Leone. *Mineral. Mag.* **2018**, *82*, 223–246. [CrossRef]

98. Bowles, J.F.W.; Suárez, S.; Prichard, H.M.; Fisher, P.C. Inclusions in an isoferroplatinum nugget from the Freetown Layered Complex, Sierra Leone. *Mineral. Mag.* **2018**, *82*, 577–592. [CrossRef]

99. Mardock, C.L.; Barker, J.C. Theories on the transport and deposition of gold and PGM minerals in offshore placers near Goodnews Bay, Alaska. *Ore Geol. Rev.* **1991**, *6*, 211–227. [CrossRef]

100. McClenaghan, M.B.; Cabri, L.J. Review of gold and platinum group element (PGE) indicator minerals methods for surficial sediment sampling. *Geochem. Explor. Environ. Anal.* **2011**, *11*, 251–263. [CrossRef]

101. Zhmodik, S.M.; Nesterenko, G.V.; Airiyants, E.V.; Belyanin, D.K.; Kolpakov, V.V.; Podlipsky, M.Y.; Karmanov, N.S. Alluvial platinum-group minerals as indicators of primary PGE mineralization (placers of southern Siberia). *Russ. Geol. Geophys.* **2016**, *57*, 1437–1464. [CrossRef]

102. Nesterenko, G.V.; Zhmodik, S.M.; Airiyants, E.V.; Belyanin, D.K.; Kolpakov, V.V.; Bogush, A.A. Colloform high-purity platinum from the placer deposit of Koura River (Gornaya Shoriya, Russia). *Ore Geol. Rev.* **2017**, *91*, 236–245. [CrossRef]

103. Reith, F.; Zammit, C.-M.; Shar, S.S.; Etschmann, B.; Bottrill, R.; Southam, G.; Ta, C.; Kilburn, M.; Oberthür, T.; Ball, A.S.; et al. Biological role in the transformation of platinum-group-mineral grains. *Nat. Geosci.* **2016**, *9*, 294–298. [CrossRef]

104. Yudovskaya, M.A.; Kinnaird, J.A.; Grobler, D.F.; Costin, G.; Abramova, V.D.; Dunnett, T.; Barnes, S.-J. Zonation of Merensky-Style Platinum-Group Element Mineralization in Turfspruit Thick Reef Facies (Northern Limb of the Bushveld Complex). *Econ. Geol.* **2017**, *112*, 1333–1365. [CrossRef]

105. McCreesh, M.J.G.; Yudovskaya, M.A.; Kinnaird, J.A.; Reinke, C. Platinum-group minerals of the F and T zones, Waterberg Project, Far Northern Bushveld Complex: Implication for the formation of the PGE mineralisation. *Mineral. Mag.* **2018**, *82*, 539–575. [CrossRef]

106. Tistl, M. Geochemistry of platinum-group elements of the zoned ultramafic Alto Condoto Complex, Northwest Colombia. *Econ. Geol.* **1994**, *89*, 158–167. [CrossRef]

107. Burgath, K.P.; Salinas, R. The Condoto Complex in Chocó, Colombia: A Pt-bearing Alaskan-type intrusion. *Zeitschrift für Angewandte Geologie* **2000**, *Sonderheft SH 1*, 163–170.

108. Cabri, L.J. The platinum-group minerals. In *Canadian Institute of Mining, Metallurgy and Petroleum, CIM Special Volume*; Cabri, L.J., Ed.; CIM: Westmount, QC, Canada, 2002; Volume 54, pp. 13–129.

109. Hattori, K.; Cabri, L.J.; Hart, S.R. Osmium isotope ratios of PGM grains associated with the Freetown Layered Complex, Sierra Leone, and their origin. *Contrib. Mineral. Petrol.* **1991**, *109*, 10–18. [CrossRef]

110. Cabral, A.R.; Beaudoin, G.; Choquette, M.; Lehmann, B.; Polônia, J.C. Supergene leaching and formation of platinum in alluvium: Evidence from Serro, Minas Gerais, Brazil. *Mineral. Petrol.* **2007**, *90*, 141–150. [CrossRef]

111. Cabral, A.R.; Lehmann, B.; Tupinambá, M.; Schlosser, S.; Kwitko-Ribeiro, R.; De Abreu, F.R. The platiniferous Au-Pd belt of Minas Gerais, Brazil, and genesis of its botryoidal Pt-Pd aggregates. *Econ. Geol.* **2009**, *104*, 1265–1276. [CrossRef]

112. Fleet, M.E.; De Almeida, C.M.; Angeli, N. Botryoidal platinum, palladium and potarite from the Bom Sucesso stream, Minas Gerais, Brazil: Compositional zoning and origin. *Can. Mineral.* **2002**, *40*, 341–355. [CrossRef]

113. Cabral, A.R.; Radtke, M.; Munnik, F.; Lehmann, B.; Reinholz, U.; Riesemeier, H.; Tupinambá, M.; Kwitko-Ribeiro, R. Iodine in alluvial Pt-Pd nuggets: Evidence for biogenic precious-metal fixation. *Chem. Geol.* **2011**, *281*, 125–132. [CrossRef]

![minerals logo] *minerals*

MDPI

Review

Variations of Major and Minor Elements in Pt–Fe Alloy Minerals: A Review and New Observations

Andrei Y. Barkov [1,*] and Louis J. Cabri [2]

[1]   Research Laboratory of Industrial and Ore Mineralogy, Cherepovets State University,
      5 Lunacharsky Avenue, Cherepovets 162600, Russia
[2]   Cabri Consulting Inc., 514 Queen Elizabeth Drive, Ottawa, ON K1S 3N4, Canada; lcabri@sympatico.ca
*   Correspondence: ore-minerals@mail.ru; Tel.: +7-8202-51-78-27

Received: 5 December 2018; Accepted: 27 December 2018; Published: 4 January 2019

check for updates

Abstract: Compositional variations of major and minor elements were examined in Pt–Fe alloys from various geological settings and types of deposits, both lode and placer occurrences. They included representatives of layered intrusions, Alaskan-Uralian-(Aldan)-type and alkaline gabbroic complexes, ophiolitic chromitites, and numerous placers from Canada, USA, Russia, and other localities worldwide. Pt–Fe alloy grains in detrital occurrences are notably larger in size, and these are considered to be the result of a special conditions during crystallization such as temperature, pressure, geochemistry or time. In addition, the number of available statistical observations is much greater for the placer occurrences, since they represent the end-product of, in some cases, the weathering of many millions of tonnes of sparsely mineralized bedrock. Typically, platinum-group elements (PGE) present in admixtures (Ir, Rh, and Pd) and minor Cu, Ni are incorporated into a compositional series $(Pt, PGE)_{2-3}(Fe, Cu, Ni)$ in the lode occurrences. Relative Cu enrichment in alloys poor in Pt implies crystallization from relatively fractionated melts at a lower temperature. In contrast to the lode deposits, the distribution of Ir, Rh, and Pd is fairly chaotic in placer Pt–Fe grains. There is no relationship between levels of Ir, Rh, and Pd with the ratio $\Sigma(Pt + PGE):(Fe + Cu + Ni)$. The compositional series $(Pt, PGE)_{2-3}(Fe, Cu, Ni)$ is not as common in the placer occurrences; nevertheless, minor Cu and Ni show their maximums in members of this series in the placer grains. Global-scale datasets yield a bimodal pattern of distribution in the Pt–Fe diagram, which is likely a reflection of the miscibility gap between the ordered Pt_3Fe structure (isoferroplatinum) and the disordered structure of native or ferroan platinum. In the plot Pt versus Fe, there is a linear boundary due to ideal Pt $\leftrightarrow$ Fe substitution. Two solid solution series are based on the Ir-for-Pt and Pd-for-Pt substitutions. The incorporation of Ir is not restricted to Pt_3Fe–Ir_3Fe substitution (isoferroplatinum and chengdeite, plus their disordered modifications). Besides, Ir^0 appears to replace Pt^0 in the disordered variants of (Pt–Ir)–Fe alloys. There is a good potential for the discovery of a new species with a Pd-dominant composition, $(Pd, Pt)_3Fe$, most likely in association with the alkaline mafic-ultramafic or gabbroic complexes, or the mafic units of layered intrusions. The "field of complicated substitutions" is recognized as a likely reflection of the crystallochemical differences of Pd and Ir, extending along the Ir-Pd axis of the Ir–Pd–Rh diagram. The inferred solid solution extends approximately along the line Ir–(Pd:Rh = 2:3). Minor Pd presumably enters the solid solution via a coupled substitution in combination with the Rh. An Ir-enrichment trend in Pt–Fe alloys typically occurs in the Alaskan-type complexes. The large size of the Pt–Fe nuggets associated with some of these complexes is considered to be related to an ultramafic-mafic pegmatite facies, whereas significant Pd-enrichment is characteristic of gabbroic source-rocks (e.g., Coldwell Complex), resulting in a markedly different trend for the Pt versus Fe (wt.%). However, based on our examination of a large dataset of Pt–Fe alloys from numerous origins, we conclude that they exhibit compositional overlaps that are too large to be useful as reliable index-minerals.

Keywords: platinum-group elements; platinum-group minerals; Pt–Fe alloys; compositional variations; element substitutions; placer deposits; ore mineralization; ultramafic-mafic complexes

1. Introduction

Natural Pt–Fe alloys are an economically important species of platinum-group minerals (PGM), which are especially abundant in the zoned Alaskan-Uralian-(Aldan)-type ultramafic-mafic complexes and placers of PGE (platinum-group elements) that are derived from these complexes [1–4]. Native platinum (0–17 at.% Fe) is cubic (*Fm3m*), with a = 3.9231 Å (synthetic equivalent). Increasing Fe content reduces the cell dimension [1,5–7]. In the range of 17–20 at.% Fe, the structure is disordered or represents a mixture of native platinum and isoferroplatinum. The latter species, ideally the Pt_3Fe (25 at.% Fe), has a cubic structure (*Pm3m*, 20–36 at.% Fe) with a = 3.86 Å [1]. Variations in the Fe cause the appearance of a partial disorder in this mineral. Within the interval 36–41 at.%, the alloy structures are disordered in the Pt–Fe system; isoferroplatinum likely coexists with tetraferroplatinum [8], i.e., tetragonal PtFe (41–65 at.% Fe; ideally 50 at.% Fe) with a = 3.84 Å and c = 3.71 Å; and its space group is *P4/mmm* [1–3]. The observed difference between the compositions of isoferroplatinum and tetraferroplatinum leads to a small reduction in the cell dimension to accommodate a greater content of Fe [1,5,6]. Commonly, natural members of the tetraferroplatinum–tulameenite [Pt_2FeCu] series are somewhat nonstoichiometric and extend towards (Pt, PGE)$_{1+x}$ (Fe, Cu, Ni)$_{1-x}$ [9].

The goals of this article are: (1) To evaluate the compositional ranges observed in Fe-bearing platinum (or ferron platinum) and isoferroplatinum with respect to the main components and admixtures on the basis of compositions from a large number of Pt–Fe alloys reported worldwide in the literature for various geological settings and types of deposits (note that the prefix "ferroan" (used in ferroan platinum) does not imply the valence state of Fe); (2) To compare the compositional characteristics of Pt–Fe alloys that are present in lode deposits versus the ones present in placer deposits derived from source rocks; and (3) To evaluate whether Pt–Fe alloys may be used as useful indicator minerals to infer the type or character of the lode source for detrital grains or placer concentrations of PGM.

2. Materials and Methods

In the present review, we used a total of 2430 data-points accumulated from the literature, including our own data and from hitherto unpublished results in internal reports, e.g., [10]. We believe this large dataset of electron-microprobe compositions (EMP) to be representative because it also reflects a large diversity of localities, geologic settings, and ore zones worldwide. The different types of lode and placer deposits reviewed for this study are listed in Table 1.

Table 1. Worldwide occurrences of Pt–Fe alloy minerals involved in the present review.

		Number of Data-Points (N)	Reference
	Lode Deposits		
1	Dunite pipes and Merensky Reef; Bushveld Layered Complex; South Africa	9	[2,11,12]
2	Kapalagulu Layered Intrusion; western Tanzania	9	[10,13,14]
3	Tulameen Alaskan-type Complex; British Columbia; Canada	6	[15]
4	Coldwell Alkaline Complex; Ontario; Canada	10	[16]
5	Gal'moenan Mafic-Ultramafic Complex; Koryak Upland; Kamchatka krai; Russia	70	[17]
6	Kondyor concentrically zoned Alkaline Ultramafic Complex, northern Khabarovskiy krai; Aldan Shield; Russia	19	[18]
7	Kachkanar Alaskan -Uralian-type Ultramafic Complex; Urals; Russia	7	[19]
8	Nizhniy Tagil Alaskan -Uralian-type Ultramafic Complex; Urals; Russia	13	[19]
9	Ophiolitic Chromitites; Uktus and Kytlym areas; central Urals; Russia	6	[20]
10	Saprolite after mineralized dunite weathered; Yubdo; Ethiopia	83	[21]

Table 1. *Cont.*

		Number of Data-Points (N)	Reference
	Placer Deposits		
11	Placers of North Saskatchewan River; Alberta; Canada	295	[4]
12	Similkameen-Tulameen River System; British Columbia; Canada	18	[4,15]
13	Placers of Liard River; Northwest Territories; Canada	196	[4]
14	Placers of Saskatchewan River; Saskatchewan; Canada	18	[4]
15	Florence Creek; Yukon; Canada	35	[4,22]
16	Au-PGE placer deposits; British Columbia; Canada	77	[9,23]
17	Detrital grains from McConnell Stream; Dease Stream; Birch Stream; British Columbia; Canada	99	[24]
18	Detrital grains; Burwash Creek; Kluane area; Yukon; Canada	15	[25]
19	Fox Gulch; Alaska; USA	2	[4]
20	Salmon River Placer; Goodnews bay; Alaska; USA	29	[26]
21	Detrital grains from Trinity County; California; USA	4	[4,27]
22	Syssert Placer Zone; Omutnaya River; Urals; Russia	6	[4]
23	Placers from Nizhniy Tagil; Kushvinskiy and Nevyansk areas; Urals; Russia	137	[4]
24	Placers; western Chukotka; Russia	47	[4,28]
25	Placers; Anabar basin; northeastern Siberian Platform; Russia	3 (mean of 105)	[29]
26	Placers derived from Filippa clinopyroxenite-dunite complex; Kamchatka; Russia	43	[30]
27	Placer of River Bolshoy Khailyk; western Sayans; Russia	10	[31]
28	Placer derived from Kondyor Alkaline Ultramafic Complex; northern Khabarovskiy krai; Russia	14	[32]
29	Placers associated with Kondyor, Inagli and Guli concentrically zoned Complexes; northeastern Russia	13	[33]
30	Placer at Pustaya River; Kamchatka; Russia	15	[34]
31	Sisim Placer Zone; eastern Sayans; Russia	19	[35]
32	Placer at River Ko; eastern Sayans; Russia	17	[36]
33	Placers of southern Siberia; Russia	20	[37]
34	Placers associated with concentrically zoned Uktus complex; central Urals; Russia	4	[38]
35	Placers from Rio Condoto area; Chocó; Colombia	461	[4]
36	Placers from Santiago River area; Esmeraldas Province; Ecuador	104	[4]
37	Placers at Yubdo; Ethiopia	5	[4]
38	Placers at Riam Kanan; South Kalimantan; Indonesia	51	[4]
39	Placers of Borneo; Sabah Province; Malaysia	20	[4]
40	Placers from Papua New Guinea	6	[4]
41	Placer of Ortakale River; Kars Province; Turkey	3	[4]
42	Placers of Transvaal and Orange Free State; South Africa	24	[4]
43	Placers of Tasmania; Australia	9	[4]
44	Placers from Itabira; Brazil	5	[4]
45	Placers of Chindwin River; Burma	315	[4]
46	Durance River; France	5	[39]
47	Placers of eastern Madagascar	18	[40]
48	Placers of rivers draining Great Dyke; Zimbabwe	3	[41]
49	Placers of rivers draining eastern Bushveld complex; South Africa	7	[42,43]

3. Results

3.1. Variations in the Pt–Fe Alloys from Lode Deposits

The Pt–Fe alloys are associated with several lode sources, specifically from the different types of ore zones within magmatic complexes (Table 1). Layered intrusions are represented by the dunite pipes (Mooihoek and Onverwacht) and the Merensky Reef of the Bushveld Layered Complex (South Africa), and by the Kapalagulu Layered Intrusion of western Tanzania. Alaskan-Uralian-(Aldan)-type zoned complexes include the Tulameen complex in British Columbia (Canada), the Kondyor Complex in northern Khabarovskiy krai, the Kachkanar and Nizhniy Tagil Complexes of the Urals, and the Gal'moenan Complex in the Koryak Upland, Kamchatka krai (Russia). The Coldwell Complex represents a giant Alkaline Gabbro-Syenite intrusion located in Ontario (Canada); Ophiolitic Chromitites are represented by occurrences from the Uktus and Kytlym areas, central Urals (Russia). In addition, Pt–Fe alloys were studied in situ in a PGM-bearing saprolite developed after mineralized dunite at Yubdo (Ethiopia) (Table 1).

Two compositional trends are generally observed in Figure 1. First, the Yubdo trend is extended and consistent with most of the data-points plotted. Compositions of the Pt–Fe alloys from the Alaskan-type complexes, i.e., Tulameen, Kondyor, Kachkanar, and Nizhniy Tagil, are similar and broadly overlap to form a small field close to the central portion of the Yubdo trend. The second, the

Coldwell trend of compositions, is uncommon. It is discordant to the Yubdo trend because of the strong Pd-for-Pt substitution occurring in these alloys (Figure 1).

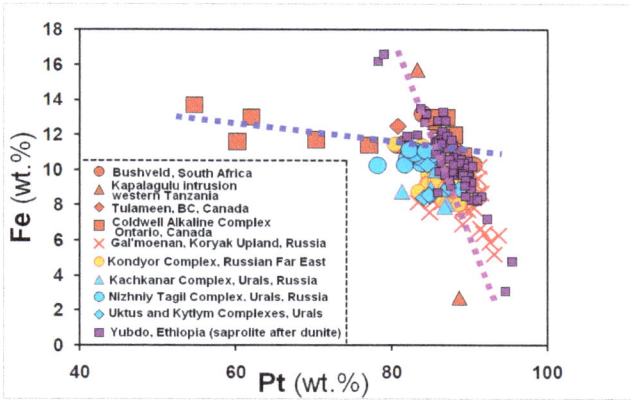

Figure 1. Compositional variations in the Pt–Fe alloy minerals from lode deposits associated with various complexes; contents of Pt plotted versus Fe are expressed in weight percent.

Pt–Fe alloys from the layered intrusions (Bushveld and Kapalagulu) and from the Coldwell Alkaline Complex are consistently poor in iridium, containing from "not detected" (n.d.) to 0.3 wt.% Ir (see References quoted in Table 1). In contrast, the Pt–Fe alloys from Alaskan-type complexes are relatively enriched in Ir (Figure 2): 0.24–3.6 (mean 1.0) wt.% Ir at Tulameen; n.d.–5.28 (1.32) wt.% at Kondyor; 0.56–5.67 (1.97) wt.% at Kachkanar; 0.65–8.85 (2.36) wt.% Ir at Nizhniy Tagil; and n.d.–8.78 (1.86) wt.% Ir at Gal'moenan. The observed maximums are notably similar in compositions of the Pt–Fe alloys from the latter two occurrences (Figure 1). The alloys analyzed in the chromitites at Uktus and Kytlym and in the Yubdo saprolite contain 0.24–2.04 (1.14) wt.% Ir and n.d. to 4.2 (mean 0.43) wt.% Ir, respectively.

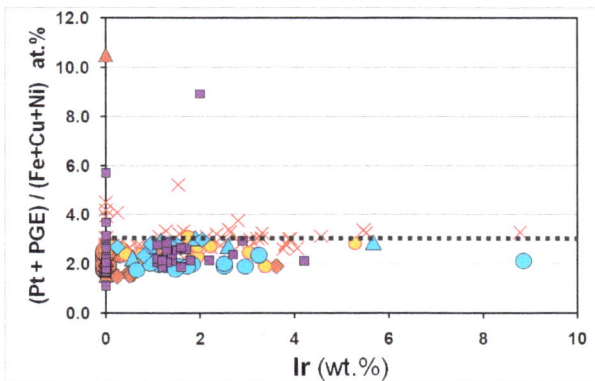

Figure 2. Plot of the values of the Σ(Pt + PGE):(Fe + Cu + Ni) ratio (in atomic %) versus the contents of Ir expressed in weight % in compositions of the Pt–Fe alloy minerals from various lode deposits. The symbols and sources used are the same as the ones shown in Figure 1 and listed Table 1. The dotted line displays the stoichiometry (Pt, PGE)$_3$(Fe, Cu, Ni).

The Rh admixtures (Figure 3) are below the level of detection (EMP) in the Pt–Fe alloys at Kapalagulu and Coldwell. The Bushveld Pt–Fe alloys contain variable amounts of Rh, from n.d. to

3.6 wt.% (mean 0.58) wt.% Rh. Pt–Fe alloys of the Alaskan-type complexes typically contain substantial levels of Rh: 0.52–0.75 (0.62) wt.% Rh at Tulameen; n.d.–1.66 (0.67) wt.% at Kondyor; 0.41–1.37 (0.92) at Kachkanar; and 0.45–1.37 (0.91) at Nizhniy Tagil, the values of which are close to 0.36–1.35 (mean 0.87) wt.% Rh in the Pt–Fe alloys analyzed at Uktus and Kytlym. Pt–Fe alloys at Gal'moenan are reportedly poor in Rh: n.d.–0.43 (mean 0.04) wt.% Rh.

Maximum Pd contents are characteristic of the Pt–Fe alloys at Coldwell, with compositions ranging from n.d. to 29.7 (mean 10.98) wt.% Pd (Figure 4). Occasionally, the Pt–Fe alloys analyzed from the layered intrusions also display elevated levels of Pd: n.d.–3.8 (mean 0.57) wt.% Pd at Bushveld and n.d.–5.6 (mean 0.62) wt.% Pd at Kapalagulu. These mean values compare well with the values observed for the Pt–Fe alloys from some of the Alaskan-type complexes: n.d.–1.07 (mean 0.59) wt.% at Kachkanar; 0.2–0.5 (0.43) wt.% Pd at Tulameen; and n.d.–1.51 (mean 0.45) wt.% Pd at Kondyor. The corresponding levels are lower at Nizhniy Tagil (n.d.–0.64, mean 0.19 wt.% Pd) and at Uktus and Kytlym (n.d.–0.86, mean 0.26 wt.% Pd).

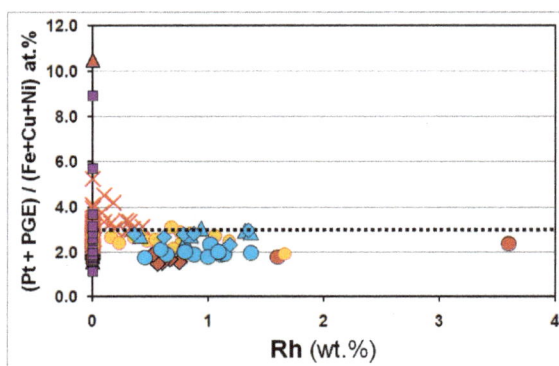

Figure 3. Plot of the values of the Σ(Pt + PGE):(Fe + Cu + Ni) ratio (in atomic %) versus contents of Rh expressed in weight % in compositions of the Pt–Fe alloy minerals from various lode deposits. The symbols and sources used are the same as the ones shown in Figure 1 and listed in Table 1.

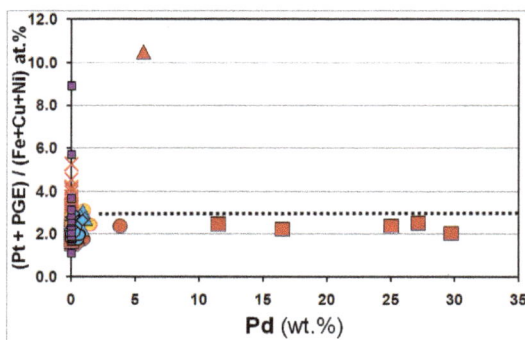

Figure 4. Plot of the values of the Σ(Pt + PGE):(Fe + Cu + Ni) ratio (in atomic %) versus contents of Pd expressed in weight % in compositions of the Pt–Fe alloy minerals from various lode deposits. The symbols and sources used are the same as the ones shown in Figure 1 and listed in Table 1.

Compositional variations of Cu in the Pt–Fe alloys are presented in Figure 5. Note that the observed levels of Cu are generally increased with decreasing values of the atomic Σ(Pt + PGE):(Fe + Cu + Ni) ratio.

Pt–Fe alloys at Tulameen (0.5–2.1, mean 1.17 wt.% Cu), Kondyor (0.12–1.75, mean 0.82 wt.% Cu), and Nizhniy Tagil (0.4–1.3, mean 0.98 wt.% Cu) are notably rich in admixtures of Cu. Alloys of Pt–Fe at Kachkanar are less enriched in Cu: 0.33–0.87 (mean 0.59) wt.%. These levels are fairly close to the ones observed at Bushveld (n.d.–1.4, mean 0.5 wt.% Cu) and Coldwell (n.d.–2.0, mean 0.64 wt.% Cu). The other lode deposit occurrences show low values of the admixtures of Cu in the Pt–Fe alloys at: Kapalagulu (n.d.–0.6, mean 0.17 wt.% Cu); Gal'moenan (n.d.–0.83, mean 0.06 wt.%); Uktus and Kytlym (0.19–0.45, mean 0.31 wt.% Cu); and Yubdo (n.d.–1.90, mean 0.30 wt.% Cu).

Elevated values of Ni (Figure 6) were found in the Pt–Fe alloys in the Bushveld samples: n.d. to 3.9 (mean 1.0 wt.% Ni). The more unexpected observation is that a substantial Ni enrichment also exists in the Pt–Fe alloys in the Tulameen (1.2–3.2, mean 2.53 wt.% Ni) and Nizhniy Tagil (0.27–2.29, mean 1.27 wt.% Ni) Alaskan-type complexes. Some of the Pt–Fe alloys at Yubdo are also enriched in Pt (n.d.–2.6, mean 0.28 wt.% Ni). The other occurrences display low contents of Ni in the Pt–Fe alloys at: Kondyor (n.d.–0.83, mean 0.24 wt.% Ni); Kapalagulu (n.d.–0.6, mean 0.17 wt.% Ni); Uktus and Kytlym (0.23–0.39, mean 0.3 wt.% Ni); Kachkanar (n.d.–0.13 wt.% Ni); Gal'moenan (n.d.–0.09 wt.% Ni); and Coldwell (n.d.).

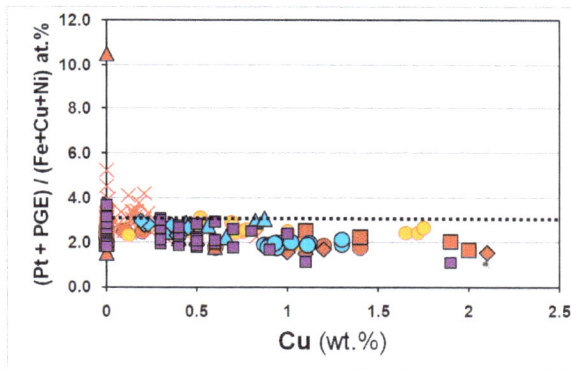

Figure 5. Plot of the values of the Σ(Pt + PGE):(Fe + Cu + Ni) ratio (in atomic %) versus contents of Cu expressed in weight % in compositions of the Pt–Fe alloy minerals from various lode deposits. The symbols and sources used are the same as the ones shown in Figure 1 and listed in Table 1.

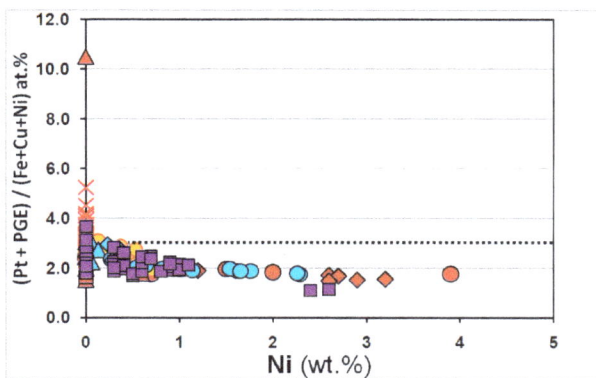

Figure 6. Plot of the values of the Σ(Pt + PGE):(Fe + Cu + Ni) ratio (in atomic %) versus contents of Ni expressed in weight % in compositions of the Pt–Fe alloy minerals from various lode deposits. The symbols and sources used are the same as the ones shown in Figure 1 and listed in Table 1.

3.2. Variations in the Compositions of the Pt–Fe Alloys from Placer Deposits

Variations in the contents of Pt versus Fe, shown in Figure 7, pertain to grains of the Pt–Fe alloys from the placer localities summarized in Table 1. Four sets of EMP data were grouped conditionally for the large territories (Canada and the USA; Russia; and "Various placers worldwide") to reduce the extensive overlaps existing amongst the compositional sets of regional scale.

Figure eight-shaped compositional fields are observed, which is indicative of a bimodal distribution pattern and consistent with the four plotted data sets that are mutually overlapped (Figure 7). Note that the multiple points of the 8-shaped clouds are clustered approximately around the central compositions $Pt_{74}Fe_{26}$ (very close to ideal Pt_3Fe, i.e., isoferroplatinum) and $Pt_{83}Fe_{17}$ (native or ferroan platinum); thus, these fields may reflect the existing miscibility gap.

As displayed in Figure 8, most of the Pt–Fe alloys contain <5–10 wt.% Ir. The observed enrichment in Ir does not necessarily occur in $(Pt, Ir)_3Fe$-type compositions related to a solid solution of isoferroplatinum and chengdeite, which is the Ir-dominant analogue of isoferroplatinum [44]. Thus, the incorporation of Ir is not controlled by the $\Sigma(Pt + PGE):(Fe + Cu + Ni)$ ratio, and presumably, can occur in both types of structures: ordered and disordered. Interestingly, the Pt–Fe alloys from the lode deposits (Figure 2) display an Ir-enrichment trend that extends along a narrow range of compositions $(Pt, PGE)_{2-3}(Fe, Cu, Ni)$.

The observed distribution of admixtures of Rh (Figure 9) displays a similar character as Ir. There is no clear relationship between the levels of Rh and the values of the atomic $\Sigma(Pt + PGE):(Fe + Cu + Ni)$ ratio in these minerals of placer alloys. In contrast, and uniform to Ir, the Rh-enrichment is related to the compositional series $(Pt, PGE)_{2-3}(Fe + Cu + Ni)$ in the lode deposits (Figure 3). However, the behavior of the Ir and Rh is not coherent in the Pt–Fe alloys. In the Rh–Ir plot, these elements formed separate trends of enrichment and were not involved in the mutual schemes of substitutions.

The Pd distribution (Figure 10) seems to be rather chaotic with respect to the ratio $\Sigma(Pt + PGE):(Fe + Cu + Ni)$ in compositions of the placer grains. The observed extent of the Pd enrichment is modest in comparison to the high-Pd compositions of the Pt–Fe alloy from the Coldwell alkaline gabbro-syenite complex (Figure 4). The latter type of lode source was unlikely to have been involved as a contributor to the studied zones of the placer PGM. In contrast, similar to Ir–Rh, the trend of Pd enrichment corresponds to the series $(Pt, PGE)_2(Fe + Cu + Ni)$ in the lode deposits (Figure 4).

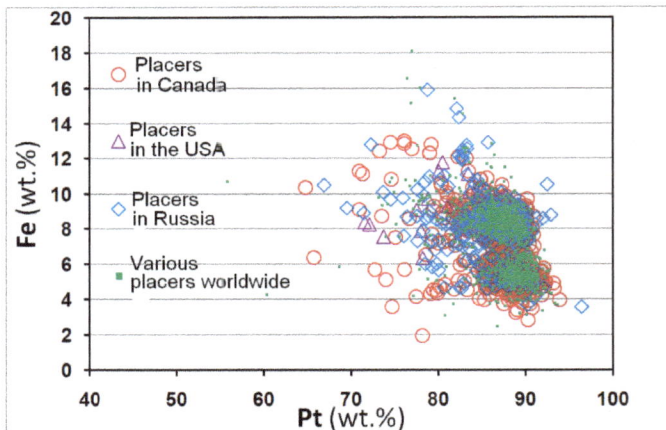

Figure 7. Pt versus Fe diagram (in weight %) showing the compositional distribution of the Pt–Fe alloy minerals from placer deposits worldwide, which are listed in Table 1.

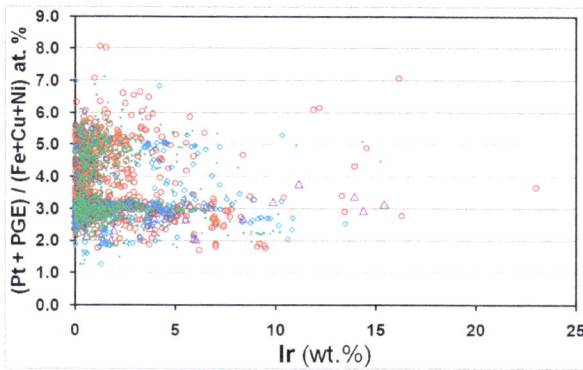

Figure 8. Plot of the values of the Σ(Pt + PGE):(Fe + Cu + Ni) ratio (in atomic %) versus contents of Ir expressed in weight % in compositions of the Pt–Fe alloy minerals from various placer deposits. The symbols and sources used are the same as the ones shown in Figure 7 and listed in Table 1.

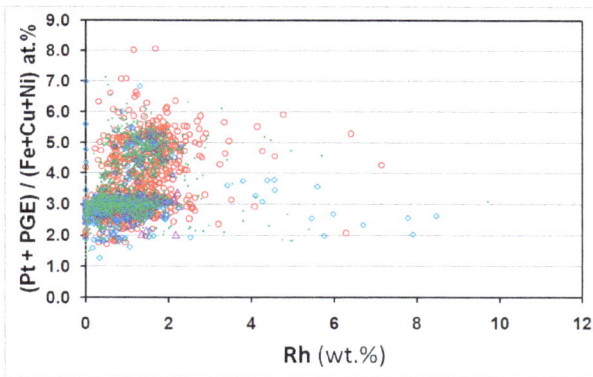

Figure 9. Plot of the values of the Σ(Pt + PGE):(Fe + Cu + Ni) ratio (in atomic %) versus contents of Rh expressed in weight % in compositions of the Pt–Fe alloy minerals from various placer deposits. The symbols and sources used are the same as the ones shown in Figure 7 and listed Table 1.

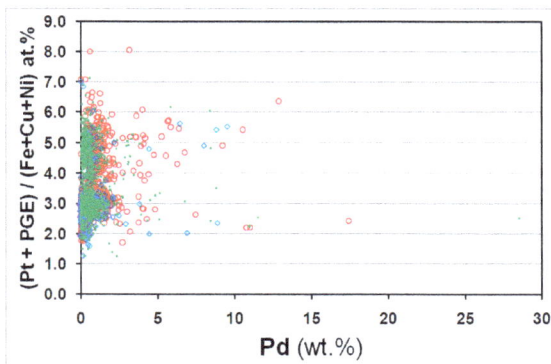

Figure 10. Plot of the values of the Σ(Pt + PGE):(Fe + Cu + Ni) ratio (in atomic %) versus contents of Pd expressed in weight % in compositions of the Pt–Fe alloy minerals from various placer deposits. The symbols and sources used are the same as the ones shown in Figure 7 and listed in Table 1.

Similar to the Pt–Fe alloys from lode sources (Figure 5), the placer Pt–Fe grains display a trend of Cu enrichment that extends along the compositional series (Pt, PGE)$_{2-3}$(Fe + Cu + Ni) (Figure 11). However, the majority of the plotted data-points are distributed more or less chaotically in the range n.d.–4 wt.% Cu (Figure 11).

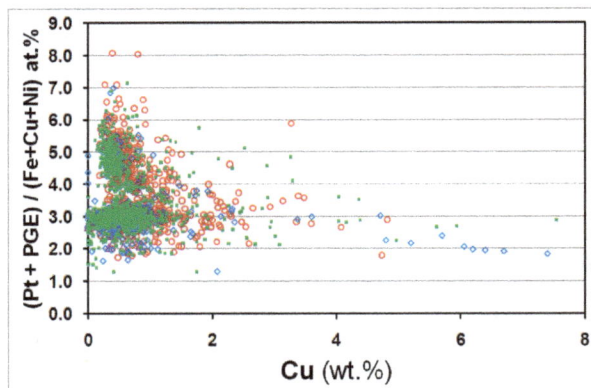

Figure 11. Plot of the values of the Σ(Pt + PGE):(Fe + Cu + Ni) ratio (in atomic %) versus contents of Cu expressed in weight % in compositions of the Pt–Fe alloy minerals from various placer deposits. The symbols and sources used are the same as the ones shown in Figure 7 and listed in Table 1.

Unlike the other elements, the distribution of minor Ni is most consistent in the Pt–Fe alloys from the lode deposits (Figure 6) and from the placers (Figure 12), showing similar trends of Ni-enrichment in the compositional series (Pt, PGE)$_{2-3}$(Fe + Cu + Ni). Additionally, the maximum contents of Ni are close in the Pt–Fe alloys in the lode and placer occurrences (≤4 wt.% Ni).

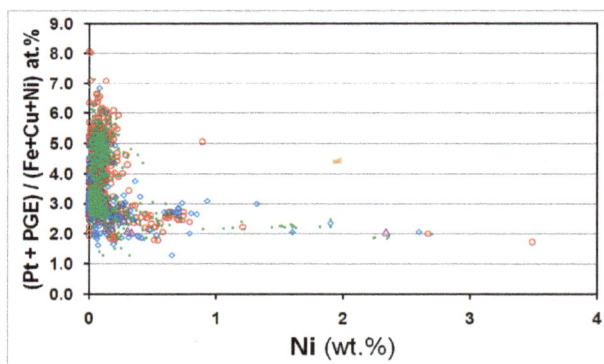

Figure 12. Plot of the values of the Σ(Pt + PGE):(Fe + Cu + Ni) ratio (in atomic %) versus contents of Ni expressed in weight % in compositions of the Pt–Fe alloy minerals from various placer deposits. The symbols and sources used are the same as the ones shown in Figure 7 and listed in Table 1.

3.3. Overall Variations in the Compositions of the Pt–Fe Alloys on a Global Scale

In Figure 13, the overall variations are shown in compositions of the Pt–Fe alloys in all types of occurrences, involving the examined lode and placer deposits (Table 1). In the Pt versus Fe diagram, the observed field of solid solutions displays a linear boundary due to the ideal scheme of Pt ↔ Fe substitution, with the observed equation of linear regression y = −x + 100 (Figure 13a).

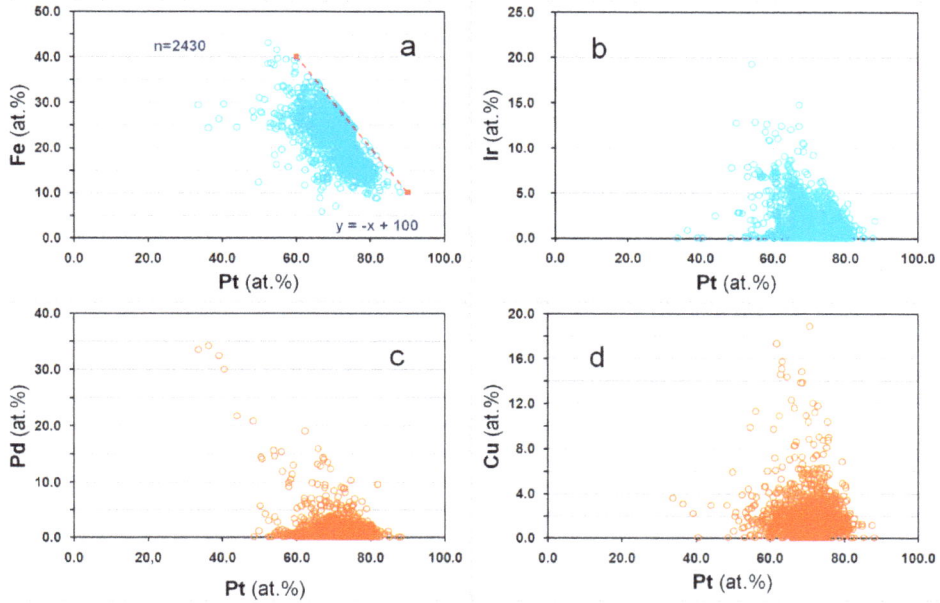

Figure 13. Plot of the contents of Pt. versus Fe (**a**), Pt. versus Ir (**b**), Pt versus Pd (**c**), and Pt versus Cu (**d**), all expressed in values of atomic %, in EMP compositions of Pt–Fe alloy minerals. A total of 2430 point analyses (n = 2430) are plotted, which pertain to a large variety of occurrences (Table 1).

The two main series of solid-solutions (Figures 13b,c and 14a) are based on the incorporation of high levels of Ir ($\leq$20 at.%), and an even greater content of Pd ($\leq$35 at.%), which are both negatively correlated with Pt. In contrast, Cu ($\leq$20 at.%) did not correlate with Pt, where the bulk of the EMP compositions had fairly low amounts of $\leq$6 at.% Cu (Figure 13d).

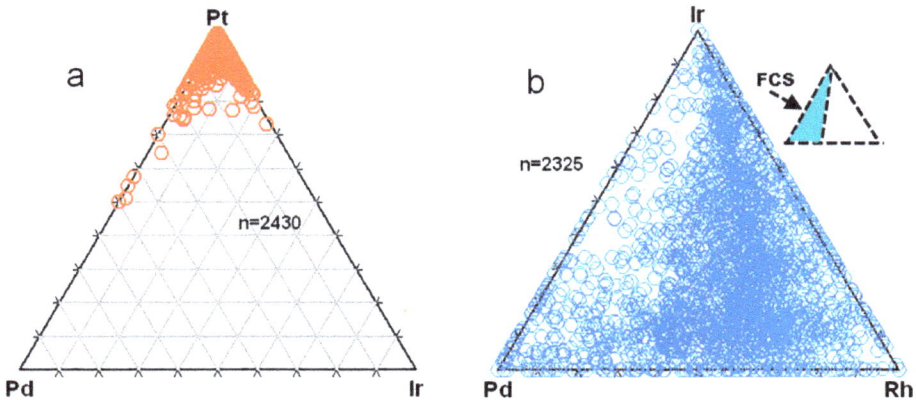

Figure 14. Ternary diagrams of Pt–Pd–Ir (**a**) and Ir–Pd–Rh (**b**) showing atomic proportions of these elements in EMP compositions of the Pt–Fe alloy minerals from numerous localities (Table 1). The total number of point analyses is 2430 (**a**) and 2325 (**b**). FCS is the inferred "field of complicated substitutions"; see text for discussion.

Thus, two series of extensive solid-solutions exist: $Pt_{3-x}Fe$ (isoferroplatinum and ferroan platinum) $\leftrightarrow$ $Ir_{3-x}Fe$ (chengdeite, and possibly, its disordered modification), and a series extending toward Pd_3Fe (unnamed), which was documented at Coldwell (Figure 14a). There is a good potential for the discovery of a new species of PGM with a Pd-dominant composition, $(Pd, Pt)_3Fe$, in relation to the latter or another complex. The ordered and disordered variants of Pd_3Fe exist in the Pd-Fe system, e.g., [45].

In the compositional Ir–Pd–Rh space, the presence of a "field of complicated substitutions" is proposed, which generally extends along the Ir–Pd axis (Figure 14b). In addition, there is a slightly rarefied area along the Ir–Rh axis, which could be ascribed to the lack of sufficient statistics. On the other hand, the observed solid-solution appears to extend, approximately, along the line Ir–(Pd:Rh = 2:3) in the Ir–Pd–Rh space, as inferred on the basis of a total of 2325 point analyses (Figure 14b).

In addition, Ru, Os, and Sb occur as traces or minor elements in the Pt–Fe alloys. The bulk of the compositions of alloy grains gave ≤ 0.5 wt.% Ru, with a maximum ~5 wt.% and mean 0.13 wt.% Ru (per n = 2430). The observed background of Os was notably higher. The majority of the Pt–Fe grains contained ≤ 0.5–2 wt.% Os, with episodic maximums ~6–8 wt.% and mean 0.53 wt.% Os (n = 2430). Typically, Sb appears to only occur in trace amounts (n.d. by EMP methods). Nevertheless, some grains of the Pt–Fe alloys yielded essential levels varying from ~0.4 to 1.4 wt.% Sb.

4. Discussion and Conclusions

In summary, we note the following observations:

(1) PGE admixtures (Ir, Rh, and Pd) are typically incorporated into the alloys of composition $(Pt, PGE)_{2-3}(Fe, Cu, Ni)$ in a variety of lode occurrences associated with ore zones and complexes of different types and geologic settings worldwide (Figures 2–4). These compositional series extend along and below somewhat the line $\Sigma(Pt + PGE):(Fe + Cu + Ni) = 3$, which corresponds to isoferroplatinum.

(2) Generally, minor Cu follows the same pattern of distribution in the Pt–Fe alloys in the lode occurrences (Figure 5), showing a clear tendency of decrease in the values of the ratio $\Sigma(Pt + PGE):(Fe + Cu + Ni)$, while the Cu content increased. The trend of Cu enrichment in the minerals that are relatively poor in Pt implies their crystallization from relatively fractionated melts at lower temperatures. This trend agrees with observations noted below for the Pt–Fe alloys from placer associations.

(3) Similarly, the Ni admixture enters preferentially in the Pt–Fe alloy, nominally $(Pt, PGE)_2(Fe, Cu, Ni)$ in the lode deposits reviewed (Figure 6). One of the suggested possibilities is that Pt_2Fe may in fact represent "invisible" mixtures of Pt_3Fe and PtFe (≤ 1 μm in size) that are exsolved upon cooling within the miscibility gap [46]. In contrast, examples of homogeneous Pt_2Fe with a disordered *fcc* structure were documented in [33].

(4) The global-scale sets of EMP data generally display Figure 8-shaped compositional fields in the Pt–Fe diagram, which are based on a large number of compositions of the Pt–Fe minerals from a large variety of placer deposits worldwide (Figure 7). This distribution is likely a reflection of the miscibility gap existing between the ordered Pt_3Fe structure (isoferroplatinum) and the disordered structure of native or ferroan platinum.

(5) The patterns of distribution of the admixtures of PGE (Ir, Rh, and Pd) appear to scatter chaotically in the plots of EMP compositions of placer grains of the Pt–Fe alloy minerals. There is no clear relationship observed between the amount of these elements and their values in the ratio $\Sigma(Pt + PGE):(Fe + Cu + Ni)$. Minerals of the type $(Pt, PGE)_{2-3}(Fe, Cu, Ni)$ were not observed as the distinctive series (Figures 8–10).

(6) In contrast to the PGE, minor amounts of Cu and Ni admixtures yielded their maximums in alloys of the compositional series $(Pt, PGE)_{2-3}(Fe, Cu, Ni)$ in the placer occurrences (Figures 11 and 12).

(7) The global-scale set of EMP compositions plotted in the Pt–Fe diagram indicates that the linear boundary observed in the overall compositional field is due to an ideal scheme of Pt $\leftrightarrow$ Fe substitution (Figure 13a).

(8) Pairs of minor elements, Ir–Rh and Cu–Ni, despite an internal similarity in the observed character of distribution (Figures 2, 3, 8 and 9), displayed incoherent behavior and formed separate trends in terms of the plots of Rh versus Ir and Cu versus Ni.

(9) The two series of solid solutions in the Pt-Fe alloys are based on the Ir-for-Pt and Pd-for-Pt substitutions. The incorporation of Ir is not restricted by the Pt_3Fe–Ir_3Fe substitution involving components of isoferroplatinum and chengdeite (also, likely their disordered modifications). Besides, Ir^0 appears to replace Pt^0 in disordered variants of the Pt–Fe alloys, as is implied by the observed pattern (Figure 8). There is a good potential for discovery of a new species of PGM having a Pd-dominant composition, $(Pd, Pt)_3Fe$, most expectedly in association with the mineralized zones of mafic units of alkaline mafic-ultramafic or gabbroic complexes.

(10) The presence of a "field of complicated substitutions" is proposed, which generally extends along the Ir–Pd axis in the Ir–Pd–Rh diagram (Figure 14b). The inferred solid solution extends approximately along the line Ir–(Pd:Rh = 2:3), suggesting that minor Pd enters the solid solution via a coupled scheme of substitution in the combination with Rh.

The "field of complicated substitutions, FCC" (Figure 14b) may likely be a reflection of the crystallochemical differences existing between Pd and Ir, which, at least under normal crystallization conditions, seem to avoid each other, and they do not participate in mutual substitution schemes in all of the PGM species presently known. Only minor substitutions are known, e.g., ~1 wt.% Ir in laflammeite $Pd_3Pb_2S_2$ [47].

The Pd-for-Rh substitution is not so common; nevertheless, it occurs in the series of solid solution of palladodymite $(Pd, Rh)_2As$—rhodarsenide $(Rh, Pd)_2As$ [48,49]. Among the PGE, Pd has the smallest size of the calculated atomic radius, which is a consequence of its outermost electron density deriving predominantly from *d*-levels, instead of *s*-levels [50].

(11) Pt–Fe alloys analyzed in situ in the lode deposits display some distinctive features of their compositions relative to the Pt–Fe grains examined in placer associations. As noted, the compositional series $(Pt, PGE)_{2-3}(Fe, Cu, Ni)$ clearly dominates in the lode occurrences of the Pt–Fe alloys from different sources (Figures 2–4). In contrast, this series was not as important in the placer occurrences (Figures 8 and 9).

In addition, placer grains of the Pt–Fe alloys generally have large grain-sizes due to crystallization under special conditions in their now-eroded source rocks, a subject extensively debated in the literature that is well-documented and discussed in [51], and references therein.

A large grain-size is known for some Pt–Fe alloy nuggets derived from placers related to Alaskan-type complexes, such as the ~1.5 kg specimens of native ferroan platinum at Kondyor [32] or the crystals exceeding 10 cm across at Nizhniy Tagil [52]. These examples are consistent with the mineral-forming environments rich in volatiles and are likely related to ultramafic-mafic pegmatite facies. For example, at Kondyor, phlogopite-rich late zones rich in PGE are known in the Anomal'nyi area [53].

(12) As an indication of the very large quantity of host Pt-bearing rocks needed to be weathered and eroded to produce tenors found in commercially-viable placers, two publications on the Goodnews Bay placers are relevant [54,55]. Detailed study of the geology and geomorphology showed that Red Mountain is the source rock that has been eroded by 2000 vertical feet over a period of about 20 million years or longer. The amount of eroded rock in this case represents a volume of about 2.4 billion m^3 and underlines the huge quantities of weathered and eroded source rocks involved and why large grains are statistically nearly impossible to find microscopically.

(13) There are some compositional features that are generally characteristic of the Pt–Fe alloy minerals hosted by the different types of source rocks, e.g., the trend of Ir-enrichment typically occurs in Alaskan-type complexes, and the strong Pd-enrichment is related with the mineralized rocks of gabbroic compositions in the Coldwell complex.

Based on our observations, we conclude that the Pt–Fe alloys of different origins (Table 1); [51,54–60] exhibit compositional overlaps that are too large to represent reliable index-minerals to define a provenance or to infer a source rock for detrital grains of Pt–Fe alloys. Nevertheless, the strong Pd enrichment in the Pt–Fe alloys from the lode source related to alkaline gabbroic deposits (e.g., Coldwell), or potentially, to another type of mineralized mafic rock (e.g., Pd-Pt zones in mafic units of layered intrusions), is distinctive and may presumably represent an indicator of provenance if found in detrital deposits.

Author Contributions: The authors wrote the article together.

Funding: A.Y.B. gratefully acknowledges the partial support of this investigation by the Russian Foundation for Basic Research (project # RFBR 16-05-00884 and # RFBR 19-05-00181).

Acknowledgments: We thank M. Economou-Eliopoulos, the editorial staff and three anonymous referees for their suggestions.

Conflicts of Interest: There are no conflicts of interest.

References

1. Cabri, L.J.; Feather, C.E. Platinum-iron alloys: A nomenclature based on a study of natural and synthetic alloys. *Can. Mineral.* **1975**, *13*, 117–126.
2. Cabri, L.J.; Rosenzweig, A.; Pinch, W.W. Platinum-group minerals from Onverwacht. I. Pt-Fe-Cu-Ni alloys. *Can. Mineral.* **1977**, *15*, 380–384.
3. Cabri, L.J. (Ed.) *The Geology, Geochemistry, Mineralogy, Mineral Beneficiation of the Platinum-Group Elements*; Canadian Institute of Mining, Metallurgy and Petroleum: Montreal, QC, Canada, 2002; Volume 54, p. 852.
4. Cabri, L.J.; Harris, D.C.; Weiser, T.W. Mineralogy and distribution of platinum-group mineral (PGM) placer deposits of the world. *Explor. Min. Geol.* **1996**, *5*, 73–167.
5. Bowles, J.F.W. Platinum-iron alloys, their structural and magnetic characteristics in relation to hydrothermal and low-temperature genesis. *Mineral. Petrol.* **1990**, *43*, 37–47. [CrossRef]
6. Nosé, Y.; Kushida, A.; Ikeda, T.; Nakajima, H.; Tanaka, K.; Numakura, H. Re-examination of Phase Diagram of Fe–Pt System. *Mater. Trans.* **2003**, *44*, 2723–2731. [CrossRef]
7. Bowles, J.F.W.; Suárez, S.; Prichard, H.M.; Fisher, P.C. Weathering of PGE sulfides and Pt–Fe alloys in the Freetown Layered Complex, Sierra Leone. *Miner. Depos.* **2017**, *52*, 1127–1144. [CrossRef]
8. Evstigneeva, T.L. Phases in the Pt-Fe system. *Vestnik Otdeleniya nauk o Zemle RAN* **2009**, *1*, 1–2.
9. Barkov, A.Y.; Fleet, M.E.; Nixon, G.T.; Levson, V.M. Platinum-group minerals from five placer deposits in British Columbia, Canada. *Can. Mineral.* **2005**, *43*, 1687–1710. [CrossRef]
10. Cabri, L.J. *Mineralogical Study of Five Samples from a PGE Regolith, Kapalagulu Intrusion, Tanzania*; Unpublished Report; Cabri Consulting Inc.: Ottawa, ON, Canada, 2004.
11. Mostert, A.B.; Hofmeyr, P.K.; Potgieter, G.A. The platinum-group mineralogy of the Merensky Reef at the Impala platinum mines, Bophuthatswana. *Econ. Geol.* **1982**, *77*, 1385–1394. [CrossRef]
12. Rudashevsky, N.S.; Avdontsev, S.N.; Dneprovskaya, M.B. Evolution of PGE mineralization in hortonolitic dunites of the Mooihoek and Onverwacht pipes, Bushveld Complex. *Mineral. Petrol.* **1992**, *47*, 37–54. [CrossRef]
13. Wilhelmij, H.R.; Cabri, L.J. Platinum mineralization in the Kapalagulu Intrusion, western Tanzania. *Miner. Depos.* **2016**, *51*, 343–367. [CrossRef]
14. Cabri, L.J.; Wilhelmij, H.R.; Eksteen, J.J. Contrasting mineralogical and processing potential of two mineralization types in the platinum group element and nickel-bearing Kapalagulu Intrusion, western Tanzania. *Ore Geol. Rev.* **2017**, *90*, 772–789. [CrossRef]
15. Nixon, G.T.; Cabri, L.J.; Laflamme, J.H.G. Platinum-group element mineralization in lode and placer deposits associated with the Tulameen Alaskan type complex, British Columbia. *Can. Mineral.* **1990**, *28*, 503–535.
16. Good, D.J.; Cabri, L.J.; Ames, D.E. PGM Facies variations for Cu–PGE deposits in the Coldwell Alkaline Complex, Ontario, Canada. *Ore Geol. Rev.* **2017**, *90*, 748–771. [CrossRef]
17. Sidorov, E.G.; Kozlov, A.P.; Tolstykh, N.D. *The Gal'moenan Basic-Ultrabasic Massif and Its Platinum Potential*; Nauchnyi Mir Publisher: Moscow, Russia, 2012; p. 288. (In Russian)
18. Nekrasov, I.Y.; Lennikov, A.M.; Zalishchak, B.L.; Oktyabrsky, R.A.; Ivanov, V.V.; Sapin, V.I.; Taskaev, V.I. Compositional variations in platinum-group minerals and gold, Konder alkaline-ultrabasic massif, Aldan Shield, Russia. *Can. Mineral.* **2005**, *43*, 637–654. [CrossRef]

19. Augé, T.; Genna, A.; Legendre, O.; Ivanov, K.S.; Volchenko, Y.A. Primary Platinum Mineralization in the Nizhny Tagil and Kachkanar Ultramafic Complexes, Urals, Russia: A Genetic Model for PGE Concentration in Chromite-Rich Zones. *Econ. Geol.* **2005**, *100*, 707–732. [CrossRef]

20. Zaccarini, F.; Garuti, G.; Pushkarev, E.; Thalhammer, O. Origin of Platinum-Group Minerals (PGM) Inclusions in Chromite Deposits of the Urals. *Minerals* **2018**, *8*, 379. [CrossRef]

21. Cabri, L.J. *Electro-Hydraulic (EH) Crushing and Hydroseparation (HS) Concentration of Three Samples from Yubdo, Ethiopia*; Unpublished CNT-MC report 2007-3; CNT-MC Inc.: Ottawa, ON, Canada, 2007; 40p.

22. Barkov, A.Y.; Martin, R.F.; LeBarge, W.; Fedortchouk, Y. Grains of Pt-Fe alloy and inclusions in a Pt-Fe alloy from Florence creek, Yukon, Canada: Evidence for mobility of Os in a Na-H$_2$O-Cl-rich fluid. *Can. Mineral.* **2008**, *46*, 343–360. [CrossRef]

23. Barkov, A.Y.; Martin, R.F.; Fleet, M.E.; Nixon, G.T.; Levson, V.M. New data on associations of platinum-group minerals in placer deposits of British Columbia, Canada. *Mineral. Petrol.* **2007**, *92*, 9–29. [CrossRef]

24. Laflamme, J.H.G. *Mineralogical Study of Platinum-Group Minerals from Au-Pt-Bearing Placer Samples from British Columbia*; Report MMSL 02-038(CR): Appendix "Electron Microprobe Data"; CANMET Mining and Mineral Sciences Laboratories: Ottawa, ON, Canada, 2002; pp. A-1–A-19.

25. Fedortchouk, Y.; LeBarge, W.; Barkov, A.Y.; Fedele, L.; Bodnar, R.J.; Martin, R.F. Platinum-group minerals from a placer deposit in Burwash creek, Kluane area, Yukon Territory, Canada. *Can. Mineral.* **2010**, *48*, 583–596. [CrossRef]

26. Tolstykh, N.D.; Foley, J.Y.; Sidorov, E.G.; Laajoki, K.V.O. Composition of the platinum-group minerals in the Salmon River placer deposit, Goodnews bay, Alaska. *Can. Mineral.* **2002**, *40*, 463–471. [CrossRef]

27. Barkov, A.Y.; Martin, R.F.; Shi, L.; Feinglos, M.N. New data on PGE alloy minerals from a very old collection (probably 1890s), California. *Am. Miner.* **2008**, *93*, 1574–1580. [CrossRef]

28. Gornostayev, S.S.; Crocket, J.H.; Mochalov, A.G.; Laajoki, K.V.O. The Platinum-Group Minerals of the Baimka Placer Deposits, Aluchin Horst, Russian Far East. *Can. Mineral.* **1999**, *37*, 1117–1129.

29. Airiyants, E.V.; Zhmodik, S.M.; Ivanov, P.O.; Belyanin, D.K.; Agafonov, L.V. Mineral inclusions in Fe-Pt solid solution from the alluvial ore occurrences of the Anabar basin (northeastern Siberian Platform). *Russ. Geol. Geophys.* **2014**, *55*, 945–958. [CrossRef]

30. Sidorov, E.G.; Tolstykh, N.D.; Podlipsky, M.Yu.; Pakhomov, I.O. Placer PGE minerals from the Filippa clinopyroxenite-dunite massif (Kamchatka). *Russ. Geol. Geophys.* **2004**, *45*, 1080–1097.

31. Barkov, A.Y.; Shvedov, G.I.; Silyanov, S.A.; Martin, R.F. Mineralogy of Platinum-Group Elements and Gold in the Ophiolite-Related Placer of the River Bolshoy Khailyk, Western Sayans, Russia. *Minerals* **2018**, *8*, 247. [CrossRef]

32. Cabri, L.J.; Laflamme, J.H.G. Platinum-group minerals from the Konder massif, Russian Far East. *Mineral. Rec.* **1997**, *28*, 97–106.

33. Malitch, K.N.; Thalhammer, O.A.R. Pt-Fe nuggets derived from clinopyroxenite-dunite massifs, Russia: A structural, compositional and osmium-isotope study. *Can. Mineral.* **2002**, *40*, 395–418. [CrossRef]

34. Tolstykh, N.D.; Sidorov, E.G.; Laajoki, K.V.O.; Krivenko, A.P.; Podlipskiy, M. The association of platinum-group minerals in placers of the Pustaya River, Kamchatka, Russia. *Can. Mineral.* **2000**, *38*, 1251–1264. [CrossRef]

35. Barkov, A.Y.; Shvedov, G.I.; Martin, R.F. PGE–(REE–Ti)-rich micrometer-sized inclusions, mineral associations, compositional variations, and a potential lode source of platinum-group minerals in the Sisim Placer Zone, Eastern Sayans, Russia. *Minerals* **2018**, *8*, 181. [CrossRef]

36. Krivenko, A.P.; Tolstykh, N.D.; Nesterenko, G.V.; Lazareva, E.V. Types of mineral assemblages of platinum metals in auriferous placers of the Altai-Sayan region. *Russ. Geol. Geophys.* **1994**, *35*, 58–65.

37. Zhmodik, S.M.; Nesterenko, G.V.; Airiyants, E.V.; Belyanin, D.K.; Kolpakov, V.V.; Podlipsky, M.Y.; Karmanov, N.S. Alluvial platinum—Group minerals as indicators of primary PGE mineralization (placers of southern Siberia). *Russ. Geol. Geophys.* **2016**, *57*, 1437–1464. [CrossRef]

38. Zaccarini, F.; Pushkarev, E.; Garuti, G.; Krause, J.; Dvornik, G.P.; Stanley, C.; Bindi, L. Platinum–group minerals (PGM) nuggets from alluvial—Eluvial placer deposits in the concentrically zoned mafic-ultramafic Uktus complex (Central Urals, Russia). *Eur. J. Miner.* **2013**, *25*, 519–531. [CrossRef]

39. Johan, Z.; Ohnenstetter, M.; Fischer, M.; Amossé, J. Platinum-Group Minerals from the Durance River Alluvium, France. *Mineral. Petrol.* **1990**, *42*, 287–306. [CrossRef]

40. Augé, T.; Legendre, O. Pt-Fe nuggets from alluvial deposits in eastern Madagascar. *Can. Mineral.* **1992**, *30*, 983–1004.

41. Oberthür, T.; Weiser, T.W.; Melcher, F.; Gast, L.; Wöhrl, C. Detrital platinum-group minerals in rivers draining the Great Dyke, Zimbabwe. *Can. Mineral.* **2013**, *51*, 197–222. [CrossRef]

42. Oberthür, T.; Melcher, F.; Gast, L.; Wöhrl, C.; Lodziak, J. Detrital platinum-group minerals in rivers draining the eastern Bushveld complex, South Africa. *Can. Mineral.* **2004**, *42*, 563–582. [CrossRef]

43. Melcher, F.; Oberthür, T.; Lodziak, J. Modification of detrital platinum-group minerals from the eastern Bushveld complex, South Africa. *Can. Mineral.* **2005**, *43*, 1711–1734. [CrossRef]

44. Yu, Z. Chengdeite—An ordered natural iron-iridium alloy. *Dizhi Xuebao* **1995**, *69*, 215–220. (In Chinese)

45. Bose, S.K.; Kudrnovsky, J.; van Schilfgaarde, M.; Blöchl, P.; Jepsen, O.; Methfessel, M.; Paxton, A.T.; Andersen, O.K. Electronic structure of ordered and disordered Pd$_3$Fe. *J. Magn. Magn. Mater.* **1990**, *87*, 97–105. [CrossRef]

46. Zhernovsky, I.V.; Mochalov, A.G.; Rudashevsky, N.S. Phase inhomogeneity of isoferroplatinum enriched in iron. *Dokl. Akad. Nauk SSSR* **1985**, *283*, 196–200. (In Russian)

47. Barkov, A.Y.; Martin, R.F.; Halkoaho, T.A.A.; Criddle, A.J. Laflammeite, Pd$_3$Pb$_2$S$_2$, a new platinum-group mineral species from the Penikat layered complex, Finland. *Can. Mineral.* **2002**, *40*, 671–678. [CrossRef]

48. Tarkian, M.; Krstić, S.; Klaska, K.-H.; Ließmann, W. Rhodarsenide, (Rh, Pd)$_2$As, a new mineral. *Eur. J. Mineral.* **1997**, *9*, 1321–1326. [CrossRef]

49. Britvin, S.N.; Rudashevskiy, N.S.; Bogdanova, A.N.; Shcherbachev, D.K. Palladodymite (Pd,Rh)$_2$As—A new mineral from placer in Miass River, Urals. *Zap. Vseross. Mineral. Obshch.* **1999**, *128*, 39–42. (In Russian)

50. Rahm, M.; Hoffmann, R.; Ashcroft, N.W. Atomic and Ionic Radii of Elements 1–96. *Chem. Eur. J.* **2016**, *22*, 14625–14632. [CrossRef]

51. Oberthür, T. The Fate of Platinum-Group Minerals in the Exogenic Environment—From Sulfide Ores via Oxidized Ores into Placers: Case Studies Bushveld Complex, South Africa, and Great Dyke, Zimbabwe. *Minerals* **2018**, *8*, 581. [CrossRef]

52. Weiser, T.W. Platinum-Group Minerals (PGM) from placer deposits in the mineral collection of the Museum of Natural History, Vienna, Austria. *Ann. Naturhist. Mus. Wien* **2004**, *105A*, 1–28.

53. Barkov, A.Y.; Shvedov, G.I.; Polonyankin, A.A.; Martin, R.F. New and unusual Pd-Tl-bearing mineralization in the Anomal'nyi deposit, Kondyor concentrically zoned complex, northern Khabarovskiy kray, Russia. *Mineral. Mag.* **2017**, *81*, 679–688. [CrossRef]

54. Mertie, J.B., Jr. The Goodnews platinum deposits Alaska. *U.S. Geol. Survey* **1940**, *918*, 97.

55. Mertie, J.B., Jr. Platinum deposits of the Goodnews Bay District Alaska. *U.S. Geol. Survey* **1976**, *938*, 42.

56. O'Driscoll, B.; González-Jiménez, J.M. Petrogenesis of the Platinum-Group Minerals. *Rev. Mineral. Geochem.* **2015**, *81*, 489–578. [CrossRef]

57. Garuti, G.; Pushkarev, E.V.; Zaccarini, F. Composition and paragenesis of Pt alloys from chromitites of the Uralian Alaskan-type Kytlym and Uktus complexes, northern and central Urals, Russia. *Can. Mineral.* **2002**, *40*, 357–376. [CrossRef]

58. Ohnenstetter, M. Platinum group element enrichment in the upper mantle peridotites of the Monte Maggiore ophiolitic massif (Corsica, France): Mineralogical evidence for ore-fluid metasomatism. *Mineral. Petrol.* **1992**, *46*, 85–107. [CrossRef]

59. Tolstykh, N.; Kozlov, A.; Telegin, Yu. Platinum mineralization of the Svetly Bor and Nizhny Tagil intrusions, Ural Platinum Belt. *Ore Geol. Rev.* **2015**, *67*, 234–243. [CrossRef]

60. Stepanov, S.Y.; Malitch, K.N.; Kozlov, A.V.; Badanina, I.Y.; Antonov, A.V. Platinum group element mineralization of the Svetly Bor and Veresovy Bor clinopyroxenite–dunite massifs, Middle Urals, Russia. *Geol. Ore Depos.* **2017**, *59*, 244–255. [CrossRef]

MDPI
St. Alban-Anlage 66
4052 Basel
Switzerland
Tel. +41 61 683 77 34
Fax +41 61 302 89 18
www.mdpi.com

Minerals Editorial Office
E-mail: minerals@mdpi.com
www.mdpi.com/journal/minerals

www.ingramcontent.com/pod-product-compliance
Lightning Source LLC
Chambersburg PA
CBHW051840210326
41597CB00033B/5719